Handbook of Open Source Tools

Sandeep Koranne

Handbook of Open
Source Tools

 Springer

Sandeep Koranne
2906 Bellevue Ct
West Linn, Oregon 97068
USA
Sandeep.Koranne@gmail.com

QA
76.76
.S46
K67
2011

ISBN 978-1-4419-7718-2   e-ISBN 978-1-4419-7719-9
DOI 10.1007/978-1-4419-7719-9
Springer New York Dordrecht Heidelberg London

Library of Congress Control Number: 2010938855

Printed on acid-free paper

Springer is part of Springer Science+Business Media (www.springer.com)

Preface

The constant and speedy progress made by humankind in the industrial revolution, and more recently in the information technology era can be directly attributed to sharing of knowledge between various disciplines, reuse of the knowledge as science and technology advanced, and inclusion of this knowledge in the curriculum. The phrases "do not reinvent the wheel" and "to stand upon the shoulders of giants" come to our mind as representative of this thought process of using existing solutions and building upon existing knowledge, but at the same time contributing to the society as a whole.

It was with this intention of documenting existing (circa 2010) Open Source Tools for Scientists and Engineers, that I set about to write this book. Computer technology has progressed at such a fast pace that it is difficult (nary impossible) to catalog all of the existing software systems which are available to us. To simplify our task I have chosen a representative software to solve a class of problem. Where space and time permitted I have provided alternatives as well.

A key benefit of using open-source applications is that the code can be compiled on a system which is non-standard. Or, it can be compiled with CPU specific optimizations which a general purpose binary released from an ISV cannot assume. As CPU technology advances rapidly, and software has a longer lifespan, the ability to recompile the source code becomes more and more important. The same can be said with open-source implementations of data-standard in image processing, and documentation retrieval. In this media focused era, more and more content is being stored as digital data. Unlike, paper, whose archival properties have been refined over centuries, digital media has not gone through the same process of archival management. In situations where data archival is necessary, a key component is the persistence of the key software components which read and write the digital data files. Since no one can predict the computers of the next century, open-source software is essential to long term archival of information.

Rather than duplicate the fine documentation for each package, this book is organized in sections related to solve a particular problem. This book should be treated as an "existential quantifier" $\exists$, rather than $\forall$, on the information provided for each task. Once the existence of a solution or software tool to address the problem is known, more details about the solution can be researched. Each software system or tool is presented in a simple to read manner describing the main problem the system addresses and the tasks performed. Each chapter is presented in a similar manner to ease referencing.

Although many of the software mentioned in this book are routinely used in science and engineering tasks, more and more I have found that students and general practitioners from other fields, such as liberal arts, music, statistics, are using these in their work and study. This book contains references to artificial intelligence programs and tools which are being widely used in cognitive sciences. Many of the software tools use libraries and development tools which are themselves open-source; this synergy is representative of the open-source concept, and a key driver to its proliferation. As such, any large open-source software is a good learning example to study the use of its components. For example the GRASS GIS software (presented in Section 14.9.1) is itself developed using a number of libraries such as (i) PROJ4, (ii) HDF5, (iii) MySQL, (iv) FFTW, and many others. To learn how to use these libraries in a real world example, one only has to study the GRASS GIS source code. This is a key advantage of open-source tools.

Another argument (made mostly in the context of mathematical proofs) stems from the scientific validity and acceptance of computer generated, or computer assisted proofs. For such proofs to be included as standard material, the software system used to arrive at the result must also be available to researchers, as well as its own correctness be verified. These goals are readily achieved by open-source mathematical software as presented in Chapter 16.

I present a short summary of the book contents:

This book is divided in six parts. The first part describes the open-source operating systems and user interaction as well as introspection tools. Chapter 1 includes a discussion on Bash shell, POSIX compliant libraries and open-source programming languages (including Erlang, Lua and Smalltalk). External utilities (such as `tar`, `find` and `rsync`) as well as OpenSSH are discussed to ease the users interaction with a modern GNU/Linux type operating system. Chapter 2 presents several text processing and document creation and management tools. These include OpenOffice and various LaTeXprocessing tools. Software for Wiki management as well as graphical page layout are also described.

Part II of this book focuses on the process of open-source software creation. For the reader who wants to know more about the methods and systems used by the authors to create the open-source tools, this part provides information on the GNU build system, version control, compilers, APIs and much more. In Chapter 3 I present the GNU Compiler Collection. Commonly used command-line options, pragmas, pre-processor defines as well as GCC intrinsics are explained. Examples of using GCC to compile Java and Ada are presented, as well as recent features of GCC including OpenMP support and C++ advice features. Source code version control with CVS and SVN is presented with the help of examples; GUI front-ends for version control (TkCVS) is presented and used in many examples. The GNU Build system is discussed with the help of examples in Section 3.3. GNU Make as well as SConstruct are described in Section 3.4. Both GNU make and SConstruct are also used in many examples in the sequel of this book.

Chapter 3 also contains a description of Bugzilla for defect tracking (Section 3.5) and a section on various editors and IDEs available on GNU/Linux for editing source code. Static code checking and analysis of source code is presented in Section 3.8, while the use of GNU debugger GDB is shown with examples in Section 3.9.1. Graphical front-ends to GDB (including GDB Insight) are shown in Section 3.9.2. Code optimization using profiling and cache measurement with GNU profiler and Valgrind is discussed with examples in Section 3.12. The C standard library and the C++ standard library (including STL) are discussed in Chapter 4.

In Chapter 5 I describe the Apache Portable Runtime (APR) library. In particular APR memory pools, APR process library, APR thread and thread pool library, APR file information and memory mapping library are explained with the help of short examples. Advanced APR concepts, dealing with the use of Memcache library are also demonstrated. All examples have been compiled and run on GNU/Linux system running Fedora Core 12, and thus are known to work. Using the examples presented in this part of the book, alongwith the documentation for the library, simplifies the learning process of the API.

Chapter 6 contains a description of the most useful parts of the Boost C++ API. For lack of space, I had to choose only a small portion of Boost for demonstration. I depict the design and usage of Boost library with the help of examples including Boost smart pointer and memory pool, the Boost asynchronous IO (asio) framework, Boost data-structures. Boost Graph Library (BGL) is an almost complete graph representation library, which also includes an implementation of many graph algorithms. BGL is presented with the help of examples in Section 6.4. Boost multi-threading (like APR) is a portable and integrated (with C++) threading system. I have presented examples which the reader can contrast with APR threading and POSIX pthread examples presented in this book. Python language integration with C++ can be achieved using SWIG as well as using Boost Python integration framework, an example of which is presented in Section 6.7. Boost generic image processing library is presented in Section 6.8, while Boost parsing framework (SPIRIT) is presented in Section 13.4.1 of Chapter 13 on Compiler systems.

Performance optimization of programs using Google perftools memory allocation and profiler is shown in Chapter 7. This chapter also includes a discussion on the Boehm garbage-collection system. A related method of memory optimization is by using compression libraries. Lossless compression engines including ZLIB and BZIP2 are discussed in Chapter 8. Recent compression implementations of LZMA and XZ Utils are also discussed.

Till this point, Part II has focused on individual application programming libraries. In Chapter 9 I discuss application development frameworks which work closely together. I first describe the remote procedure call (RPC) system which provides the user, an integrated and secure method to invoke pre-defined functions and tools on remote machines. RPC usage is usually accompanies by data transmission across networks, and endian varying machines. Data integrity is checked and maintained using a combination of tools. XDR (Section 9.1.1) presents a library for data representation which is impervious to endian variation and transport across the network. Data verification using checksum can be computed using MD5 of SHA1 checksums (Section 9.2). OpenSSL implements many of the required tools for secure and verifiable transport of data, examples demonstrating the efficacy and ease of use of OpenSSL are shown in Section 9.3. Another closely related method of data representation is XML which is investigated in Section 9.4 with the help of two APIs, libXML and Expat.

Chapter 9 also contains a description and example use of the Berkeley DB database system. Section 9.5 describes the embedded database features of BerkeleyDB with the help of examples written in C++ as well as Python.

A very interesting network caching library, Memcache is presented in Section 9.6. The use of Memcache in applications is presented, and the use of the API is described. The use of the `nc` tool to check the status of a running `memcache` server is highlighted. Many of the examples presented in this book are in Python. A very common use of Python is in applications which also have compute intensive portions written in C or C++. The Simplified Wrapper Interface Generator (SWIG) tool is described in Section 9.7, which simplifies the creation of Python modules from C/C++ headers. A complete example of this interface generation is presented in this chapter.

Scientific datasets can be enormously large in size (comprising of many terabytes of data). It is thus essential that large datasets are transported and accessed in an efficient and standardized manner (otherwise multiple tools will have to reinvent the wheel of doing performance optimized IO on large datasets). Since scientific datasets also have a lot of structure to them, efficient storage policies can be adopted. An example of such standardization is the HDF5 standard as shown in Chapter 10.

Graphics and Image Processing (not to be confused with graphical interfaces, which are described separately in Chapter 19) tools and APIs, including vector graphics rendering library Cairo is shown in Chapter 11. Image processing library APIs (libPNG and JPEGlib) are described in Section 11.2, while Scalable Vector Graphics (SVG) files are described in Section 11.2.2.

Part III contains two chapters. Chapter 12 on Parallel programming deals with multi-processing, and multi-threading. With the advent of multi-core computers, parallel programming has become essential. In this chapter I discuss the POSIX threading library (pthread). User annotated compiler supported parallelism with OpenMP is described with the help of examples in Section 12.2. Examples in Fortran and C++ are presented. A comparison of various scheduling strategies in OpenMP is performed and the results are presented in Figure 12.2. OpenMP directives and control over the visibility of variables is demonstrated with the help of small and concise examples. The new features of OpenMP version 3.0 (task computing) is presented with the help of examples in Section 12.2.0.3.

In addition to multi-threading, parallel computing has been successfully deployed on cluster grids (called the Beowulf class). The most common API used in cluster computing is MPI (message passing interface). MPI is described in Section 12.3. As before, the examples presented in this chapter were actually executed on a cluster (albeit small cluster). Examples of using Boost MPI for distributed computing is also shown.

In addition to these well established parallel programming systems, the rapid rise of many-core and other forms of parallelism has also created new systems which have had less exposure. In particular the Intel Thread Building Block library (as discussed in Section 12.4.1) is described with the help of examples.

Another exciting development in parallel programming has been the emergence of GPGPU (general purpose computing on graphics processing units) computing. Using the nVIDIA CUDA and OpenCL libraries, I present several examples of writing compute kernels in CUDA. Experiments on a GeForce GT 240 card, demonstrate the power of CUDA and GPGPU for many applications.

The second chapter in Part III is on Compilers. Although many readers will think that with the availability of compilers for many programming languages, they themselves may never have to write a compiler for a new language (or an existing one for that matter); experience has shown, that even if a complete language need not be required, small components of compiler theory and the open-source tools which implement that theory is often indispensable. Running with the theme of this book, Chapter 13 is also an $\exists$ type chapter (as there have been tomes written on the theory and practice of writing compilers), I have only illustrated the open-source tools which perform the task (with small examples where warranted). The interested reader will surely have access to more detailed documentation on these tools, which include `flex`, `yacc`. More specialized tools include `m4`, `getopt` and `gperf`. Instruction set manipulation and system utilities (part of GNU binutils) are also described in this chapter.

Chapter 13 also contains a description of Low Level Virtual Machine (LLVM), which in our opinion is a very valuable tool for writing compilers or optimization framework. LLVM is described in Section 13.6. LLVM tools including Clang and the `dragonegg` GCC plugin are also described in that section.

The next part (Part IV I discuss the central theme of the book (in a very global sense), Engineering and Mathematical software (which includes many scientific software as well). In Chapter 14 I present engineering libraries such as Computer Vision, CImg and FWTools. Geospatial data abstractions are becoming very important with the rise of location aware computing, and several open-source tools such as GDAL and PROJ4 are described in this chapter. Image processing, audio processing, and computational fluid dynamics (CFD) have been part of many engineering applications. More recently, molecular dynamics and simulation programs have also become heavy contenders for the compute time on grids. Molecular dynamics programs as well molecule viewers are shown in Section 14.7.3. Geographical Information Systems (GIS) including GRASS and QGIS are described in Section 14.9.1.

Mechanical engineering, including use of mechanical CAD software in other disciplines can be accomplished using open-source tools such as QCAD. Solid modeling tools BRL-CAD are described in Section 14.11 and Blender 14.12.

VLSI CAD Tools are described in Chapter 15. VLSI tools have a rich history of open-source tools including schematic capture, Verilog simulation and synthesis.

The Alliance CAD system is described in Section 15.7. Magic VLSI system and NgSpice simulator are also discussed with examples.

A large fraction of engineering tools are based on a fixed set of mathematical libraries. The most important libraries which implement mathematical features are described in Chapter 16. In particular the mathematical libraries of BLAS, ATLAS, LAPACK, NTL, GSL, GMP and MPFR are discussed. The FFTW system for computing FFTs is also described with the help of an image-processing example. Linear programming has become an important solution method for engineering disciplines (especially financial applications). The GNU GLPK and COIN-OR systems for linear programming are discussed in Section 16.10.

In addition to mathematical libraries, there exists a number of open-source mathematical software systems which implement complete mathematical framework. These are described in Chapter 16, and they include Maxima, GNU Octave, R, PSPP, Pari/GP, Nauty, OpenAxiom, Reduce, Singular, and polymake. Other specialized math software for algebraic geometry include Macaulay2 and CoCoA. The CGAL (computer geometry algorithms library) implements many computational geometry algorithms and data-structures.

Mathematical front-ends to the above mentioned software include TeXMacs and SAGE. SAGE, in particular combines many of the features of the previously described software with uniform notation and the ability to pass data from one tool to another. SAGE is described in Section 17.14.

A closely related area of mathematical software is artificial intelligence software which implements expert systems, automatic theorem provers, genetic algorithms, simulated annealing, machine learning and artificial neural networks. These are described in Chapter 18.

Part V discusses scientific visualization software and libraries. Chapter 19 describes the many GUI libraries on GNU/Linux systems. These include, GTK, Qt, as well as wxWidget and Fox Toolkit. High-performance 3d graphics applications are synonymous with OpenGL, and the use of OpenGL for creating rich and appealing 3d graphics is shown in Section 19.2. I present OpenGL through many examples, which also present GLUT, GLUI and show example of using OpenGL from within Python. Object-oriented Graphics rendering engines (OGRE) and OpenGL helper libraries are also discussed.

In addition to 3d graphics, graphics layout are also available using the Graphviz dot tool. Plotting software Gnuplot, and vector drawing tools Xfig and Inkscape are discussed. Raytracing with PovRay is shown with the help of examples in Section 19.9. Programmatic creation of graphics is shown with the help of gd library, and the Asymptote library.

Graphics visualization with GeomView, HippoDraw is described in Section 19.15. Multi-dimensional data visualization with GGobi is discussed. High-performance scientific data visualization with ParaView and OpenDX are discussed in Section 19.17 and Section 19.18.

Chapter 20 describes the use of open-source software for Internet and Database systems. In particular, the Apache HTTP server is described in the context of the LAMP (Linux, Apache, MySQL and PHP) stack. Virtualization and cloud computing software is described in Section 20.4. The latter part of this chapter also describes the open-source database systems such as MySQL, PostgreSQL, SQLite and CouchDB. I conclude in Chapter 21.

The most glaring omission from this book are Web based software systems, advanced database concepts, and a detailed study of the programming languages which are available as Open Source. As cloud based computing and hosted platforms become more common, I plan to address these in an addendum. Moreover, open-source software which is generally useful (such as Firefox web-browser, Thunderbird, GNUcash, or, software used in non-engineering fields, such as music composition and notation, library catalog management etc, has not been covered either. The field of open-source computing is ever-expanding, and selection of which software to include was guided by pragmatic choices, but also stayed focused on mathematical and engineering fields.

One caveat, which I should mention at this time, is that the software considered for inclusion in this book has to be Open Source. But that does not imply that it is free to use, especially in a commercial product. While the code is available to look at and learn, I would advise the reader to contact the author of the software, and read the license carefully to determine the responsibility the user has prior to including the software library, or using the software system.

As with almost all software written on GNU Linux, I gratefully acknowledge the large part GNU and Linux have played in not only the writing of this text, but also my programming experience in general. This book is written in Emacs on a Linux box, using LaTeX, xfig, gnuplot and other fine pieces of free software mentioned in this book. My heartfelt thanks to all the developers of these projects for burning the midnight oil. I would also like to thank my wife, Jyoti and my 8 year old son Advay for being patient while this book was written. Its finally done!

West Linn, Oregon,   *Sandeep Koranne*
                                                                     May 2010

Contents

List of Tables

List of Figures

Listings

Part I
Fundamentals

Chapter 1
GNU/Linux Operating System

Abstract In this introductory part we describe the basic GNU/Linux system usage and the various tools and utilities which are used with it. In particular, the use of the GNU/Linux command-line, Bash shell scripting, and external programs such as `find`, `tar` etc. are explained. Secure communication tools OpenSSH and VNC are also described. Since open-source software is written using computer programming languages we describe the commonly used open-source programming languages, including C/C++, FORTRAN, Ada, Java, Python, Tcl/Tk, Perl, Common Lisp, Scheme, Erlang, Smalltalk, Ruby, Scala, X10, and Lua in this chapter. Example for each of these languages and their salient points are described.

Contents

We focus on the fundamentals of any Open Source system for Scientists and Engineers. Since this book focuses on the usage of computer tools, the Operating System (OS) which forms the primary interaction layer with the user, is particularly important. We have written the book primarily for a POSIX[1] compliant UNIX like system; but many of the applications described in this book have been ported to other OS as well.

In Section 1.1 we discuss the GNU/Linux operating system and its relevance to open source tools for scientists and engineers.

Besides the OS, the compiler used for generating executable binaries from the source code also has a fundamental impact on the quality, performance and portability of the application. Especially for Open Source tools, where availability of the

---

[1] Portable Operating System Interface for UNIX, IEEE 1003.

S. Koranne, *Handbook of Open Source Tools*,
DOI 10.1007/978-1-4419-7719-9_1, © Springer Science+Business Media, LLC 2011

source code is a given, knowing the intricacies of the compiler can be very beneficial. In Section 3.1 we talk about GCC (GNU Compiler Collection) compilers.

GNU/Linux (basically all UNIX derived OS) share a number of common features
and facilities. This include:

1. Files: the OS has several types of files, such as regular files, directories, devices,
   symbolic links, named pipes, and sockets. Interaction with kernel, and devices
   takes place using special device files. Similarly, a *program* is a binary file re-
   siding on a file-system file which is *loaded* into memory, combined with other
   shared libraries, to form an executable object which the kernel then schedules as
   a process,
2. Multi-user: the OS has user id (uid), group id (gid), and process id (pid). These
   are used for *accounting*, where the OS keeps tracks of system usage based on
   user accounts, and group quotas,
3. Resource permissions: GNU/Linux is one of the most secure operating systems
   (when configured correctly), and modern technologies such as SE Linux have
   improved the authentication and security even more. Every resource has permissions set on its use, be it files, directories, and executables. An user can only
   perform the action with the resource if the user id has sufficient privilege level.
   This prevents misuse of the system, and guards against malicious attempts to
   gain access,
4. Multi-tasking: since its conception GNU/Linux was a multi-tasking OS, but
   multi-core computers have only recently become commonplace (so much so, that
   now in 2010 it is difficult to mind a single core CPU). By design GNU/Linux has
   support for multiple processes, and multi-threading. The scheduler in the Linux
   kernel has been enhanced to improve the response time on regular desktop workloads, while the multi-CPU SMP kernel is already used in many supercomputers
   with large number of CPUs. In addition to supporting multiple CPUs, the Linux
   kernel also has excellent support for inter-process communication using shared
   memory, named pipes (FIFOs), and sockets. All POSIX features for IPC are supported, and Linux also supports extensions which are not in POSIX,
5. System calls: being POSIX compliant, Linux kernel supports all the POSIX system calls. A system call is a facility or functionality provided by the kernel to
   the application program through a well defined interface. The use of system
   calls provides a layer of abstraction which enables a well written program to
   be ported from one POSIX OS to another with minimal change. System calls in
   GNU/Linux are provided with a C language API, and are similar in form to the
   function APIs provided by the C library. For example a program may register a
   cleanup function to be called when the application program exits. An example is
   shown in Listing 1.1.

```
/** \file example_atexit.c
    \author Sandeep Koranne (C) 2010
    \description Example of using the atexit function
*/
#include <stdio.h>           /* program IO */
#include <unistd.h>          /* POSIX */
#include <stdlib.h>          /* atexit */
```

```
     static void CleanupAfterwards(void) {
10     fprintf( stdout, "%s: ", __FUNCTION__ );
       fprintf( stdout, "Deleting application files...\n");
     }

     int main( int argc, char* argv[] ) {
15     long max_atexit = sysconf( _SC_ATEXIT_MAX );
       int rc;
       if( max_atexit == 0 ) perror(" atexit..\n" );
       rc = atexit( CleanupAfterwards );
       if( rc != 0 ) perror( "atexit failed..\n");
20     return (0);
     }
```

Listing 1.1 Using the atexit function

atexit may even be registered for dynamic libraries, to be called when they are unloaded,

6. Sysconf functionality: the system configuration of the currently running system can be introspected using the sysconf functionality. Portable programs, or application programs that attempt to retune their runtime parameters based on system configuration can inspect the state of the system. Some parameters, such as *page size* can have a large impact on the performance of the program. The most common system configuration arguments are given in Table 1.1.

Table 1.1 POSIX.1 variables for sysconf

Variable	Description
ARG_MAX	maximum length of arguments to exec functions,
CHILD_MAX	maximum number of simultaneous process per user id,
HOST_NAME_MAX	maximum length of a hostname,
LOGIN_NAME_MAX	maximum length of login name,
_SC_CLK_TCK	number of clock ticks per second,
OPEN_MAX	maximum number of open files per process,
PAGESIZE	size of page in bytes,
_SC_PHYS_PAGES	number of pages of physical memory,
_SC_AVPHYS_PAGES	number of pages of memory currently available,
_SC_NPROCESSORS_CONF	number of processors configured,
_SC_NPROCESSORS_ONLN	number of processors online.

1.1 Basic GNU/Linux Usage

Linux (Figure 1.1 shows the official Linux mascot) is the most popular open source operating system. Technically, Linux refers to the UNIX-like kernel written by Linus Torvalds, but most users don't interact with the computer's kernel, instead they rely on a host of other software which runs on top of the kernel. It is for this rea-

Fig. 1.1 Tux, the Linux mascot

son that the operating system described in this book is referred to as GNU/Linux. It includes the kernel, the C runtime library, GNU utilities and tools, shells, and graphical windowing systems.

Today GNU/Linux runs on mainframes, personal computers, and even embedded systems such as cellphones and gaming consoles. GNU/Linux is also a favorite development platform for open source systems relevant to scientists and engineers, and is the epitome of the open source movement. This book was designed and authored on a number of GNU/Linux systems, using open source tools.

The advantages of GNU/Linux can be summarized as follows:

1. Linux has a stable and efficient kernel which is optimized (and undergoes constant improvement),
2. Complete source code of the kernel and system utilities is available,
3. Its POSIX compliant and thus easy for application developers to port,
4. Since it has the UNIX philosophy, it is resilient against malicious software (to a large extent), has good security model, is multiuser, and separates the graphical windowing system from the underlying OS,
5. GNU/GCC and other high quality development tools are readily available,
6. More and more high-performance supercomputers are running GNU/Linux, and thus systems such as MPI, OpenMP, cluster computing are first available on GNU/Linux.

GNU/Linux is often available as a *distribution* which is a packaged, tested and complete operating system. The distribution can be commercial (open source does not imply free-of-cost), or it can be free (for example Fedora or Ubuntu). But all GNU/Linux systems are fundamentally the same, the differences mostly are in the application packaging, updates, and themes. Some distributions tend to focus on vertical segments such as education, electronics. There is also a distribution dedicated for scientists and engineers, but many of the software tools are available on all distributions, or can be compiled on the available distribution. The source code of the kernel can be found in `/usr/src/kernels` if the kernel development package was installed.

1.1.1 System Calls

An useful tool to find out which system calls are being issued by a program is
strace, which intercepts the system calls made by its argument program and prints
them either to standard error or to a file (with the -o option). Running strace
uptime gives the following data on a Linux 2.6.31 system:

```
execve("/usr/bin/uptime", ["uptime"], [/* 86 vars */]) = 0
brk(0)                                  = 0x816c000
mmap2(NULL, 4096, PROT_READ|PROT_WRITE
...
access("/etc/ld.so.preload", R_OK)
mprotect(0xaa7000, 8192, PROT_READ)     = 0
mprotect(0x934000, 4096, PROT_READ)     = 0
munmap(0xb787d000, 133998)              = 0
uname({sys="Linux", node="celex", ...}) = 0
open("/proc/stat", O_RDONLY|O_CLOEXEC)  = 3
read(3, "cpu  230207 235 59156 1265206 32"..., 8192) = 2856
close(3)                                = 0
time(NULL)                              = 1277426073
brk(0)                                  = 0x816c000
brk(0x818d000)                          = 0x818d000
open("/etc/localtime", O_RDONLY)        = 3
...
open("/proc/uptime", O_RDONLY)          = 3
lseek(3, 0, SEEK_SET)                   = 0
read(3, "119176.07 12652.06\n", 2047)   = 19
..
open("/proc/loadavg", O_RDONLY)         = 4
lseek(4, 0, SEEK_SET)                   = 0
read(4, "0.77 1.00 1.17 2/301 5624\n", 2047) = 26
write(1, " 17:34:33 up 1 day,  9:06,  4 us"..., 69) = 69
exit_group(0)                           = ?
```

Even without consulting the code for the uptime utility we can garner that it
produces its output by reading various system files in /proc, namely, /proc/stat
and /proc/loadavg.

1.1.1.1 Usage Limits

The functions getrlimit and setrlimit are used to get and set resource limits
respectively. These functions prevent a single rogue process on the machine from
consuming all of the available resources (such as CPU time, memory, files, etc).
Using sensible limits on the resource in your programs is advised. Consider the
small program in the listing below:

```
/////////////////////////////////
// limitexample.cpp
// (C) Sandeep Koranne, 2010
// Example showing how to put limits on resources
/////////////////////////////////
```
5

```
       #include <iostream>
       #include <sys/resource.h>
       #include <stdio.h>
       void AllocateMemory() {
10       while( true ) {
           int *n = new int[1024];
           if( !n ) return;
         }
       }
15
       int main( ) {
         struct rlimit mem;
         mem.rlim_cur = mem.rlim_max = 1024000;
         int rc = setrlimit( RLIMIT_AS, &mem );
20       if( rc ) perror("setrlimit:");
         AllocateMemory();
         return( 0 );
       }
```

Listing 1.2 Example of using resource limits

By using the `setrlimit` function we cap the maximum memory usage by this program to 10240 bytes. Without this usage restriction, the program would go into swap, and the whole system becomes unresponsive, until the operating system kills the offending program, which could be a long time as the virtual memory system is implemented using disk (which is many orders slower than physical RAM). Thus, specifying reasonable limits, even then they are computed at runtime (based on input file size), makes for a robust application software. An application program can inspect the current *environment* of the user using the `getenv` function. This provides a simple way of allowing the user to customize parameters to the program. An example is shown in Listing 1.3.

```
       /** \file example_getenv.c
          \author Sandeep Koranne, (C) 2010
          \description Example of using getenv
       */
5      #include <stdio.h>            /* for program IO */
       #include <stdlib.h>           /* for getenv */
       #include <string.h>           /* for strcmp */

       int main( int argc, char *argv[] ) {
10       enum { NO_ALGO=0, LINEAR, BINARY };
         int which_algo = NO_ALGO;
         const char* use_algo = getenv( "USE_ALGORITHM" );
         if( use_algo == NULL ) which_algo = NO_ALGO;
         else if( !strcmp( use_algo, "LINEAR" ) )
15         which_algo = LINEAR;
         else if( !strcmp( use_algo, "BINARY" ) )
           which_algo = BINARY;
         else {
           fprintf(stderr, "Unknown algorithm : %s\n", use_algo );
20         which_algo = NO_ALGO;
         }
         printf("Using algorithm %d\n", which_algo );
         return (0);
       }
```

Listing 1.3 Example of `getenv` function

Running this program gives us:

`./example_getenv`

```
Using algorithm 0
$ export USE_ALGORITHM="BINARY"
$ ./example_getenv
Using algorithm 2
$ export USE_ALGORITHM="RANDOM"
$ ./example_getenv
Unknown algorithm : RANDOM
Using algorithm 0
```

1.1.1.2 Signals

Signals in GNU/Linux (and UNIX) are event notifications. The nomenclature has roots in hardware implementations of control, and even today, actions such as keypress interrupt of Ctrl-C sends a signal to the application program. The computation itself may raise a signal from the CPU such a arithmetic divide-by-zero. Signals can be responding to an external condition, or based on timer intervals (alarm signal). The key concept to be noted for signals is that they are asynchronous, and a program using signals should be cognizant that a signal can arrive at any time, and during any computation. A common use of signals is to intercept the SIGINT signal (which is the result of Ctrl-C) to perform cleanup actions. We should note that `atexit` does not cleanup when the program terminates due to signal. An example of intercepting Ctrl-C (or the equivalent SIGINT key on the OS) is shown in Listing 1.4.

```
   /** \file example_signal.c
       \author Sandeep Koranne, (C) 2010
       \description example of using SIGNALS in GNU/Linux
   */
5  #include <stdio.h>           /* program IO */
   #include <signal.h>          /* for sigaction */
   #include <stdlib.h>          /* for exit */
   #include <unistd.h>          /* for _exit */
   #include <string.h>          /* for memset */
10
   static void SecondChance( int signal_number ) {
     int yes_or_no;
     fprintf( stderr, "Are you sure ?\n" );
     fscanf( stdin, "%d", &yes_or_no );
15   if( yes_or_no ) _exit( 0 );
   }

   int main( int argc, char *argv [] ) {
     int i;
20   struct sigaction action;
     memset( &action, 0, sizeof( sigaction ) );
     action.sa_handler = SecondChance;
     if( sigaction( SIGINT, &action, NULL ) ) {
       perror( "sigaction failed....\n" );
25     exit( 1 );
     }
     while( 1 ) ;
     return 0;
   }
```

Listing 1.4 Example of signal handling in GNU/Linux

For an user a GNU/Linux is like any other multiuser UNIX system with login shells, hierarchical file systems, text configuration files, and UNIX development tools. The shell (which is the program responsible for interaction with the user) supports job control, scripting, input redirection and other facilities. In the next section we have discussed briefly the Bash shell.

1.1.2 GNU/Linux introspection

GNU/Linux has a number of system commands which can be used to gather information about the system. These commands can be useful when discussing or reporting problems about particular software installations, or runtime problems. We discuss some of these commands below:

- uname: with the -a option prints detailed information about the system, as:

```
Linux celex 2.6.31 Wed Dec 9 11:14:59 EST 2009 i686 GNU/Linux
```

  with the -m option prints shorter machine identification which can be used in compiling machine specific code,
- /proc/cpuinfo: is a /proc file which contains detailed information about the number and type of CPU(s) in the machine. The instruction set of the CPU can be checked to see if it has SIMD or advanced instructions,

```
processor : 0
vendor_id : GenuineIntel
cpu family : 6
model : 13
model name : Intel(R) Pentium(R) M processor 1.60GHz
stepping : 6
cpu MHz : 600.000
cache size : 2048 KB
...
```

- /proc/meminfo: similarly contains information about installed memory, also see the free command.

1.1.3 GNU coreutils

For any GNU/Linux system there is a set of command known as *coreutils*, these can be thought of as fundamental user commands which perform common and useful actions. We list some of these below:

- man: format and display the manual pages. This command can be used to find more details about the commands listed below. In fact, man man gives you information about man itself,

- Commands to list output of files: includes `cat`, `od`, and `nl`,
- Formatting: `fmt`, `fold`,
- Output parts of files: `head`,`tail`, and `split`,
- Summarizing files: `wc`, `sum`, `cksum`, `md5sum`, and `sha1sum`,
- Sorting: `sort` and `uniq`,
- Operating on fields within a line: `cut`, `paste`, and `join`,
- Operating on characters: `tr`, and `expand`,
- Directory listing: `ls`, and `dir`,
- Basic operations: `cp`, `mv`, `rm`, `shred`, `dd`, and `install`,
- Changing file attributes: `chown`, `chmod`, `chgrp`, and `touch`,
- Disk usage: `df`, `du`, `stat`, `truncate`, and `sync`,
- Printing text: `echo`, `printf`, and `yes`,
- Conditions: `false`, `true`, `test`, and `expr`,
- Redirection: `tee`,
- File name manipulation: `basename`, `dirname`, and `pathchk`,
- Working context: `pwd`, `tty`, and `printenv`,
- User information: `id`, `logname`, `whoami`, `groups`, `users`, and `who`,
- System context: `date`, `arch`, `uname`, `hostname`, `hostid`, and `uptime`,
- Process control: `kill` and `sleep`.

1.2 Bash shell

The Bash shell is an UNIX shell for the GNU system and is often the default shell on GNU/Linux. Bash is actually an acronym for *Bourne-again shell*, which is a pun on the *Bourne* shell. Bash is POSIX compliant, but also has a number of extensions. We list some of the important features of Bash below:

- Brace expansion: running:

```
[skoranne@celex ~]$ echo a,{b,c,d},e
a,b,e a,c,e a,d,e
```

- Keyboard shortcuts: supports Emacs style keyboard shortcuts,
- Startup scripts: loads and executes `/etc/profile` for login, `/.bashrc` for interactive shells,
- Redirection: supports input and output redirection.

1.2.1 Bash shell scripting

The Bash shell supports scripting and programming. This feature is very useful in automating repetitive tasks and system administration. Consider a directory with JPG files, and that you want to add a prefix to all the filenames, such as "Printed".

If there are hundreds of files doing such a task by hand is tedious, and more importantly, un-necessary on a system like GNU/Linux running Bash.

```
$for i in $(ls *.jpg);do
  mv $i "Printed_$i"
done
```

Consider a real life example:

```
for i in $(ls *.png);do
  echo "Converting image file $i"
  convert $i -colorspace Gray $(basename $i .png)_bw.png
done
```

This simple program will do the required action. Using the `basename` command, we can separate a file name from its designated extension (.png) in this case. The external command `convert` is then called with the two file names, the input and the output (which has the suffix 'bw'). The conversion is a grayscale conversion. In addition to the above mentioned `for-do-done` loop, there are also `if-then-else` statements.

A more detailed example of using Bash scripting to perform an experiment is shown in Listing 1.5.

```
#!/bin/bash
# Sandeep Koranne, (C) 2010
#
# Shell script to sweep configuration parameters for OpenMP
5  # we run the program multiple times with different options
# and varying the THREAD count

# Scalability plot for program on 2-CPU
CMD=./collatz_omp
10 for num_threads in $(seq 1 2); do
export OMP_NUM_THREADS=$num_threads
NUM_TH=$(echo $OMP_NUM_THREADS)
for algo in $(seq 0 5);do
echo "Running algorithm $algo"
15 datFileName="$algo"_"$NUM_TH".dat
echo "Data goes in $datFileName"
for i in $(seq 19 21);do
echo "Running N = " $i
/usr/bin/time -f "%e %U" -o $datFileName -a $CMD $i $algo
20 done
done
done
```

Listing 1.5 Example of using Bash scripting

1.3 External commands and programs in GNU/Linux

In addition to the GNU coreutils commands, there are a number of external commands which are readily available on GNU/Linux and perform many useful tasks.

The regular expression feature is common to many of these commands, and it is instructive to list the regular expression syntax.

1.3.1 GNU regular expression syntax

A regular expression is built up from blocks that match single characters, or patterns. Single characters are patterns which match themselves, any meta-character can be matched by preceding it with a backslash. The '.' character (period) matches any single character. Characters listed inside *brackets* form a list, from which any single character can be matched. For example:

```
[school]
[^home]
```

the first pattern list matches any character in 's', 'c', 'h', 'o', and 'l'; the second list has a '^' symbol as the first character, implying that characters *not* in the list can be matched.

Character ranges can be specified by a hyphen. A + after the list bracket implies one or more repetition of that class can be matched, while implies a zero or more match. For example:

```
[0-9]+ matches any non-empty number
[a-z]* matches any word, even empty
[0-9]{9} matches 9 numbers
```

Now we discuss some useful commands below:

1. wget: utility to download files from the Internet in a non-interactive manner. It supports HTTP, HTTPS as well as FTP. wget can work in the background, as well as resume downloading partially downloaded files. It can also be used for recursive downloads of sites. Another related program is curl, which is used to transfer a single URL (see an example of using curl in Section 20.8),

2. find: GNU find searches the directory tree routed at the given path, example:

```
find /home/skoranne -name '*.cpp'
find /home/skoranne -name '*.tar' | xargs ls -l
```

find can search for files based on their type, date of modification, etc. Actions can also be specified for files which match the pattern,

3. grep: program to find regular expression based pattern in files, or standard input. Some of the important options for grep are:

   - -v: negate the matching,
   - -m: specify number of matches per file,
   - -A: print number of lines after context,
   - -B: print number of lines before context,
   - -w: match complete words only,

- `-i`: case independent matching,
- `-n`: print line number with match,
- `-H`: print file name for each match.

4. `sed`: Sed is a stream editor. A stream editor is used to perform basic text transformations on an input stream, for example:

```
cat A.txt | sed -e 's/alpha/beta/g/' > B.txt
```

5. `awk`: GNU awk is an implementation of the *awk* language. The most common use of `awk` is to parse an input file (line by line) into fields. Consider the following example:

```
$cat A.txt
ABC   WV    65
DEF   SJ    87
GHI   MN    18
$cat A.txt | awk '{print $1}'
ABC
DEF
GHI
$cat A.txt | awk '{ printf "City " $2  " # " $3 "\n"}'
City WV # 65
City SJ # 87
City MN # 18
```

There are a number of other features in `awk`, and a complete list of commandline options and features can be found by running `info awk`,

6. `rsync`: remote file copying and synchronization tool. This tool can be used to keep two computer's contents in sync, as is the case when mirroring a website (for example), or transfering documents from one workstation to another,

7. `ssh`: is an OpenSSH client and remote login program. SSH has rapidly become the remote login program of choice since it is much more secure and flexible than `telnet`, and `rlogin`. It provides secure and encrypted communication capability between two untrusted hosts over an insecure network. It can also be used to *tunnel* other protocols such as VPN, X11 and VNC (see Section 1.7.1). `ssh` can be configured to use *authentication keys* in lieu of passwords. A related program is *secure copy*, `scp`. A more detailed description of OpenSSH is given in Section 1.5,

8. `nc`: nc (short form of netcat) is a TCP/UDP analysis tool. It can open TCP/UDP connections, send and listen on ports and deal with IPv4 as well as IPv6,

9. `screen`: GNU screen is a full-screen window manager that multiplexes a physical terminal between several processes. When `screen` is first called it creates a single window with a shell. To switch between terminals a key combination can be used, such as `Ctrl-a` n. A new window can be created using "C-a c". Using `screen -x` a remote second user can attach to a non-detached screen session (which is a way to share a session),

10. `xargs`: the `xargs` programs reads input separated by blanks, and executes the specified command for each input token. `xargs` is used in combination

with `find` or other file name listers to perform system administration and disk cleanup tasks. For example, to check the time stamp for all files which match the regular expression `m[a-e]k` we can use the following *oneliner*[2]

```
$ ls | grep -i m[a-e]k | xargs ls -l
   2915 2010-06-27 01:26 make.aux
  19103 2010-07-31 22:54 make.tex
  19099 2010-07-26 00:27 make.tex~
```

In the above example we chained the output of `ls` to `grep` where we used regular expression matching. As we know, `grep` will output only those patterns which match the expression, and these are passed to `xargs`; the command to be executed by `xargs` is `ls -l`, which prints the required time stamp information. `xargs` can also take the input from a file (in addition to the standard input),

11. `tar`: GNU `tar` (acronym for Tape Archive) is used to collect directories and files into a single file for archival on tape, or backup. It supports a number of command-line options, the most common ones are:

```
-c, --create               create a new archive
-r, --append               append files to the end of an archive
-t, --list                 list the contents of an archive
-x, --extract, --get       extract files from an archive
-p, --preserve-permissions, --same-permissions
-f, --file=ARCHIVE         use archive file or device ARCHIVE
-C, --directory=DIR        change to directory DIR
-v, --verbose              verbosely list files processed
```

Thus a new archive to contain all files in directory XYZ can be done as `tar cvf xyz.tar XYZ/`. Moreover, the archive can be compressed using the following compression schemes. The compression schemes included in GNU `tar` are:

a. gzip: using the `z` option,
b. bzip2: using the `j` option,
c. lzma: using the `J` option,

12. `diff`: the `diff` command can be used to find differences between files. Especially for source code, `diff` is used for version control management and change control.

1.4 Next steps

As familiarity with GNU/Linux grows, the user can write useful shell scripts, or *one liners*, to perform administration work. The software systems presented in the remainder of this book assumes that the user is able to operate the operating system to perform the input, output, and file system tasks. Before closing this chapter let me

---

[2] A oneliner in UNIX terminology is a combination of UNIX utilities and shell constructs, which is created to solve a particular problem, and is often comprised of only one line.

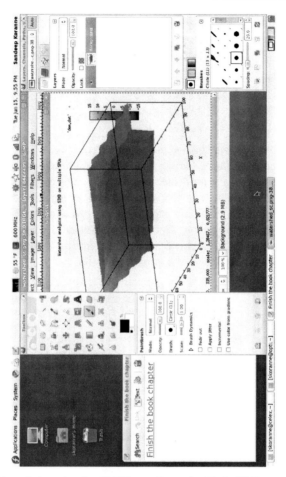

Fig. 1.2 GNU/Linux desktop showing GNU Image Processing Tool (GIMP)

present the graphical user interface of GNU/Linux as this may be the more common interaction paradigm for most users today.

Consider the GNU/Linux screenshot of a computer running Fedora 12 as shown in Figure 1.2. The graphical system is GNOME, and the system menus have a number of applications for tasks such as system administration, office applications, Internet and networking, and software development. These are shown in Figure 1.3. I am not listing programs such as Firefox, Thunderbird since these are in popular use.

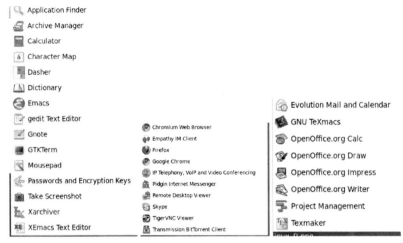

(a) Applications, Internet and Office

(b) Software development and System

Fig. 1.3 GNU/Linux with GNOME Menus

1.5 OpenSSH: OpenBSD Secure Shell

OpenSSH (OpenBSD Secure Shell) is an open source implementation of the SSH connectivity tools. As compared to Telnet, Rlogin, and FTP (which transmit their password and all user data in clear text on the network), SSH encrypts all traffic. This eliminates the risk of snooping and hijacking, as well as other attacks. OpenSSH also provides tunneling capabilities over which other protocols such as X11, and VNC can be used with the same security advantages.

OpenSSH is comprised of a number of tools:

1. sshd: the SSH services daemon (can be compared to 'telnetd' daemon),
2. ssh: rlogin and telnet like login client,
3. ssh-keygen: key generator tool,
4. ssh-config: daemon configuration file,
5. scp: secure copy program (like rcp).

1.5.0.1 The ssh client

The ssh client is the primary login and remote execution client for SSH. The usual manner of using ssh is:

```
$ssh [options] remote-server
```

When logging into a remote system, ssh supports a number of command line arguments; the most common and useful ones are listed below:

- -C: requests compression of data using zlib compression,
- -c: selects cipher for encryption, supported ciphers include *blowfish*, des,
- -F :specify alternative configuration file,
- -f: request ssh to go background just before command execution,
- -i: specify identity file,
- -L: specifies port forwarding,
- -l: specify remote login name,
- -o: specify option,
- -p: specify port on remote server,
- -v: verbose mode (useful for debugging a connection problem),
- -X: enables X11 forwarding,
- -Y: enables trusted X11 forwarding.

ssh exits with the exit status of the remote command or 255 if an error occurred.

Prior to a successful connection, sshd verifies that the client machine is a trusted party (by checking the name of the host in /etc/hosts.equiv), and that the user names are same on both machines. Moreover the server must be able to verify the client's host key. The file /.ssh/authorized_keys on the remote server lists the public keys that are permitted to login. The ssh program on the client informs the server of the key pair it is going to be using for the session, and then the sshd server checks whether the corresponding public key is authorized.

The user can create a key-pair by running `ssh-keygen`. If using OpenSSH protocol 2 with DSA the generated keys are stored in the file `/.ssh/id_dsa`, when using OpenSSH protocol 2 with RSA the generated keys are stored in the file `/.ssh/id_rsa`. The corresponding public keys (which must be transmitted to the server and placed in `/.ssh/authorized_keys`) are generated in `/.ssh/id_dsa.pub` (for DSA) and `/.ssh/id_rsa.pub` (for RSA). Once the public keys are placed on the server, the `ssh` client (and any program which uses `ssh`) can login to the remote server without giving the password. Although, if the public keys do not exist, then `ssh` falls back to password based login.

The `ssh` client also maintains a database of identifications of all hosts it has connected to; if the host key changes for some reason (or more commonly if the host IP address changes), `ssh` warns about this change.

External tools such as CVS (see Section 3.2.2) and Subversion SVN (see Section 3.2.3) also use SSH to perform remote version control through remote repositories.

1.6 Programming Languages

Since this book is about open-source software, and all software is written in a computer programming language, it behooves us to describe some of the languages and their implementations which are available as open-source systems themselves. In this section we discuss some of the popular programming languages available on GNU/Linux system. Particular programming syntax for each language is not presented, however the particular strength of the language and its application domain is mentioned in case the reader wants to investigate the use of a particular language for a specific task.

1.6.1 C and C++

C and C++ are used as system programming, as well as application programming languages. Many of the libraries presented in the sequel of this book (see Chapter 4) are written in C/C++. C is especially popular for system programming, writing system kernels and tools. The function calling conventions of C language are sufficiently common so as to become the de-facto standard for new languages. Consequently, a number of existing libraries in C can be easily used from other language using *foreign function interface*.

A number of compilers for C and C++ are available. These include the GNU GCC Compiler suite. GCC is discussed in depth in Section 3.1. LLVM (low level virtual machine) (discussed in Chapter 13, section 13.6) also implements a C and C++ compiler `clang`. Open64 group also has an open-source compiler for C and C++. In recent years, GCC has made promising strides in optimization, better instruction generation on modern processors, automatic parallelization, compiler optimizations and error checking, threading support, and support for OpenMP. The upcoming C++ standard is also supported by GCC's C++ compiler and standard library.

1.6.2 GNU FORTRAN

On GNU systems, there exist a number of FORTRAN compilers and source translators, including g77, f2c and GNU FORTRAN (part of GCC). GNU FORTRAN has support for FORTRAN 95 and uses the same back end as GCC C/C++ compilers thus benefiting from portability and integration with libraries written in C and C++. GNU FORTRAN can be run using the gfortran command and accepts many of the same general command line arguments as GCC (see Section 3.1 for a detailed list of GCC's options). GNU FORTRAN accepts the OpenMP directives as specified in OpenMP version 3.0. See Section 12.2 for a detailed explanation of OpenMP support and programming. An example of a short FORTRAN program to print the number of OpenMP threads is shown in Listing 1.6.

```
  ! \file fomp.f
  ! \author Sandeep Koranne, (C) 2010
  ! \description Example of using OpenMP in GNU Fortran
        PROGRAM FOMP
5       IMPLICIT NONE
        INTEGER OMP_GET_MAX_THREADS
        INTEGER OMP_GET_THREAD_NUM
        write( 6, "(a,i3)") " MAX THREADS : ", OMP_GET_MAX_THREADS()
        write( 6, "(a,i3)") " Thread num  : ", OMP_GET_THREAD_NUM()
10      END PROGRAM
```

Listing 1.6 GNU FORTRAN example with OpenMP

We can compile this program as:

```
$gfortran fomp.f -o fomp -lgomp
$./fomp
 MAX THREADS :   16
 Thread num  :    0
```

1.6.3 Ada

The GNU GCC compiler collection also has a compiler for the ADA programming language. Consider the simple hello, world style program in Ada.

```
-- this is hello world in Ada
with Text_IO; use Text_IO;
procedure HelloWorld is
begin
```

```
   Put_Line( "Hello from Ada!" );
end HelloWorld;
```

We create a file `helloworld.adb` with these content, then we can compile this Ada program using the GNU gnat system.

```
$gnatmake helloworld.adb
gcc -c helloworld.adb
gnatbind -x helloworld.ali
gnatlink helloworld.ali
$./helloworld
Hello from Ada!
```

Later (in Section 15.6) when we discuss the VHDL hardware description language, its similarities to ADA will be apparent.

1.6.4 Java

Although many Java software development kits are available on GNU/Linux, we discuss the GNU GCC compiler suite's `gcj` Java compiler. GNU `gcj` is an *ahead-of-time* compiler for the Java language. Using the `-C` option `gcj` can generate the Java bytecode, otherwise it will generate native instructions (for the architecture GCC was compiled for). Consider the small Java program in Listing 1.7.

```
   // \file hw.java
   // \author Sandeep Koranne, (C) 2010
   // \description Hello world in Java compiled with GNU gcj

5  class hw {
       public static void main( String args[] ) {
       System.out.println(" Hello, World!\n" );
       }
   }
```

Listing 1.7 Compiling Java program with GNU `gcj`

The Java language has restrictions on the names of classes present in files, for example the file 'hw.java' has **class** hw which has the `main` function. Compiling this file using the GNU `gcj` compiler we get:

```
$gcj --main=hw hw.java
$ldd a.out
linux-gate.so.1 =>  (0x00f80000)
libgcc_s.so.1 => /lib/libgcc_s.so.1 (0x00603000)
libgcj.so.10 => /usr/lib/libgcj.so.10 (0x0381f000)
```

Compiling the file using the `--main` command line argument specifies the class which contains the `main` function. This creates an executable file which is linked with the `libgcj` runtime library. We can also compile the file to Java bytecode using the `-C` argument as:

```
$gcj -C hw.java
$ls -l hw.class
                        413 2010-07-16 20:09 hw.class
$java hw
Hello, World!
```

The bytecode is executed through the Java Virtual Machine.

1.6.5 Python

The Python programming language implemented on most GNU/Linux is the CPython reference implementation. The version used for many of the examples in this book is Python version 2.6.2. Many of software libraries discussed in this book have been integrated with Python. And even complete tools such as SConstruct (see Section 3.4.2) have been constructed around Python. Due to its interactive nature Python has also been used to develop glue software integrating diverse tools into a complete package. In fact the SAGE mathematical software system has been written in Python (see Subsection 17.14), and is developed as a collection of several independent math software tool, all connected using Python.

Although Python is a general purpose programming language, some features of Python are especially suited for high-performance scientific computing and solving engineering problems.

1.6.5.1 NumPY

NumPy is a Python package developed for scientific computing with Python. It contains:

1. N-dimensional array: *n*-dimensional array object,
2. sophisticated broadcasting functions
3. tools for integrating C/C++ and FORTRAN code
4. linear algebra, Fourier transform: integration with external library codes for linear algebra, FFT and random number generation.

As an example of NumPy consider:

```
    Python 2.6.2 (r262:71600, Aug 21 2009, 12:22:21)
    [GCC 4.4.1 20090818 (Red Hat 4.4.1-6)] on linux2
    Type "help", "copyright", "credits" or "license" for more information.
    >>> from numpy import *
 5  >>> a = array( [ 1,2,3,4])
    >>> a
    array([1, 2, 3, 4])
    >>> b = array( [5,6,7,8] )
    >>> a+2*b
10  array([11, 14, 17, 20])
```

In the above listing, the use of NumPy arrays as well as functions operating on array elements can be observed.

NumPy is used effectively with SciPy (a package for scientific computing with Python). These packages can be imported in Python as:

```
>>> import numpy as np
>>> import scipy as sp
>>> import matplotlib as mpl
>>> import matplotlib.pyplot as plt
```

The `matplotlib` is a general purpose plotting library for Python. SciPy contains many mathematical functions including optimization, linear interpolation, special functions and more. Consider an example shown below:

```
  import numpy as np
  from scipy import interpolate
  import matplotlib.pyplot as plt
  x = np.arange(0,20)
5 y = np.exp(-x/3.0)
  f = interpolate.interpolate1d(x,y)
  xnew = np.arange(0,20,0.1)
  plt.plot(x,y,'o',xnew,f(xnew),'-')
  plt.title(``Example of interpolation'')
10 plt.show()
```

The resulting plot is shown in Figure 1.4.

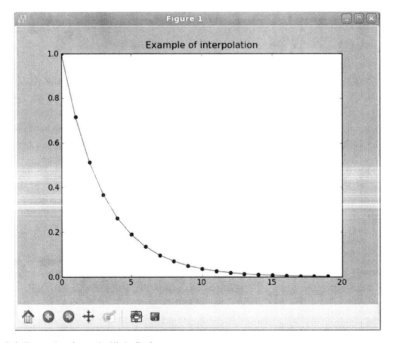

Fig. 1.4 Example of matplotlib in Python

Matplotlib has many other features which are shown in Section 19.1.5.5, where we discuss building graphical user interfaces (GUI) with Python and Qt.

1.6.5.2 PyCUDA

Python has well designed facilities for multi-processing and multi-threading. An example of calling external processes through Python is shown in Section 19.1.5.5. In this section we describe a more recent development in the parallel programming arena, GPGPU (general computing on graphics processing units). A leading software library for GPGPU computing is nVidia's CUDA software. PyCUDA is a Python module which makes the integration with CUDA simple. Consider the following example:

```
   import numpy as np
   import pycuda.driver as cuda
   cuda.init()
   assert cuda.Device.count() >= 1
5  dev = cuda.Device( 0 )
   contxt = dev.make_context()
   func = mod.get_function(``myf'')
   a_on_cpu = np.arange(0,10)
   a_on_gpu = cuda.mem_alloc( a_on_cpu.nbytes )
10 cuda.memcpy_htod( a_on_gpu, a_on_cpu )
   func( a_on_gpu, block=(1,1,1) )
   cuda_memcpy_dtoh( a_on_cpu, a_on_gpu )
```

1.6.5.3 PuLP

SciPy has a number of general-purpose optimization routines including: (i) Nelder-Mead Simplex algorithm, (ii) Quasi-Newton method, and (iii) Minimize sum-of-squares. But for general linear problems (LP) formulated using variables and constraints, or combinatoric problems modeled as integer linear programs (ILP) the Python LP modeler (PuLP) can be used. Once the model has been defined in PuLP a number of linear solvers can be called to solve the problem. See Subsection 16.10 for more details on PuLP, however a simple example is presented below:

```
   x = LpVariable(``x'',0,10)
   y = LpVariable(``y'',5,20)
   problem = LpProblem(``example'', LpMinimize )
   prob += x + 3*y <= 20
5  prob += 2*x+y
   status = prob.solve( GLPK( msg = 0 ) )
   LpStatus[status]
```

The above listing creates a LP problem with two variables x and y which has one constraint, $x + 3y \leq 20$, while the objective function is $2 * x + y$ and the goal is to *minimize* the LP. The solver chosen in GLPK (see Section 16.9 for more details on GNU GLPK and other linear solvers).

1.6.6 Tcl/Tk

Tcl (Tool control language) and Tk GUI Kit are interactive programming languages which were designed for embedding into applications. Tcl, however, is a general purpose programming language which is used for rapid prototyping and has been used to develop complete applications (see TkCVS in Section 3.2.5). Tk is the GUI toolkit which is used to design and deploy GUIs written in Tcl and other languages (such as Python using the Tkinter library). Tcl is designed to be extensible by C, C++ and other Tcl modules. All expressions in Tcl are written as *commands* which are variadic and written in prefix notation. Example:

```
% set sum [expr 1+2]
3
% puts "Hello"
Hello
pack [button .b -text "Hello"]
```

Tk provides a number of *widgets*, including: (i) buttons, (ii) message-box, (iii) tool-bars, and (iv) drawing canvas. Tcl/Tk are commonly used in providing scripting capabilities in engineering applications; however, Python is also being used in more modern applications for the same purpose.

1.6.7 Perl

Perl is a general purpose, high-level computer programming language. It was originally designed for efficient and easy report generation and processing. Its design was heavily influenced by C, shell scripting, AWK, and `sed`. Perl's handling of regular expressions and capabilities to process large amounts of textual data, combined with its rapid use in initial CGI scripting for the Internet caused its exponential rise. Perl is often used to automate system administration tasks which are too complex for shell scripting, while not compute intensive enough to warrant a full C/C++ application. Moreover, Perl has integrated support for hash-tables, regular-expressions, string support, and file-handling on most operating systems. With Perl version 5.0, the language was optimized with a new interpreter, lexical variables, and modules. Extension modules to Perl (often written in Perl, but possibly in C) are available for many tasks such as XML processing. Modules are available on CPAN (Comprehensive Perl Archive Network). It is easy to recognize Perl code since it uses symbols to distinguish major data types. Fundamental data types included in Perl are: (i) scalars ($), (ii) array (), and (iii) associative array (%). Other data-types are: (iv) globs, (v) file-handles, and (vi) sub-routines. Implementation of Perl on GNU/Linux are version 5.10.0. A short example of Perl code using an associative array is shown below:

```
#!/usr/bin/perl -w
my %days = ("mon" => 1, "tue" => 2, "wed" => 3 );
```

```
foreach my $day (keys %days) {
    print "$day has $days{$day}\n";
}
```

1.6.8 Common Lisp

Common Lisp is general purpose programming language which is best associated
with the development of large and complex applications dealing with symbolic
logic and artificial intelligence. Symbolic math software (such as Maxima, see Section 17.1) are usually written in Common Lisp. Lisp is an acronym of List Processing, although the Common Lisp language has first class support for many data-
structures including arrays, hash-tables, strings, and streams. Common Lisp was
one of the first standardized languages to have a complete object-oriented (OO)
methodology as part of the language. The Common Lisp Object System (CLOS) is
an advanced OO framework which supports inheritance and message dispatch. Lisp
(like Python) has an interpreter (called the REPL (read-eval print loop)), but Common Lisp implementations have sophisticated compilers which often produce object
code rivaling C and FORTRAN in their number crunching speeds. Lisp also introduced the concept of *garbage collected* storage, and dynamic type. Open-source
implementations of Common Lisp include GNU Common Lisp (clisp), CMU Lisp,
and SBCL. Consider a Common Lisp function as shown in Listing 1.8.

```
   (defun compute-function ( a b )
     (if (= 0 (logxor a b)) -1 1))
   (defun ck (A n k)
     (let ((ans 0)(temp_j 0)
5         (n_temp_k 0)(ub (- n k )))
       (dotimes (j ub)
         (setf temp_j (aref A j))
         (setf n_temp_k (logxor (aref A (+ j k)) 1))
         (incf ans (compute-function temp_j n_temp_k)))
10     ans))
   (defun cs (A n)
     (let ((sum 0))
       (dotimes (i (- n 1))
         (let ((val (ck A n (1+ i))))
15         (incf sum (* val val))
           (format t "~%c_~D = ~D" (1+ i) val)))
       sum))
   (cs #(1 0  1 0 1) 5)
   (quit)
20 (defun merit-factor (A)
     (let ((n (length A)))
       (* 1.0 (/ (* n n) (* 2 (cs A n))))))
   (defun log2 (n)
     (dotimes (j 64)
25     (when (> (ash 1 j) n) (return-from log2 j))))
   (defun eb (n pos)
     (let ((SHIFT (1- (log2 n))))
       (ash (logand
       n (ash 1 (- SHIFT pos)))
30     (- pos SHIFT)))))
```

Listing 1.8 Common Lisp functions

1.6.9 Scheme

A closely related language to Common Lisp is the Scheme language which shares many of the functional programming roots with Lisp. GNU's Scheme implementation is called Guile as it is the official scripting language for application extension of the GNU project. The idea behind using interactive scripting languages for application extension comes from separating the compute intensive portion of the application and writing it in a language such as C or C++, while the bulk of the application can be written (and customized, even by the end-user) using a language such as Scheme. Towards this end, GNU Guile has a simple integration method where Guile can be linked with an application and used as the scripting interface. The wrapper generator SWIG can be used to automate this process; more details on using SWIG for wrapper interface generation are given in Section 9.7. Python, Tcl and Lua are also commonly used for application scripting interface. An example of Scheme code is:

```
(define sum (lambda (a b ) (+ a b)))
guile> (sum 1 2)
3
```

Scheme, although superficially similar to Lisp, had (prior to Common Lisp, atleast) fundamental differences surrounding the rules for variable scoping (lexical scoping for Lisp was first introduced in Scheme), namespaces for functions and implementation of iterative loops. Even today, Scheme is generally regarded as an elegant and more research oriented language than Common Lisp, which arguably has many more features.

1.6.10 Erlang

On GNU/Linux, the Erlang implementation can be run using the `erl` command-line which starts the Erlang shell:

```
Erlang R13B04 (erts-5.7.5) [source] [rq:1]
                           [async-threads:0]
                           [kernel-poll:false]

Eshell V5.7.5 (abort with ^G)
1> % this is a comment
1> 3+3
1> .
```

6

The advantages of using Erlang are manifest when programming on multi-processor machines, or when writing applications for fault-tolerance, as Erlang has *single assignment variables*.

```
X = 1234.
1234
3> X = 5678.
** exception error: no match of right hand side value 5678
```

In the above example the variable *X* was assigned the value 1234; then later when we try to assign another value 5678 to *X*, Erlang raises an exception. Since a variable can be set only once in Erlang, the debugging of programs as well as analyzing the program for execution on a parallel machine is simplified (as Erlang has no mutable state in variables, it does not need shared memory locking).

Erlang has enumerated types called *atoms* defined by the use of lowercase variable names, and a fixed number of items can be collected into a single entity using *tuples*. However, unlike C language `struct`, tuples in Erlang are anonymous, so it is common to use an atom as the first element in a tuple to denote its intended use in the program:

```
Point = {point, {x,10}, {y,20}}.
{point,{x,10},{y,20}}
8> Point.
{point,{x,10},{y,20}}
```

Erlang uses *pattern matching* operator to satisfy the equality operator:

```
{point,CX, CY} = Point.
{point,{x,10},{y,20}}
13> CX.
{x,10}
14> CY.
{y,20}
{another, DX, DY } = Point.
** exception error: no match of right hand
    side value {point,{x,10},{y,20}}
```

By checking the pattern `{point,X,Y}` against the `Point` tuple, we assign the variable *CX,CY* to the corresponding elements in the tuple, as the first element in both LHS and RHS is `point`. In the second example, when we try to match a pattern with `another` as the first element, we get an exception.

Erlang has *lists* which are very similar to Common Lisp lists:

```
AL = [1,2,3,4,5].
[1,2,3,4,5]
17> [CAR_AL| CDR_AL ] = AL.
[1,2,3,4,5]
18> CAR_AL.
```

```
1
19> CDR_AL.
[2,3,4,5]
```

The [X|Y] notation denotes the *head* and *tail* of the list respectively, and like all Erlang pattern matchings, can be used to extract the head and tail of the list respectively.

Functions can be defined in Erlang as shown below. The listing is placed in a file called 'number_theory.erl':

```
%% \file number_theory.erl
%% \author Sandeep Koranne, (C) 2010
%% \description Examples in Erlang, for number theory

-module( number_theory ).
-export( [calc_collatz/1] ).
-export( [fibonacci/1] ).

%% Actual functions
calc_collatz( {collatz,1 } ) -> 1; % simple case
calc_collatz( {collatz,X } ) when ( X rem 2 ) == 0
      -> calc_collatz( {collatz,X div 2} );
calc_collatz({collatz,X}) -> calc_collatz({collatz,3*X+1}).

fibonacci(0) -> 0;
fibonacci(1) -> 1;
fibonacci(X) -> fibonacci( X-1 ) + fibonacci( X-2 ).
```

The *module* declares the number_theory module. It *exports* the two functions calc_collatz and fibonacci, both accepting a single input (the *arity* of the function is part of the function declaration in Erlang). We can bring this module into a running Erlang shell by:

```
c(number_theory).
{ok,number_theory}
>number_theory:fibonacci(6).
8
```

Similar to Perl, Erlang also has a Comprehensive Erlang Archive Network (CEAN) which attempts to gather the major applications for Erlang in one common place. Erlang is commonly used in designing fault-tolerant communication software. An example of fault-tolerant software is CouchDB (see Section 20.8 for information).

1.6.11 Smalltalk

Smalltalk-80 is a general purpose programming language which introduced many concepts in programming language design such as *object oriented design, messages,* and the *model-view-controller* method of GUI design. Smalltalk-80 is also noted for its use of a workspace environment for interacting with the user, as compared to the syntactical text in, object code out view of programming languages. Smalltalk-80 is implemented on GNU/Linux system using the Squeak software system. A picture of an actual Smalltalk-80 session in GNU/Linux is shown in Figure 1.5.

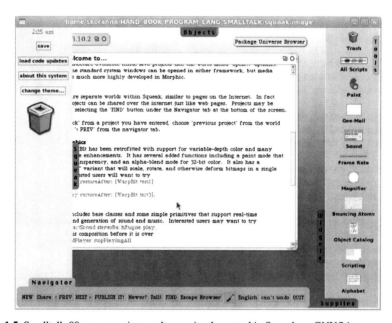

Fig. 1.5 Smalltalk-80 programming workspace implemented in Squeak on GNU/Linux

1.6.12 Scala

The design of the Scala programming language started as a research project to develop better language support for component software, where components can be libraries, modules, classes, frameworks, processes, or even web services. As such, Scala is by design an amalgamation of object-oriented and functional language design, with the over riding goal of *scalability*, that it should be possible to describe large as well as small systems using the same concepts. Scala has been implemented on the Java Virtual Machine on GNU/Linux, and Scala programs resemble Java programs in many ways. However, Scala introduces the concept of singleton class objects using the `Object` keyword. Scala also does not distinguish between statements and expressions, and every function returns a value.

Scala allows arguments to be passed by name, which can be used to implement *short-circuiting* as below. Consider the following user-defined class in Scala:

```
abstract class Boolean {
  def && (x: => Boolean): Boolean;
  def || (x: => Boolean): Boolean;
}
```

The use of => in passing the formal argument allows the parameter to be passed without evaluation into the function.

Functions in Scala are values (they can be passed as first-class objects) as well as Objects. The map method applies a given function to every object in a sequence. Scala's library defines many types of sequences including: (i) arrays, (ii) lists, (iii) streams, and (iv) iterators. Like Python, Scala too has *list comprehensions* which can succinctly represent code which operates on elements of a sequence:

```
def sqrts(xs: List[double]): List[double] =
  for( val x <- xs; 0 <= x ) yield Math.sqrt(x)
```

Here, the constraint $0 \le x$ acts as a filter to remove negative values from the *xs* list of doubles. The values which pass through this filter are used as a generator and passed to the variable *x* which is then *yielded* to form the return sequence (which is also of type List[double]).

Scala has been used to create interactive and high traffic sustaining websites which require scalability to 100 million or more transactions per hour. In addition to the facilities described in this short section, Scala has other features such as pattern recognition, compositions and XML processing which are also very powerful.

1.6.13 Google's GO Programming Language

Go is an expressive, concurrent, garbage-collected systems programming language developed by Google. Consider the following canonical example:

```
// this is a simple Go Program
package main
func main() {
   print("Hello, World! from Go\n")
}
```

We can compile this program using the 8g Go compiler on an i386 machine running GNU/Linux as:

```
$ echo $GOOS
linux
$ echo $GOARCH
386
$ ../bin/8g -h
gc: usage: 8g [flags] file.go...
```

```
$ ../bin/8g -V
8g version 5917
$ ../bin/8g helloworld.go
# this produces the file helloworld.8 file
# which can be linked using the 8l linker
$ ../bin/8l -h
usage: 8l [-options] [-E entry] [-H head] [-L dir]
          [-T text] [-R rnd] [-r path] [-o out] main.8
$ ../bin/8l helloworld.8
# which produces the file called 8.out
$ ./8.out
Hello, World! from Go
```

The Go programming language is already being used for system programming tasks.

1.6.14 X10 Language

As parallel programming on a large scale (clusters of millions of computers) becomes a reality, the existing programming paradigms of shared-memory, or message-passing have shown the limits of their scalability. Towards this end, a new language, the X10 Programming Language, has been designed with the concept of partitioned global address space (PGAS). X10 is a modern object-oriented programming language in the PGAS family with the goal of enabling scalable and high-performance programming for next-generation computer systems.

In addition to supporting PGAS semantics, X10 introduces asynchronous actions and distinguishes between the two views of synchronous execution (defined as executing within a *place*) and asynchronous execution. Each *place* runs lightweight *activities* which are synchronous and atomic in the memory space of the place they execute on. Furthermore, data is classified as immutable (in which case no consistency management is necessary, and it can be freely copied) or data consistency management may require one or more *clocks* to order the execution. The array datatype supports parallel collective operation on data.

X10 is object oriented, and like C++ programmers write X10 code by defining containers for data and behavior called *interfaces*, e.g.:

```
interface Diameter {
   def compute_diameter() : Double;
}
class SimpleGraph implements Diameter {
   var x : Double = 0;
   public def isConnected() { }
   public def compute_diameter() { }
}
struct DirectedGraph implements Diameter {
   public def compute_diameter() {}
}
```

In the above listing, both `SimpleGraph` as well as `DirectedGraph` implement the `Diameter` interface, in that they both implement the `compute_diameter` method which returns a `Double`. When the programmer states that `SimpleGraph` implements `Diameter`, it must implement all the method that the interface demands.

In X10 there is a distinction between **class** and **struct**, **struct** are headerless values and cannot have mutable fields, like the **class** `SimpleGraph` has field *x*. The advantage of **struct** is the methods can be inlined; moreover, **struct** are immutable. Although, X10 has no primitive class, the X10 standard library defines classes for: (i) Boolean, (ii) Byte, (iii) Short, (iv) Integer, (v) Long, (vi) Float, (vii) Double, (viii) Complex, and (ix) String. X10 is also a functional language, with functions being first-class objects (functions can be stored and passed to other functions as values). The GNU/Linux implementation of X10 (which is still considered a research language) is available and implements version 2.0 of the X10 language specification.

1.6.15 Lua

Lua is a general purpose programming language designed to be used as a scripting language for application extension. It is a lightweight language, and is used in extending game development engines and providing interactive control to users of the application. In many respects, Lua is similar to Scheme (see Section 1.6.9) as both languages are used for application extension and scripting. Lua is a dynamically typed language and supports a small number of fundamental data-types, including: (i) Boolean, (ii) numbers, and (iii) strings. The native data structure in Lua is the *table* from which other data-structures such as list, set, arrays, and records can be devised. The Lua table is basically an associative array. Like Scheme, Lua also has support for *first class functions*, garbage collection, closures, proper tail recursion, and co-routines. The use of Lua tables is essential to programming in Lua, consider the following program fragment:

```
person={name="Jack", age=15,
        marks={physics=80, math=75, english=70}}
print(person)
table: 0x84bc578
print(person.marks.physics)
80
```

Tables in Lua are always passed by reference, so if a table is assigned and modified, both tables are modified. Tables can be used to implement record structures as shown in the above example.

Lua is used by video game developers as a scripting language due to its small embedded footprint, fast execution speed, and relatively small learning curve. As an application extension language Lua competes with Scheme and Python.

1.7 Miscellaneous Topics

In addition to the programming languages discussed above, software implementation can be done using database processing languages such as SQL (structured query language) (see Chapter 20 for a discussion on open-source database systems).

1.7.1 VNC : Virtual Network Computing

In Section 1.5 we saw the use of OpenSSH for tunneling X11 connections securely over SSH protocol. VNC provides additional facility for remote X11 connections. It allows for tunneling using SSH through a trusted intermediary node (often a designated VPN gateway), allows for sharing of the display (allowing collaborative work), and persistent display workspace which can be connected from any machine.

VNC comprises of two tools, the VNC server `vncserver` and the VNC client viewer `vncviewer`. The server has to be launched on the remote machine and is setup to ask for a connection password. The display size, and color depth can be specified on the `vncserver` command-line. The display then starts off with a display number 1 larger than the maximum X11 display. To connect to a VNC server we can use the `vncviewer` application. The viewer application also has a number of command-line options such as: (i) full-screen mode, (ii) JPEG quality level, (iii) gateway tunnel server, (iv) shared display, (v) viewonly, and (vi) zlib compression level. Another remote desktop sharing application on GNU/Linux is `rdesktop` (remote desktop).

1.8 Conclusion

In this chapter we have provided an overview of the GNU/Linux operating system and its shell interface. Commonly used external tools were categorized, as well as most of the open-source languages in which tools are written. In particular we discussed GNU FORTRAN, GNU Ada, Java, Python, Tcl/Tk, Perl, Common Lisp, Scheme, Erlang, Smalltalk, Scala, X10, and Lua. PHP and Ruby are discussed in Chapter 20 (on web technologies). We have also presented an overview of remote computing with VNC.

Chapter 2
Text processing

Abstract Text processing and document creation are an important part of the open-source world. Most open-source tools come with documentation which is written using SGML or TeXinfo (both described in this chapter). Complete tools such as OpenOffice, devote significant portion to word processing. In this chapter we describe open-source tools for text document processing. We discuss OpenOffice, TeXand LaTeX, SGML, and TeXinfo. Page layout tool, Scribus, and document classification/citation tools are described.

Contents

Even with all the advent in graphical user interfaces, electronic document processing and production remains an important part of any product development project. There are many open source document preparation and markup systems; we discuss some of them in this chapter.

2.1 OpenOffice.org Suite

The OpenOffice.org system is an open-source office software suite which includes (i) word processing, (ii) spreadsheets, (iii) presentation, (iv) graphics and diagrams, and (v) database. It is available in many languages and on many platforms. It can read-write data files from many different software packages. Example of OpenOffice.org tools is shown in Figure 2.1.

S. Koranne, *Handbook of Open Source Tools*,
DOI 10.1007/978-1-4419-7719-9_2, © Springer Science+Business Media, LLC 2011

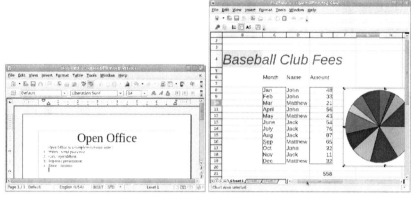

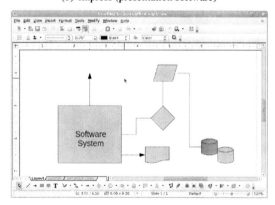

(a) Writer (word processor) and Calc (spreadsheet)

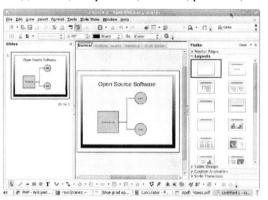

(b) Impress (presentation software)

(c) OpenOffice Draw

Fig. 2.1 OpenOffice.org System

2.2 TeX and LaTeX

Consider the example of an article written in LaTeX:

```
%% Simple document in LaTeX
\documentclass{article}[12pt]
\usepackage{a4wide}
\usepackage{amsmath}
\usepackage{times}
\title{Introduction to Writing with \LaTeX}
\author{Sandeep Koranne}
\begin{document}
\maketitle
\begin{abstract}
This short document demonstrates the use of \LaTeX in writing
professional quality typeset articles. In
Section~\ref{section:math} we give an example
of using \TeX for typesetting mathematics, which was the
reason for its invention. We conclude in
Section~\ref{section:conclusion}.
\end{abstract}

\section{Introduction}

\section{Typesetting Mathematics}
\label{section:math}
Consider the following quadratic equation in $x$:
\begin{equation}
ax^2 + bx + c = 0 \label{eqn:quadratic}
\end{equation}
Equation~\ref{eqn:quadratic} has exactly two roots
$x = \frac{-b \pm \sqrt{ b^2 - 4ac}}{2a}$. The same
solution can be written as:
\begin{eqnarray}
x_1 & = & \frac{-b + \sqrt{ b^2 - 4ac}}{2a} \\
x_2 & = & \frac{-b - \sqrt{ b^2 - 4ac}}{2a}
\end{eqnarray}

\section{Conclusion}
\label{section:conclusion}
The use of references within the article is made
easier by using labels and references are automatically
generated.
\end{document}
```

We process this article with LaTeX to generate a device independent file (DVI) as shown in Figure 2.2.

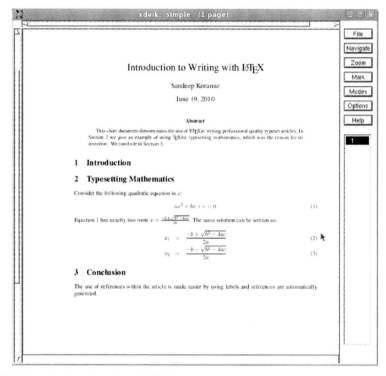

Fig. 2.2 Example of article written in LaTeX

2.2.1 *Lout Typesetting System*

Lout is a typesetting software similar in concept to TeXbut implemented quite differently. It has a markup language, and can generate PostScript from Lout documents.

2.2.2 *SGML Processing*

Standard Generalized Markup Language. Most GNU manuals are written using markup language as opposed to WYSIWYG word processors. Markup language is identified by the following:

- structure as opposed to formatting: the markup should describe the structure of the material as opposed to any specific formatting to be applied to it at any given time,
- automatic processing: the structure imposed on the source document should be sufficient to allow automated processing of data using software tools. We discuss such an automated document generator for source code in Section 3.10.

As a documentation system SGML was designed to enable the sharing of machine readable large corpus of documents from various agencies of the government, law, and business applications. SGML is mostly associated with textual data, and XML and HTML are closely related to SGML.

2.2.3 Texinfo : GNU Documentation System

Texinfo is the official documentation format of the GNU project. It is based on the concept of markup language, and from a single source document many different formats, including 'dvi', 'html', 'info', and 'pdf' can be automatically generated. An example of a '.texi' file from GCC is shown below:

```
@node Compatibility
@chapter Binary Compatibility
@cindex binary compatibility
@cindex ABI
@cindex application binary interface

Binary compatibility encompasses several related concepts:

@table @dfn
@item application binary interface (ABI)
The set of runtime conventions followed by all of
the tools that deal with binary representations of a program
```

Texinfo files can be converted to output formats using the `makeinfo` program. The output format can be specified using the following options:

```
--docbook   : docbook in XML
--html      : output in HTML
--xml       : Texinfo XML
--plaintext : plain text
```

The related tool `troff` generated `man` pages from descriptions formatted in a similar manner.

2.2.4 LyX Frontend

LyX is a word-processor front end to the TEXand LATEXtypesetting systems. Modeled after the WYSIWYG (what you see is what you get) principle, authors who are not familiar with TEXcan use the GUI and commands of LyX to produce high quality manuscripts. An example of LyX in action is shown in Figure 2.3.

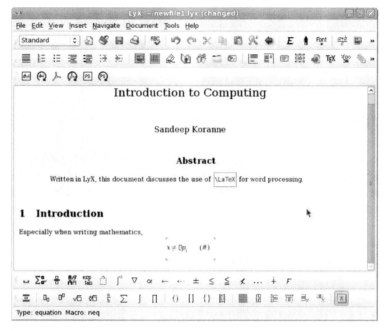

Fig. 2.3 LyX word processing software using LaTeX

2.2.5 *Texmaker LaTeX Editor*

Another frontend to LaTeXis TexMaker; but unlike LyX TexMaker is not a WYSI-WYG word processor. The author using TexMaker has to be more familiar with LaTeX, although like a good GUI, TexMaker does provide menus and toolbars to automate and simplify many tasks. An example of TexMaker editing a chapter of this book is shown in Figure 2.4.

Fig. 2.4 TexMaker LaTeXfrontend

2.2.6 PostScript and PDF Support

The output of LATEXis a DVI file, but more and more, PDF (portable document format) and PostScript output is also expected from the tools. There are a number of open-source programs which can generate and process PostScript and PDF files. Document viewers such as (i) ghostscript, (ii) evince, and (iii) xpdf are available. Ghostscript in particular has PostScript processing tools which can manipulate PostScript data for efficient printing, double-sided printing, and even format conversion from PostScript to PDF.

In a similar vein, the text to PostScript converter program a2ps can be used to convert TEXT files (including source code listing) to neatly formatted PostScript data.

2.3 Scribus

Unlike Lyx and TexMaker, Scribus is a *desktop publication* (DTP) software. Such software is often used when designing and printing brochures, pamphlets, menus, and other promotional material. An example of Scribus in designing a library information brochure is shown in Figure 2.5.

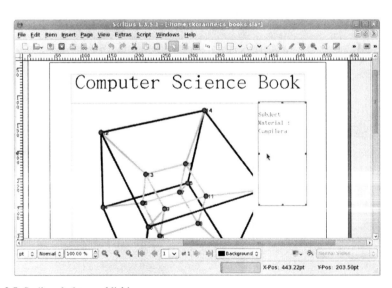

Fig. 2.5 Scribus desktop publishing program

2.3.1 Citation management

If using TEXor LATEX, the appropriate text citation tool is BibTex, which accepts an input '.bib' file containing citations of the form:

```
@ARTICLE{karger,
    author = {David R. Karger and Clifford Stein},
    title = {A new approach to the minimum cut problem},
    journal = {Journal of the ACM},
    year = {1996},  volume = {43}, pages = {601--640}
}
```

This article can be cited in a LATEXsource as:

```
In his paper Karger~\cite{karger} presented a
```

which produces output such as "paper Karger [18] says", where the number 18 is automatically calculated based on the citation order. The software also produces the list of used references for inclusion in the main text.

2.4 Document classification software

In addition to text document creation tools, engineers and scientists also need document classification, citation, and search/index tools. We discuss document file searching tools on GNU/Linux systems. We have already seen the GNU `find` tool for searching for file names in a hierarchical path. Since `find` actually traverses the file system its performance can be slow for large file systems.

2.4.1 GNU `locate`

GNU tools `locate` and its security enhanced version `slocate` provide an effective command line file searching tool for GNU/Linux systems. The command line for `locate` is:

```
$locate [options] file-pattern
```

GNU `locate` uses database prepared by `updatedb`, and thus is not bottlenecked by the file system access speed or size, and is significantly faster than using `find` when searching for files. By default `locate` does not check whether the file reported still exists in the system or not; moreover, `locate` reports on files which were present when the last database update was performed. Like many other UNIX tools, `locate` exits with status 0 if a match was found, 1 otherwise. The options accepted by `locate` are:

1. `-b`: match basename only (instead of wholename),

2. `-c`: print count of matching files (instead of names),
3. `-e`: report only files which still exist,
4. `-L`: follow symbolic links,
5. `-l <n>`: limit search to the first n matches only,
6. `-S`: print statistics about the database,
7. `-r`: use regex (regular expression) pattern matching.

The default database is located in `/var/lib/mlocate/mlocate.db`, running `locate -S` returns:

```
Database /var/lib/mlocate/mlocate.db:
65,147 directories
927,777 files
59,636,635 bytes in file names
20,767,916 bytes used to store database
```

A security enhanced version of `locate` also exists, and is called `slocate`. It only allows searches for files which have appropriate permission for the user who is performing the search.

2.5 Wiki

A Wiki is described as a collaborative software which is implemented as a website. A wiki allows multiple users to edit any page or to create new pages which can be used to maintain collaborative information between groups of users. An example of Wiki is Wikipedia, which is an encyclopedia maintained by a group of users. Wiki updates are written with markup languages using web browser. A starting wiki web page simply contains an Edit as well as a Save button. New pages in the wiki set can be created by the users and added as a link to existing pages. The singular point of the Wiki is that it allows any user to edit any page of the wiki website (of course, this feature can be selectively switched off). This allows for a Wiki to grow as more and more users add to the collective knowledge. Many wikis are completely open to the general public, while others may have certain restriction to allow rudimentary quality control on the content. Another very recent open-source addition is the Eureka Streams open social networking project.

2.6 Conclusion

Information dissemination is an integral part of scientific research and information science also benefits from the generation of high-quality documents. The software presented in this chapter can help the scientist present the research in a standardized and attractive manner. Above, we discussed text processing and document creation tools including OpenOffice, Wiki, Scribus, SGML, LaTeXand its various front-ends (including LyX and Texmaker). Document management and searching tools, along-with citation tools were presented with examples.

Part II
Software Engineering and Libraries

Chapter 3
Software Engineering

Abstract In this chapter we discuss software construction tools including the venerable GNU Compiler Collection (GCC) compilers, source code configuration systems (CVS, SVN and git), as well as the GNU Build Tools. Automatic build management tools GNU `gmake` and SCons are described with examples. The Bugzilla defect tracking system is described; incidentally, this book had a Bugzilla page during its development for issue tracking. Source code editors and IDEs (including Emacs, Kdevelop and Eclipse) are shown as well as debugging, documentation and profiling tools.

Contents

3.1 GCC : GNU Compiler Collection

GCC stands for the GNU Compiler Collection (and not the GNU C Compiler), as today it supports many languages in addition to C. In this Section we describe GCC (as of version 4.5.0), and its command-line options, extensions, and optimizations. The key features of GCC in the recent years have been support of modern architectures with super-scalar instruction issue, parallelism (auto-SIMD as well as support for OpenMP) and more complete support of C++ features, link-time optimization, and with version 4.5.0, support of *plugins*. GCC is the official compiler of the GNU project and is thus used to build the GNU tools, as well as the Linux kernel. It can generate cross-platform executables, and is instrumental in getting a new chip, or machine architecture off the ground. Once GCC is self-hosting on the new architecture, many other GNU tools are easy to port to the new architecture.

Currently (as of version 4.5.0), GCC supports the following programming languages:

1. C/C++/Objective-C
2. GNU FORTRAN
3. Java
4. Ada

For the purposes of this section we are not focusing on the internals of GCC, although the interested reader is strongly advised to read the 'GCC Internals' document as well as the source code, as well as refer to Chapter 13 for more information on compiler construction tools and techniques. Instead, we present the common options for GCC command-line, extensions and new features which are of interest in building high-performance, robust applications. It also supports pre-compiled headers, automatic link time instantiation of C++ templates, both features come in handy when compiling large C++ programs.

3.1.1 GCC Command-line Options

The commonly used options for GCC (which are used in almost all non-trivial compilations) are:

- `-v`: print version number of GCC,
- `-I /directory/`: location of additional paths to search for include files present in the program,
- `-L /library/path`: location of additional paths to search for library locations,
- `-l libname`: additionally link to this library also,
- `-o filename`: write output to this filename,
- `-c`: compile only, do not perform linking,
- `-D foo`: define a pre-processor macro on the command-line,
- `-static`: link against static libraries,
- `-fPIC -shared`: link against shared libraries,
- `-ggdb`: include debugging symbols in the output,
- `-On`: perform optimization (using `-O0` switches off optimization),
- `-ansi`: support all ANSI standards,
- `-pedantic -Wall` :turn on warnings,
- `-mtune`: specify machine architecture for optimizations,
- `-fsyntax-only` :quick check of syntax only,

- `-combine`: pass multiple files to compiler at same time,
- `-E`: pre-process only,
- `-S`: produce assemble output,
- `-frepo`: C++ automatic template instantiation at link time,
- `-pipe`: use pipes instead of temporary disk files,
- `-fprofile-arcs`: generate profile data for PGO (profile guided optimization),
- `-fbranch-probabilities`: use PGO runtime data to optimize code,
- `-fomit-frame-pointer`: do not generate stack frames,
- `-pg`: instrument code for profiling,
- `--coverage`: instrument code for code coverage,
- `-fopenmp`: enable handling of OpenMP directives,
- `-fstats`: generate statistics about compilation,
- `-Q`: prints each function name during compilation,
- `-fverbos-asm`: add more text to assembly output,
- `-fWeffc++`: warn about C++ stylistic problems,
- `-fplugin`: load the specified plugin.

GCC also has a number of machine target specific command-line options:

- `-m128bit-long-long`: size of `long long` is 16,
- `-m32`: generate 32-bit code,
- `-m64`: generate 64-bit code,
- `-malign-loops`: align loop code,
- `-mtune`: tune code for a specific architecture.

GCC can also be used in *defensive programming*, by using its options for warning on suspect code. Some options are:

- `-Waddress`: warn about suspicious use of memory addresses,
- `-Wbuiltin-macro-redefined`:
- `-Wwrite-strings`:

Run `gcc --help=warn` for a complete list of command-line options for warning and error generation.

3.1.2 GCC Preprocessor

The preprocessor reads in source code files and performs textual transformations including text replacement, inclusion, conditional expansion and inclusion, before passing the source to the compiler proper. The GNU GCC preprocessor is `cpp` and operates on *directives* placed in the source file. We enumerate the important directives in Table 3.1.

Table 3.1 Directives in GNU GCC for the preprocessor

Directive	Description
`#include`	inclusion of file into current
`#define`	define pre-processor symbol or macro
`#undef`	undefine the macro or symbol
`#ifdef`	conditional compilation
`#else`	based on defined macro
`#endif`	-do-
`#if`	conditional based on value of variable
`#pragma`	additional hints to the compiler
`#warning`	generate warning message
`#error`	generate error message
`#line`	replace current line number
`##`	text concatenation operator

3.1.2.1 GCC Pragmas

The following GCC pragma are useful:

```
#pragma GCC poison malloc
// produces message whenever malloc is used
#pragma GCC dependency "a.inc"
// adds a.inc to dependency list of current file
#pragma GCC system_header
// treats file as a GCC system header
```

3.1.2.2 Predefined Macros

A list of the predefined macros in GCC is given below in Table 3.2.

3.1.3 GCC Support of OpenMP

The OpenMP support in GCC is part of the GOMP project. It is an implementation of OpenMP for GCC supported languages, i.e., C, C++ and GNU FORTRAN. GCC has supported OpenMP since version 4.2. OpenMP code is added to otherwise standard C, C++, FORTRAN code using *pragmas*, or directives to the compiler. The compiler then uses these hints, and the code to produce parallel programs under the semantics of the OpenMP specification. GCC OpenMP uses a runtime library `libgomp` to facilitate the runtime parallelism requirements of the compiled program. The pragmas are written in the form:

```
#pragma omp parallel for
```

Table 3.2 Directives in GNU GCC for the preprocessor

Predefined Macro	Description
__BASE_FILE__	quoted string (full file name)
__cplusplus	defined when compiling C++
__DATE__	current compilation date
__TIME__	current compilation time
__FILE__	current translation file
__LINE__	current line number
__FUNCTION__	name of function
__PRETTY_FUNCTION__	for C++ more decorated name
__GNUC__	major version number of GCC
__GNUC_MINOR__	minor version of GCC
__VERSION__	complete version number
__OPTIMIZE__	defined when optimization is on
__STDC__	compiler is conformant to C++

```
for( i = 0; i < N; ++i )
```

OpenMP is a shared memory model, and thus the user is responsible for variable sharing, and avoiding concurrent access situations, called *race conditions*. Further details about parallel programming with OpenMP and GNU `libgomp` are given in Section 12.2.

3.1.4 GCC Advice Mode

Starting with version 4.5.0 GCC has an *advice* mode which gives performance improvement advice for suboptimal usage of C++ data structures and algorithms from the C++ standard library. It is an non-intrusive solution, the application code does not need to be changed, and the advice is call context sensitive. However, to take benefit of this advice mode the application has to be recompiled and executed on representative workloads (akin to profile guided optimization, and branch probability analysis). Consider the following program fragment from a discrete optimization application:

```
     std::ifstream ifs( file_name );
     std::vector< LONG_WORD* > GraphCollection;
     while( ifs ) {
       LONG_WORD* graph = ParseGraph( ifs, dimension );
5      if( graph == NULL ) break;
       GraphCollection.push_back( graph );
       AnalyzeGraph( graph, dimension );
     }
     while( graph_display_mode == 1 ) { // simple printing
10     int which_one = 0;
       std::cerr << std::endl << "Which graph to print ?";
       std::cin >> which_one;
```

```
       if( which_one < 0 ) break;
       LONG_WORD* graph = GraphCollection[ which_one ];
15     ConstructIntoGraph( graph, dimension );
       PrintAdjacencyGraph( which_one, dimension );
     }
```

Listing 3.1 GCC advice

We compile the full program as follows:

```
g++ -ggdb -D_GLIBCXX_PROFILE -O0 f.cpp -o f_dbg
$./f_dbg <representative input>
cat libstdcxx-profile.txtvector-size: improvement = 2:
 call stack = 0x804abfa 0x94ebb6 0x80499a1 :
 advice = change initial container size from 0 to 108
vector-to-list: improvement = -2:
 call stack = 0x804abfa 0x94ebb6 0x80499a1 :
 advice = change std::vector to std::list
```

Using the `addr2line` program (see Section 13.5.1.1 for more details on using this tool) we can convert the call stack addresses to program line numbers, these correspond to the definition of `GraphCollection` as a `std::vector< LONG_WORD*>`. The representative input we ran on this program to calculate the advice did not include the code in the second **if** clause which uses the data-structure as a random-access container. Thus, GCC's profile based advice informs us that (i) an initial size of 108 for the vector is recommended and (ii) replacing the vector (we only use `push_back`) with a `std::list` (this will actually degrade performance as indicated by the negative score of -2, but can save memory).

3.1.5 GCC Attributes

GCC allows for *attributes* (additional information) to be added to functions and data variables. The attributes are defined using __attribute__ keyword. There are a number of pre-defined attributes in GCC and these are listed below:

- __noreturn__: specifies that the function does not return to the caller, e.g.,

  ```
  void fatal_error() __attribute__ ((noreturn));
  ```

- __align__: specifies alignment of data variables, e.g.,

  ```
  struct GraphNode {
      int    value;
      void* data __attribute__ ((align(4)));
  };
  ```

- alias: defines a function to be a weak alias of another function,
- always_inline: if a function has been declared inline, then with this attribute, it will be expanded inline even without any optimization,
- noinline: will never be expanded inline,

- pure: a function with this attribute has no side effects whatsoever (no changes of global variables, but can read global variables),
- const: a *pure* function with the additional constraint that it does not read global variables also (function output is dependent on input parameters only),
- constructor: a function which is called before main(),
- destructor: a function called after main() exits,
- format: a function which has a single va_arg style argument,
- malloc: a function which can be treated as malloc,
- deprecated: a warning will be issued for every *call*, or place of reference of this variable or function,
- packed: a variable with this attribute has the smallest possible alignment (improves memory footprint of structures).

3.1.6 GCC : Inline Assembly

Inline assembly refers to the ability of the GCC compiler to copy and transform assembly code from the source C/C++ file and insert direct assembly code into the generated assembler output. This facility is useful when optimizing loops, calling CPU instructions which have no equivalent in the language, and calling vector CPU instructions directly. The major differences between writing assembly code versus writing inline assembly are the following:

1. Operand size is determined by the op-code name,
2. Register naming: register names in inline assembly are prefixed by % symbol,
3. variables references: to C/C++ code variables can be made in the inline code.

It should be noted that GCC uses the AT&T assembler syntax as opposed to the Intel x86 syntax. The source-destination ordering in At&T is source, destination. Thus

```
mov %ebx, %eax ; eax <- ebx
movb %al, %bl  ; byte move
movw %ax, %bx  ; word move
movl %eax, %ebx; longword move
movl $0xAABB, %ecx; immediate operand
movb (%eax), %bl; bl <- byte pointed to by eax
```

To write inline assembly, the following syntax has to be observed:

```
asm ( assembler template
    : output operands
    : input operands
    : list of clobbered registers
    );
```

example:

```
#include <stdio.h>
int func(void) {
  int a=1,b=2,c;
  __asm__ ("movl %1, %%eax\n\t"
   "movl %2, %%ebx\n\t"
   "add %%eax,%%ebx\n\t"
   "movl %%ebx,%0\n\t"
   : "=r"(c)
   : "r"(a),"r"(b)
   : "%eax","%ebx"
   );
  return c;
}

int func2(void) {
  int a=1,b=2,c;
  __asm__ ("add %1,%2\n\t"
   "movl %2,%0\n\t"
   : "=r"(c)
   : "r"(a),"r"(b)
   : /* none */
   );
  return c;
}

int main( ) {
  printf("%d %d", func(), func2() );
  return 0;
}
```

The relevant part of the assembler output is shown below:

```
# func
# 4 "inl.c" 1
movl %edx, %eax
movl %ecx, %ebx
add %eax,%ebx
movl %ebx,%edx
# func2
# 17 "inl.c" 1
add %eax,%edx
movl %edx,%eax
```

The critical difference between func and func2 is that, in func2 we did not specify to GCC which registers are to be used for variables a and b. In register strapped architectures such as x86, long instruction sequences increase the *register pressure*. Unless a specific register is required by the instruction (such as CLD), it is better to let GCC choose a register. In this case (when GCC automatically chooses a register), that register need not be added to the list of clobbered registers.

3.1.6.1 cpuid instruction

Above we had mentioned the use of inline assembly in calling CPU instructions which do not have a C/C++ language equivalent. One such instruction is the `cpuid` instruction which returns information about the current CPU. An example of using inline assembly with this instruction is shown:

```
/* CPUID instruction,
   calling with 0 returns the maximum call value

*/
#include <stdio.h>
int get_max_cpuid_call(void) {
  int retval = 0;
  __asm__ ("xor %%eax, %%eax\n\t"
   "cpuid\n\t"
   "movl %%eax,%0\n\t"
   : "=r"(retval)
   : /* no inputs */
   : "%eax"
   );
  return retval;
}

int print_ids(void) {
  int id;
  int ax;
  for( ax=0; ax < get_max_cpuid_call(); ++ax ) {
    __asm__ ( "movl %1, %%eax\n\t"
      "cpuid\n\t"
      "movl %%eax,%0\n\t"
      : "=r"(id)
      : "r"(ax)
      : "%eax" );
    printf("ID[%d] = %d\n", ax, id );
  }
}
#define cpuid(func,ax,bx,cx,dx)    \
  __asm__ __volatile__ ("cpuid": \
 "=a" (ax), "=b" (bx), "=c" (cx), "=d" (dx) : "a" (func));


unsigned int get_cpu_information(void) {
  unsigned int info = 0x0;
  int param = 1;
  __asm__ ( "movl $1, %%eax\n\t"
    "cpuid\n\t"
    "movl %%eax,%0\n\t"
```

```
    :  "=r" (info)
    :  "r" (param)
    :  "%eax" );
  return info;
}
```

```
   void print_cpu_information(unsigned int param) {
     printf("\n CPU Information = %d", param);
     printf("\n Stepping number = %d", (param & 7) );
     printf("\n Model number = %d", (param & 0x00F0) >> 4 );
5    printf("\n Family number = %d",(param & 0x0F00) >> 8 );
     printf("\n Processor type= %d",(param & 0xA000) >> 12);
   }

   int main() {
10   printf("Max CPUID call = %d\n", get_max_cpuid_call() );
     print_ids();
     print_cpu_information( get_cpu_information() );
     printf("\n");
     return 0;
15 }
```

The output of the previous program is shown below:

```
Max CPUID call = 2
ID[0] = 2
ID[1] = 1750

 CPU Information = 1750
 Stepping number = 6
 Model number = 13
 Family number = 6
 Processor type= 0
```

One inherent problem with inline assembly is that, not only is the code now specific to GCC (or atleast to a compiler which understands GCC style inline assembly), but more importantly, the code is now specific to an instruction set. Thus, it is a good idea to encapsulate inline assembly code in functions and for equivalent C/C++ code (if it exists) to be maintained at the same time as assembly. The results of the C/C++ code should be compared against the inline assembly code as well. In the next section we discuss GCC intrinsics which in some cases remove the necessity of using inline assembly, while maintaining some portability.

3.1.7 GCC Intrinsics

GNU GCC contains a number of *extensions* to the C/C++ language, not the least of which are the *intrinsic* functions. Intrinsics are optimized versions of functions which have been tuned for a particular architecture or instruction set. The extensions supported by GCC include:

- typeof: returns the type-of (akin to sizeof) an expression,

- long long: double-word integer,
- __int128: 128-bit integers,
- Local labels: labels which are local to a block,
- Statement expressions: are permitted with this extension,
- Nested functions: are permitted with this extension,
- Intrinsic functions: discussed below,
- Empty structures: are allowed,
- Thread local variables: using the __thread storage class,
- Return address: getting the return address from within the function,
- Offset of: calculates the offset of a particular data member within a structure.

3.1.7.1 X86 builtin functions: intrinsics

The major categories of builtins are:

1. Mathematical functions: such as sin, cos,
2. Compilation hints: include functions __builtin_types_compatible_p, __builtin_expect, __builtin_constant_p, __builtin_prefetch,
3. Atomic functions: these functions implement *atomic* operation functions. They include, fetch-and-add, compare-and-swap, synchronize, and lock-and-test. These functions are full *barrier* functions, thus no memory operand will be moved across the operation. Atomic builtins can be used to implement *spin-locks* and other synchronization facilities,
4. Vector arithmetic: SIMD on small size vectors, e.g., `fabsq`, `padd`, and `psub`,
5. Memory load/store, shuffle instructions:
6. CRC32: checksum functions,
7. AES: encryption functions,
8. Bit counting: population count for number of bits set.

3.1.8 Compiling Java using GCC

We can use the `gcj` GNU Java compiler to compile Java programs. Consider the pedantic and canonical Java program:

```
/* helloworld.java */
public class helloworld {
    public static void main( String arg[] ) {
        System.out.println( "Hello!\n" );
    }
}
```

We can compile this program as:

```
gcj --main=helloworld helloworld.java -o H
```

One key distinction between using GNU gcj and other Java compilers such as javac, is that GNU gcj produces native compiled code binaries, which do not need to be run on the Java Virtual Machine (JVM). This has advantages of performance, but at the same time the compile once, run everywhere facility of Java is discarded.

3.1.9 Compiling Ada using GCC

Consider the simple hello, world style program in Ada.

```
-- this is hello world in Ada
with Text_IO; use Text_IO;
procedure HelloWorld is
begin
   Put_Line( "Hello from Ada!" );
end HelloWorld;
```

We create a file helloworld.adb with these content, then we can compile this Ada program using the GNU gnat system.

```
$gnatmake helloworld.adb
gcc -c helloworld.adb
gnatbind -x helloworld.ali
gnatlink helloworld.ali
$./helloworld
Hello from Ada!
```

3.1.10 Conclusion

Most of the open-source software developed on GNU/Linux is compiled using some version of GCC, and thus it is an essential part of the open-source software stack. Using GCC effectively is important, and in this section we have described the important facets of using GCC and its related tools. In particular, GCC extensions, pragmas, support for OpenMP, attributes, inline assembly, and builtin intrinsics were discussed. Using GCC to compile languages other than C/C++ was also briefly discussed.

3.2 Source Code Configuration Systems

3.2.1 Introduction to Version Control Systems

Software configuration management (SCM) and version control system (VCS) are used to manage changes to program text, documents and other information stored in computer files. It is used during the process of writing and developing computer software, where a number of people working on the same software product are developing and working on the same set of files. Using SCM and VCS software, the programmers aim to collaborate on the development, while maintaining the ability to work on private copies of the shared source code files. Changes are usually referred to as a version number and are a combination of numeric version numbers and letters.

As the team continues to develop software it is common to introduce new features, and fix bugs existing in previous versions of the software. Since the life of the deployed software is many years, the software development team must be able to go back to a *snapshot* of the released version to reproduce a customer bug, or fix a problem. Moreover, during development, errors can creep in, and developers can use the version control system to do a quick analysis of the change they have made to the code files from the last known good version.

Another common feature of the version control system is the concept of *branches*, where the files at some point in time are *forked* into differing versions, presumably for experimental development, or feature freeze prior to release. We shall present the example of the software systems to support branching below. Once the feature under development has been completed, it is possible to *merge* the changes back to the main development line, which is often called the *trunk*, to take the analogy of the software development tree further.

Moder software development systems and integrated development environments (IDEs) have built in features to access version control systems through the user interface. In this chapter we discuss the common version control systems including CVS, SVN, and git. We discuss the command line utilities to access the version control repository, and also discuss some graphical tools.

3.2.2 CVS

As mentioned above CVS (concurrent version system) is a version control system. The main activities for working with a CVS *repository* can be divided into two roles (i) the repository administrator, and (ii) CVS user. CVS allows the following features:

To begin working with CVS, first the location of the CVS *repository* has to be informed to the CVS tools. A common method is to define the CVSROOT environment variable with the location of the repository. If the repository is a local file it can be specified as:

```
export CVSROOT=/usr/local/cvsroot
```

or when working with multiple repositories it can be specified on the command-line for every CVS tool as:

```
cvs -d /usr/local/cvsroot diff file.c
```

Once a local copy of the repository has been checked out, CVS maintains the current CVSROOT information in the file CVS/Root. It is possible to connect to a remote repository using the RSH protocol.

- cvs init: this command creates a repository in the named location with access controls. The choice of the machine and file system to deploy CVS depends on the expected load on the repository server, concurrent usage, and work load.
- cvs import: given an existing collection of files and directories, these can be added to the CVS repository using the import command. An example is:

```
cvs import -m "Import" POLY POLY init
```

this commands add all the files and directories in the current working directory as a child directory and files in the POLY directory of the repository. The commit log is given by the -m argument, and the init tag is specified.

- cvs checkout: copies a copy of the module or file specified on the command line. Internally, CVS also maintains hidden directories which contain the repository information for each directory. Example:

```
$cvs co POLY
cvs checkout: Updating POLY
U POLY/cano_proc.cpp
U POLY/dual.cpp
U POLY/graph_reader.cpp
```

the directory structure within POLY/CVS is shown below:

```
drwxrwxr-x. 2  4096 2010-06-13 17:52 .
drwxrwxr-x. 3  4096 2010-06-13 17:52 ..
-rw-rw-r--. 1   150 2010-06-13 17:52 Entries
-rw-rw-r--. 1     5 2010-06-13 17:52 Repository
-rw-rw-r--. 1    24 2010-06-13 17:52 Root
```

- cvs update: periodically update the local copy of the files to bring in changes from the repository.

```
RCS file: /home/skoranne/CVSROOT/POLY/cano_proc.cpp,v
retrieving revision 1.1.1.1
retrieving revision 1.2
Merging differences between 1.1.1.1 and 1.2
                              into cano_proc.cpp
```

```
rcsmerge: warning: conflicts during merge
cvs update: conflicts found in cano_proc.cpp
C cano_proc.cpp
```

The conflict markers show the difference between the local changes and the modifications made on the repository copy since the last update.

```
    static void WritePostamble( std::ofstream& ofs ) {
    <<<<<<< cano_proc.cpp
            // this function should have some comments
    =======
5           // I also want to add some comments
    >>>>>>> 1.2
      ofs << std::endl << "\\end{document}" << std::endl;
    }
```

- cvs commit: once changes in the file have been made, to make them apparent to other users of the CVS system, the changes have to be committed back to the repository. This is don't using the cvs commit command; this command requires an accompanying log for each commit. The log can be specified on the command line using the -m option, or an editor can be invoked.

```
$cvs commit -m "Add my comments" cano_proc.cpp
Checking in cano_proc.cpp;
CVSROOT/POLY/cano_proc.cpp,v  <-- cano_proc.cpp
new revision: 1.2; previous revision: 1.1
done
```

- cvs diff: this command runs the diff command on the current version of the file with the CVS copy. Results are presented in the diff format.

```
[skoranne@celex POLY]$ cvs diff cano_proc.cpp
Index: cano_proc.cpp
===================================================
RCS file: /home/skoranne/CVSROOT/POLY/cano_proc.cpp,v
retrieving revision 1.1.1.1
diff -r1.1.1.1 cano_proc.cpp
130a131
>   // I also want to add some comments
```

- cvs stat: shows the status of the current files:

```
===================================================
File: cano_proc.cpp      Status: Locally Modified

   Working revision: 1.1.1.1 Mon Jun 14 00:50:22 2010
   Repository revision: 1.1.1.1 CVSROOT/POLY/cano_proc.cpp,v
   Sticky Tag: (none)
   Sticky Date: (none)
   Sticky Options: (none)
```

- cvs release: removes file from the working copy, if specified using -d option physically deletes the file from disk as well.

A full list of CVS commands is:

```
add          Add a new file/directory
admin        Administration front end for rcs
annotate     Show last revision
checkout     Checkout sources for editing
commit       Check files into the repository
diff         Show differences between revisions
edit         Get ready to edit a watched file
editors      See who is editing a watched file
export       Export sources from CVS
history      Show repository access history
import       Import sources into CVS
init         Create a CVS repository if it doesn't exist
log          Print out history information for files
login        Prompt for password for authenticating server
logout       Removes entry in .cvspass for remote repository
pserver      Password server mode
release      Indicate that a Module is no longer in use
remove       Remove an entry from the repository
rtag         Add a symbolic tag to a module
server       Server mode
status       Display status information
tag          Add a symbolic tag
unedit       Undo an edit command
update       Bring work tree in sync
version      Show current CVS version(s)
watch        Set watches
watchers     See who is watching a file
```

As an alternative to memorizing the CVS command, the graphical tool TkCVS
can be used as shown in Subsection 3.2.5.

3.2.3 SVN

Subversion, or SVN is designed to be a better version control system than CVS, but
for ease of migration it has retained many of the same command-line functions as
CVS. In addition it has the following features:

- Commits as true atomic operations,
- Renamed/copied/moved/removed files retain full revision history,
- The system maintains versioning for directories, renames, and file metadata,
- Branching and tagging as cheap operations,
- Costs proportional to change size,
- XML log output,
- File locking for unmergeable files,
- Path-based authorization.

Running svn stat on a locally checked out revision gives:

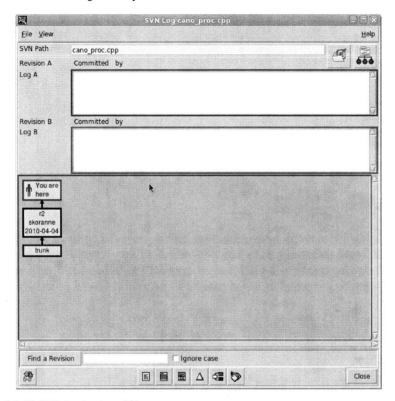

Fig. 3.1 TkCVS showing log of file

```
?        build.sh
M        graph_reader.cpp
```

The '?' symbol shows that the file 'build.sh' is not in the version control, while the file graph_reader.cpp has been locally modified. We can inspect the changes to the file using the `svn diff` command or by using the TkCVS graphical tool.

```
--- graph_reader.cpp (revision 2)
+++ graph_reader.cpp (working copy)
@@ -96,8 +96,10 @@
  GRAPH[i][j] : GRAPH[i][k]+GRAPH[k][j];
```

Some of the available commands for SVN are given below, by design many of the commands have similar name and purpose as the above mentioned CVS commands.

```
add
blame (praise, annotate, ann)
changelist (cl)
checkout (co)
cleanup
commit (ci)
```

```
copy (cp)
delete (del, remove, rm)
diff (di)
export
import
merge
move (mv, rename, ren)
revert
status (stat, st)
update (up)
```

3.2.4 GIT

Git is a distributed revision control system with an emphasis on speed. Git was initially designed and developed by Linus Torvalds for Linux kernel development. Every Git working directory is a full-fledged repository with complete history and full revision tracking capabilities, not dependent on network access or a central server.

3.2.5 TkCVS

TkCVS is a graphical tool which makes it easy to work with diverse version control systems. It can support CVS, SVN, and other version control systems. TkCVS is a Tcl/Tk-based graphical interface to the CVS and Subversion configuration management systems. It displays the status of the files in the current working directory, and provides buttons and menus to execute configuration-management commands on the selected files. Limited RCS functionality is also present. TkDiff is bundled in for browsing and merging your changes.

TkCVS also aids in browsing the repository. For Subversion, the repository tree is browsed like an ordinary file tree. For CVS, the CVSROOT/modules file is read. TkCVS extends CVS with a method to produce a browsable listing of modules. TkCVS is invoked as:

```
tkcvs [-dir directory] [-root cvsroot]
      [-win workdir|module|merge] [-log file]
```

3.2.6 Tinderbox

Tinderbox is a software suite that provides continuous integration capability. Tinderbox allows developers to manage software builds and to correlate build failures on various platforms and configurations with particular code changes.

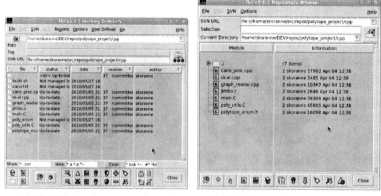

(a) TkCVS showing current branch   (b) TkCVS showing repository model

Fig. 3.2 TkCVS branch and repository

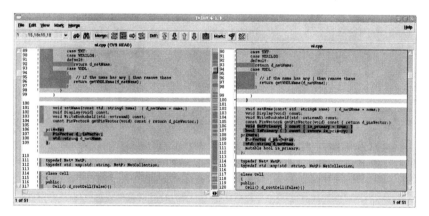

Fig. 3.3 TkCVS showing TkDiff module with CVS

Tinderbox is described as "a detective tool for software development". It allows the developer to see what is happening in the source tree. It shows who checked in what (by asking Bonsai); what platforms have built successfully; what platforms are broken and exactly how they are broken (the build logs); and the state of the files that made up the build (cvsblame). Tinderbox is composed of a server with clients running builds and reporting status via mail.

3.3 GNU Build System

As the reader probably knows already, most GNU software has a traditional way
of installation: (i) download the source TAR file, (ii) untar the archive in a local
directory, and (iii) run:

```
$./configure
$gmake
$gmake install
```

The goal of the GNU build system is to simplify the development of portable programs, as well as encourage the open-source concept by simplifying the process of
releasing products as source code which can then be compiled by the user. Moreover, using the GNU build tools allows for the VPATH build concept where the
source code is placed in a single directory while operating-system or machine dependent variations of the build are carried out in their own directories. In this section
we focus our attention on the `configure` script which is omnipresent in GNU
software. This script is generated as a part of the GNU Build Tools process which
involves the following tools (i) autoconf, (ii) automake, and (iii) libtool; these are
described below.

3.3.1 Autoconf

`autoconf` produces a configuration shell script (`configure`) which probes the
host system for portability related information. This information is then used as part
of the build, but can also be used by the application program. The most common use
of `autoconf` is to generate customized source code blocks using pre-processor
`#define` on the computed values.

   The version of `autoconf` can be checked using:

```
$autoconf --version
autoconf (GNU Autoconf) 2.63
```

3.3.2 Automake

`automake` produces make files (see Section 3.4.1 for more information on GNU
gmake). The tool produces the file Makefile.in to be used by `autoconf` from
a very highlevel specification file called Makefile.am, thus reducing the effort required.

   We use a simple example to describe the use of `autoconf` and `automake`.
Before running the GNU build tools the directory containing the example is:

```
$ ls
```

```
configure.in  hello_world.cpp  Makefile.am
```

The files 'configure.in' and 'Makefile.am' are shown below:

```
# \file configure.in
# \author Sandeep Koranne, (C) 2010
# \description Example file for autoconf

AC_INIT(hello_world.cpp)
AM_INIT_AUTOMAKE(hello_world,1.0)
AC_PROG_CXX
AC_PROG_INSTALL
AC_OUTPUT(Makefile)
```

The statements listed in the 'configure.in' file above initialize the configuration system (AC_INIT), any source code file name can passed as an argument. The AM_INIT_AUTOMAKE denotes the use of `automake` and also contains the name of the package and version number. Features on the host system, such as version of C++ compiler can be checked using AC_PROG_CXX statement. To check for `lex` (see Section 13.3) we can add AC_PROG_LEX, and AC_PROG_YACC (for `yacc`).

```
# \file Makefile.am
# \author Sandeep Koranne, (C) 2010
# \description Example file for automake

bin_PROGRAMS = hello_world
hello_world_SOURCES = hello_world.cpp
```

We proceed to run `autoconf` as:

```
$aclocal
$ls
aclocal.m4  autom4te.cache  configure.in
hello_world.cpp  Makefile.am
$autoconf
$ ls
aclocal.m4      configure      hello_world.cpp
autom4te.cache  configure.in  Makefile.am
```

Finally, we have generated our own '.configure' file, its contents being too numerous to copy here. Next we run `automake −A` to get:

```
$automake −a
configure.in:6: installing './install-sh'
configure.in:6: installing './missing'
Makefile.am: installing './INSTALL'
Makefile.am: required file './NEWS' not found
Makefile.am: required file './README' not found
Makefile.am: required file './AUTHORS' not found
```

```
Makefile.am: required file './ChangeLog' not found
Makefile.am: installing './COPYING' using GNU
General Public License v3 file
$ ls
aclocal.m4       configure      COPYING
hello_world.cpp  install-sh     missing
autom4te.cache   configure.in   depcomp
INSTALL          Makefile.am
```

To create the required files 'NEWS', 'README', we can create them in an editor, or simply call `touch NEWS` to create an empty file which we promise to fill with useful and relevant data *later on*. Thereafter, we have to run `automake -a` again, and this time it completes without any error, creating the file 'Makefile.in' which is used by 'configure'. Running './configure' gets:

```
$ ./configure
checking for a BSD-compatible install... /usr/bin/install -c
checking whether build environment is sane... yes
checking for a thread-safe mkdir -p... /bin/mkdir -p
checking for gawk... gawk
checking whether make sets $(MAKE)... yes
checking for g++... g++
checking for C++ compiler default output file name... a.out
checking whether the C++ compiler works... yes
checking whether we are cross compiling... no
checking for suffix of executables...
checking for suffix of object files... o
checking whether we are using the GNU C++ compiler... yes
checking whether g++ accepts -g... yes
checking for style of include used by make... GNU
checking dependency style of g++... gcc3
checking for a BSD-compatible install... /usr/bin/install -c
configure: creating ./config.status
config.status: creating Makefile
config.status: executing depfiles commands
```

This produces the Makefile which can be processed using GNU `gmake` as:

```
gmake
g++ -DPACKAGE_NAME=\"\" -DPACKAGE_TARNAME=\"\"
    -DPACKAGE_VERSION=\"\" -DPACKAGE_STRING=\"\"
    -DPACKAGE_BUGREPORT=\"\" -DPACKAGE=\"hello_world\"
    -DVERSION=\"1.0\" -I. -g -O2 -MT hello_world.o
    -MD -MP -MF .deps/hello_world.Tpo
    -c -o hello_world.o hello_world.cpp
mv -f .deps/hello_world.Tpo .deps/hello_world.Po
g++  -g -O2   -o hello_world hello_world.o
```

The produced Makefile conforms to the GNU Makefile standard and has the default targets of 'clean', 'distclean'. An additional target 'distcheck' produces a

TAR gzipped archive of the current sources and proceeds to test it against the listed regression suite.

3.3.3 Libtool

The `libtool` tool is a platform independent method of generating shared libraries and position independent code. It supports versioning of shared libraries. The `libtool` command-line tool can be used independently of `automake` and `autoconf`, but is used by them for compiling position independent code and generating shared libraries.

The above example may appear contrived and trivial to warrant the use of GNU Build tools, and indeed it is, as it is an expository and pedantic example. However, for large and complex software which is expected to be compiled on many different systems (some of which have not even been invented yet), the separation between source code and operating system and machine dependencies is very important. The GNU Build tools help enforce this separation, and indeed the tools are used in most open-source applications, thus even if the reader does not have to write a Makefile.am file from scratch, at the very least its purpose and semantics should be understood.

3.4 Automatic Build Dependency Management

We have already seen example of building source code using the GNU Build tools. One clear advantage of using automated build tools is the reduction of effort in making a new binary from source code changes. However, a naive approach to re-compiling the whole software suite since a single file has been modified does not scale well beyond small projects. Towards that end, automated build dependency tools have been developed. In this section we discuss two tools for automated build dependency, (i) GNU `gmake` and (ii) SCons.

3.4.1 GNU make : automatic build dependency

GNU Make (`gmake`) is a utility for automatically building executable programs and libraries from source code. An input control file (called a Makefile) specifies the dependencies of the various source files (using gcc -MD, the dependencies based on inclusion of files can be generated automatically). The `make` tool then computes the topological sort of the files and performs the required actions on the dependencies to produce the *target*.

The control file specifies the relationship between the target and the dependencies. A simple rule looks like:

```
target:      dependencies
             action
```

If the time stamp of the source dependency is more recent than the time stamp of the target, the target needs to be *rebuilt*. The procedure for the rebuilding entails the execution of the commands in the control file. Consider an example where a static library depends on hundreds of object files (each of which is a compiled C/C++ file); in turn the library contributes to a binary. This can be represented as:

```
libapp.a:    parser.o optimizer.o app.o
      $(AR) cr libapp.a parser.o optimizer.o app.o
app.exe:     libapp.a main.o
      $(CC) $(CFALGS) main.c -lapp -o app.exe
```

When specifying rules and actions GNU make understands the following features:

- Variables: variables can be defined in the Makefile using

  ```
  APP_INCLUDE := -I$(apps)
  ```

  predefined variables for the current rule are:

  ```
  $< : current list of depencies which changed
  $@ : current target
  $^ : all dependencies of current rule
  ```

  There are also targets which can be marked as .PHONY implying that they do not represent real on-disk files, so that even the presence of a real file of that name does not match the target.
- Pattern substitution:

  ```
  OBJS   = $(patsubst %.cpp,%.o,$(SOURCES))
  ```

  or pattern substitution based on extension with pre-defined actions:

  ```
  .cpp.o:
          $(CXX) $(CXXFLAGS) -c $< -o $@
  ```

- Include files: using `include` to bring in parts of a global Makefile (containing rules or dependencies),
- Conditionals: GNU make supports conditional dependencies, consider

  ```
  ifdef STATIC
  CCOPT += -static
  endif
  ```

  Then on the command-line we can say `gmake STATIC=1` to enable the variable. The second form of the conditional is based on the value of variables:

```
ifeq ($(cc), gcc )
      $(CC) -mcpu=pentium -c file.c
else
      $(CC) -c file.c
```

In the above example, if the variable cc has a value gcc then we compile the
source file with the -mcpu command line option.

- Functions for file names and others: GNU make has inbuilt functions which operate on file names. These functions include:

```
$(dir src/file.c ) : produces src
$(suffix src/file.c src/app.h) : produces .c .h
$(basename src/file.c) : produces src/file
$(join a b, .c .o) : produces a.c b.o
$(wildcard src/*.c): produces wildcard expansion
$(foreach ) example
    files := $(foreach dir,$(dirs),$(wildcard $(dir)/*))
$(call variable,param, param):
$(eval ) function
```

In addition to the Makefile syntax there are a number of commonly used command-
line options to gmake:

```
-B : unconditionally make all targets
-k : continue despite errors
-d : print lots of debugging information
-f : specify name of Makefile
-n : dry-run, dont perform any action
```

3.4.2 SCONS : A software construction tool

SCons is a computer software construction tool that automatically analyzes source
code file dependencies. Its function is analogous to the traditional GNU build system based on the make utility and the autoconf tools. SCons uses the Python general
purpose programming language as a foundation, so that all software project configurations and build process implementations are Python scripts.

Consider a simple program in C/C++ (hw.cpp):

```
#include <iostream>
int main( ) {
  std::cout << "Hello, World!\n";
  return 0;
}
```

The SCons build file for this application is extremely simple: we create a file called
SConstruct and in it we add

```
$cat SConstruct
Program('hw.cpp')
$scons
scons: Reading SConscript files ...
scons: done reading SConscript files.
scons: Building targets ...
g++ -o hw.o -c hw.cpp
g++ -o hw hw.o
scons: done building targets.
$./hw
Hello, World!
```

To build an object file, instead of `Program` use the `Object` *builder method*, and for building libraries use the Library builder. The SCons tool knows the default actions it needs to perform for the given builder object. For Program it compiled and linked the file, for Object it only performs the compilation. The Program construct can also take the name of the produced binary as the first argument, as:

```
Program( 'new_hello', 'hello.c' )
```

Multiple files can be passed as a list to the Python builder object, or by using a wildcard matching `Glob(*.c)`. Multiple Program builder objects can be placed in a single file to build more than one program. Since SCons file are Python scripts, and the Program objects are Python objects, their constructors can accept named parameters; to link a program with a library we can specify it as:

```
Library('compression', ['lz77.c', 'file_io.c'])
Program('main.c', LIBS=['compression'], LIBPATH='.')
```

3.4.2.1 The SConstruct file

SConstruct files are not like normal Python programs, in that they are not executed sequentially. It is after-all a Makefile equivalent, and the order in which the builder objects are *defined* in the SConstruct file are independent from the order in which the targets are built.

Node objects Builder methods return a list of node objects that identify the target file. These node objects themselves can be passed as arguments to other builder methods. Moreover, properties on node objects can be used to control various build options for specific files and targets.

SCons dependencies Just as GNU make can be used to calculate the minimum set of files which have to be rebuilt to build a target (make uses timestamp information to populate the change graph), SCons also calculates the set of files to be rebuilt. Instead of using the timestamp, SCons uses a MD5 *signature* or checksum for each file (although SCons can be configured to use timestamp). It is even possible to specify a custom Python function to decide if an input file has changed (imagine a tool where CVS/SVN changes on the repository are automatically inspected to calculate change information for some files). SCons MD5 mechanism is robust and supports include files and symbolic links.

Calculating which files have changed is just part of the dependency analysis. The main graph of dependencies has to be constructed as well. For this purpose, SCons uses the CPPPATH variable for every source file. For every include file of the source file located in the CPPPATH directory, SCons adds a dependency from the included file to the source file. If the included file has been modified, the source file has to be recompiled.

For Library builder object a dependency is added from every member of the list to the library itself. Scanning each source file takes extra time, but as the contents of the file are being inspected for the MD5 anyway, the include information can be cached as well. This is performed by using the implicit cache command-line discussed below.

To explicitly add dependencies, we can use the Depends('A','B') method. Additionally, we can use GCC's -MD option to generate the dependencies as:

```
A = Object ('file.c',
            CCFLAGS='-MD -MF file.d',
            CPPPATH='.')
SideEffect ('file.d',A)
ParseDepends ('file.d')
```

Dependencies (for a specific target) can be ignored using the Ignore method.

SCons Environments An *environment* defines a related set of option to be used to build a piece of the software. An environment can be created using:

```
env = Environment ()
```

method. An easy method to create a custom build environment based on operating system environment variables is:

```
import os
env = Environment (CXX = 'g++', CXXFLAGS = '-ggdb')
env.Program ('file.cpp')
```

This SConstruct file specifies that file.cpp source should be compiled using the debug option unless the CXXFLAGS variable is defined to some other value in the OS. The default environment can be set using the DefaultEnvironment function.

In addition to the control file SCONS has a number of command-line options:

```
-c : remove specified targets
-Q : reduce verbosity of messages
--implicit-cache : cache dependencies
--random : build dependencies in random order
```

A more detailed example of using SCons is shown in Listing 3.2.

```
   import os
   env = DefaultEnvironment ()
   print "Default CXX = ", env['CXX']
5  dbg = Environment (CXXFLAGS='-O0 -ggdb')
```

```
     opt = Environment(CXXFLAGS='-O3 -funroll-loops')

     if ARGUMENTS.get('opt', 0):
         env = opt
10   else:
         env = dbg

     env.Program('udp_server.cpp', LIBS=['boost_system','pthread'],
                 CPPPATH=['/home/skoranne/INCLUDE'],
15               LIBPATH=['/usr/lib'])
     Program('udp_client.cpp', LIBS=['boost_system','pthread'])
```

Listing 3.2 SCons file for Boost UDP

3.4.3 CMAKE and QMake

3.4.3.1 CMake

CMake is a cross platform, open-source make system. CMake generates native Makefiles (which can be processed with GNU make). Nokia/Trolltech Qt's qmake also generates Makefile for the platform. Like SCons, CMake maintains the control file in a named file, in this case CMakeLists.txt; an example is given below:

```
project(app)

CMAKE_MINIMUM_REQUIRED(VERSION 2.4.5 FATAL_ERROR)
if ( GCC_FOUND )

add_config_flag( CGAL_USE_GMP )
add_config_flag( CGAL_USE_MPFR )

endif()
macro( add_programs subdir target ON_OFF )

  cache_set( CGAL_EXECUTABLE_TARGETS "" )

endmacro()
```

Run man cmake for more information on CMake.

3.4.3.2 QMake

QMake is Qt's build program. An example qmake '.pro' file is shown below:

```
TEMPLATE = app
TARGET =
DEPENDPATH += .
```

```
INCLUDEPATH += .

# Input
SOURCES += qsig.cpp
```

Unlike SCons or CMake, qmake .pro (for project) file has a simpler keyword-value syntax. In addition to turning off Qt libraries, it also allows for specifying debug or release builds, location of Qt resources such as bitmaps, and additional libraries. Like CMake, qmake also generates platform specific Makefiles, which are then processed using GNU make to build the target.

3.5 Bugzilla : Defect Tracking System

Bugzilla is a defect tracking system implemented as a server software featuring optimized database structure for efficiency, security, advanced query tools, and integrated email. Bugzilla supports the complete life-cycle of a defect, from ticket initiation to final resolution. Moreover, it can be used to generate charts and reports with diverse statistics on the defects stored in its database.

After installation, the initial Bugzilla screen in the browser is shown in Figure 3.4.

Fig. 3.4 Bugzilla screen

To deploy Bugzilla the system must have:

1. compatible database management system
2. suitable release of Perl 5
3. assortment of Perl modules
4. compatible web server
5. suitable mail transfer agent, or any SMTP server

These requirements are fairly easy to meet on a modern GNU/Linux system, and deploying Bugzilla is easy, see Figure 3.4 for the main Bugzilla screen where user authentication can be performed. This book maintained its own Bugzilla database during its authoring. We have added *products* defining the scope of the book material and divided the products into *components* as shown in Figure 3.5.

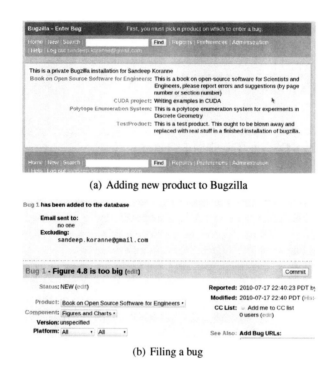

(a) Adding new product to Bugzilla

Bug 1 **has been added to the database**

>   **Email sent to:**
>   no one
>   **Excluding:**
>   sandeep.koranne@gmail.com

Bug 1 - **Figure 4.8 is too big** (edit)   Commit

>   Status: NEW (edit)   **Reported:** 2010-07-17 22:40:23 PDT by
>   **Modified:** 2010-07-17 22:40 PDT (Hist:
>   Product: Book on Open Source Software for Engineers ·   **CC List:**   Add me to CC list
>   Component: Figures and Charts ·   0 users (edit)
>   Version: unspecified
>   Platform: All   ·   All   ·   See Also: **Add Bug URLs:**

(b) Filing a bug

Fig. 3.5 Setting up and using Bugzilla

3.6 Editing Source Code

Many disciplines (if not most) of science now routinely use and develop software as part of the research. Software development systems have been designed to aid the software engineering process, of which writing the original code is an integral part. All software starts with some source code written in a file or storage system. In this section we discuss open-source code editors

3.6.1 Emacs

Emacs (acronym for Editor Macros) is an extensible text editor, however, its use goes much further than simple editing of code and text. Emacs has more than 1,000 commands for editing, and also allows the user to write new functions using Emacs Lisp, and to attach these new functions to keyboard shortcuts. Thus, the user can define keyboard macros to automate repetitive tasks (such as running LATEXon a section to preview it). An example of using Emacs to edit source code is shown in Figure 3.6.

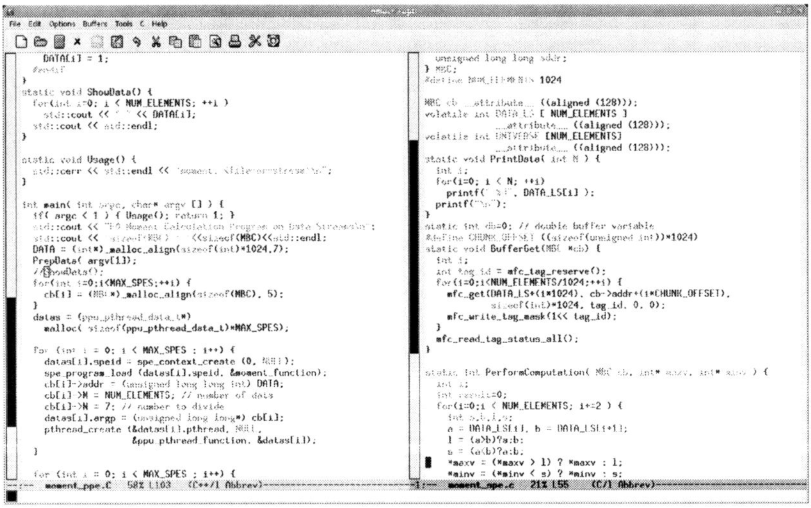

Fig. 3.6 EMACS: editor for code editing

As shown in Figure 3.6, Emacs can be used as a syntax highlighting source code editor. Using Emacs Lisp, various *modes* have been written which automate tasks such as indentation, highlighting and even using the C++ compiler to check for source code errors while typing.

Emacs can be used effectively as an IDE (integrated development environment) as it can read TAGS (see Section 3.7.1) data, as well as having customized interaction capability for source-code editing. Using the `speedbar` command, relevant information on files in the current directory of the editor can be inspected, as shown in Figure 3.7.

Fig. 3.7 'speedbar' in Emacs

Emacs also has an interactive front-end to GDB, as well as hooks to popular version control systems such as CVS (see Section 3.2.2) and Subversion (see Section 3.2.3). File merging, difference checking, and annotations can also be performed within Emacs.

3.6.2 Eclipse

Some of the tools and development environments which are integrated with Eclipse are listed below:

- CDT: C and C++ development,
- PTP: Parallel Tools Platform which contains a scalable parallel debugger,
- Cell Broadband Development: environment for developing software on the Cell Broadband Engine,

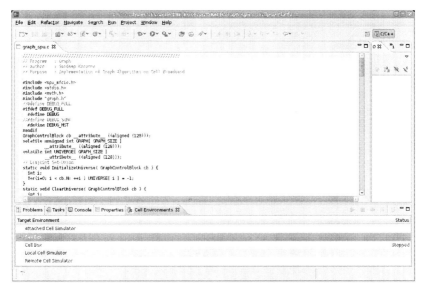

Fig. 3.8 ECLIPSE: editor for code editing

3.6.3 KDevelop

KDevelop is a free software integrated development environment for the KDE desktop environment for Unix-like computer operating systems. See Figure 3.9 for an example of a session running within kdevelop.

Its features include:

1. Source code editor with syntax highlighting and automatic indentation,
2. Project management for different project types using automake, qmake,
3. Class browser,
4. Integrated debugger (using gdb),
5. Automatic code completion (C/C++), and class definitions,
6. SCM support for CVS, SVN, Perforce.

KDevelop's dialog for creating a new project is shown in Figure 3.10.

The settings for the project can be changed within KDevelop, as shown in Figure 3.11.

Fig. 3.9 KDEVELOP: IDE for code editing

Fig. 3.10 KDEVELOP: Dialog for creating new project

Fig. 3.11 KDEVELOP: Dialog for project options

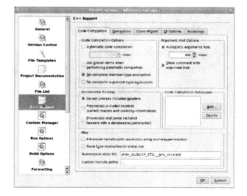

3.7 Static Checks on Source Code

In addition to run-time checks (profiling, debugging) another class of checks on source code which is very helpful in reducing errors are *static checks*. Static checking involves analysis of source code. In this section we discuss two tools which can provide analysis on source code.

3.7.1 ctags

The `ctags` and `etags` programs generate index file for a number of programming language files. The index file contains a listing of the various source-code objects detected in the input files. This index file can thereafter be used by source-code editors to quickly locate a section of source-code text which contains an object, such as variable name, or function definition.

The `ctags` program has a number of command-line options which control its output. It can be run in the root directory of the source code tree with the −R (recursive) option to generate an index for all known source code files located under the root. It can also *append* TAGS to a file. Consider the following example of `ctags` being run to produce Emacs (see Section 3.6.1) compliant TAGS index files.

```
$ctags -e -R --verbose=yes|grep OPEN
OPENING dual.cpp as C++ language file
OPENING build.sh as Sh language file
OPENING large.f as Fortran language file
OPENING jimbo.c as C language file
OPENING poly_utils.C as C++ language file
OPENING enum.scons as Python language file
OPENING graph_reader.cpp as C++ language file
```

A part of the produced TAGS file (which is a binary file) is also shown; it can be seen that `ctags` has encoded information about source-code objects in the file.

```
dual.cpp,464
typedef std::vector<int> Vector;^?Vector^A37,653
```

In Emacs, using `tags-apropos` we can perform a regular-expression search for any symbol present in the TAGS file:

```
Tags matching regexp `polytope.*hash':

[poly_utils.C]: std::set< std::string > GlobalPolytopeHash;
[poly_utils.C]: bool MarkUniquePolytope( const CBPolytope&
[poly_utils.C]: string& CBPolytope::compute_hash_code( ) const {
[cano_proc.cpp]:    PolytopeRepresentation():hash_code(""),
```

Using the −x option to `ctags` causes it to produce a tabular, human-readable cross-reference file instead of the TAGS file. Example:

```
LONG_WORD    function 19 gr_reader.cpp LONG_WORD():j1(0x0) {}
LONG_WORD    struct   17 gr_reader.cpp struct LONG_WORD {
MAX_DIM      macro    29 gr_reader.cpp #define MAX_DIM 10
MAX_N        macro    58 gr_reader.cpp #define MAX_N 128
MY_INT_MAX   macro    64 gr_reader.cpp #define MY_INT_MAX 10000
```

The cross-reference file contains the tag-name, type, location in source-code. This file can be post-processed to produce statistical data about the source-code. The use of ctags is not limited to source-code for programming, it can also be used to generate TAGS file for other types of text documents, as ctags supports many languages, including Bash scripts, and even TEXand LATEX.

3.8 GNU gcov: Test Coverage Program

GNU gcov is a test coverage program. It is used in conjunction with GNU gcc to analyze the program and to uncover untested parts of the source code. The gcov tool can be used to analyze which part of the source code is being executed to better aid and guide algorithmic optimization efforts. It can also be used to discover portions of the source code which are not covered by the program testbench. When used alongwith a profiling tool such as gprof (see Section 3.12.1), gcov can be used to calculate source code metrics such as (i) execution count of each line of source code, (ii) which source code lines are actually executed, and (iii) compute time spent in each section of code.

In this section we describe the use of gcov to calculate the test coverage of a program. We assume that a testbench for the program exists and is representative of the common usage expected of the program. Ofcourse, gcov is also used to guide the development of the said testcase, as it can be used to pinpoint source code areas which are un-stressed. To use gcov effectively, the program should be compiled without optimization and with debug symbols in place. This can be achieved using the following command-line for GNU gcc:

```
$gcc -O0 -ggdb -c <filename>
```

Moreover, since gcov computes statistics on a line-by-line basis, it works best with programs which have only one statement per source code line. Similarly, the use of complex macros causes gcov to report line usage after macro expansion, which can produce confusing results. A complex macro should be wrapped in an inline function for this purpose.

3.8.1 Compiling programs for gcov

To use gcov the source must be compiled to report the execution count and branch probabilities. This is referred to as *instrumenting* the source code, and is similar to the action of compiling with the −pg flag for profiling (see Section 3.12.1). The GNU gcc command-line options for compiling with gcov are:

```
$gcc −fprofile-arcs −ftest-coverage <file> −lgcov
```

The −lgcov argument *links* the GNU gcov library (which is required for the code instrumentation to work) with the application.

3.8.2 Running gcov

gcov has a number of command-line options as listed below:

−a −−all−blocks: write individual execution counts for every block,

−c −−branch−counts: write branch frequencies as number of branches taken rather than percentage,

−b −−branch−probabilities: write branch frequencies to the output file, and write branch summary information to standard output,

−f −−function−summaries: output summaries for each function in addition to file summary,

Consider the small program shown in Listing 3.3.

```
    // \file gcov_example.cpp
    // \author Sandeep Koranne, (C) 2010
    // \description Example of using GNU gcov program
    #include <iostream>        // for program IO
5   #include <cassert>         // for assertion checking
    #include <cstdlib>         // for exit

    static const unsigned int N = 16;

10  static int gMatrix[N][N];

    static void InitMatrix(void) {
      for( unsigned int i=0; i < N; ++i )
        for( unsigned int j=0; j < N; ++j )
15        gMatrix[i][j] = i + j;
    }

    int main( int argc, char *argv[] ) {

20    std::cout << "Example of using GNU gcov.\n";

      return (0);
    }
```

Listing 3.3 Example of using gcov

We compile this program as:

```
g++ −ggdb −O0 −fprofile-generate −fprofile-arcs \
            −ftest-coverage gcov_example.cpp
$ls
a.out   gcov_example.cpp   gcov_example.gcno
```

We then run the program on representative input:

```
$./a.out
$ls *.gcda
gcov_example.gcda
```

The execution of the program produces a binary data file .gcda, which contains the execution count, branch probabilities, and other information about the program's execution. We can now use gcov to analyze this file:

```
$gcov -a -b -c -l -f -p -u gcov_example.cpp
Function '_ZStorSt12_Ios_IostateS_'
Lines executed:0.00% of 2
No branches
No calls


Function '_ZL10InitMatrixv'
Lines executed:0.00% of 5
No branches
No calls


Function '_ZNKSt9basic_iosIcSt11char_traitsIcEE7rdstateEv'
Lines executed:0.00% of 2
No branches
No calls
....
Function 'main'
Lines executed:100.00% of 3
No branches
No calls
```

In addition to the report generated on standard output, gcov also creates detailed reports for each of the *translation units* and header files it encounters in the source file. C++ language header files are also annotated, though for the most part (for application developers) execution counts within these system files can be ignored.

What is important though, is the reported execution count per line for the application program source code, as that directly corresponds to the testability of the program. If a particular section of code has very low execution count, it means that, the code in question has not been exercised enough in the testbench. The generated annotation for our original code (from Listing 3.3) is shown in Listing 3.4.

```
        -:     0:Source:gcov_example.cpp
        -:     0:Graph:gcov_example.gcno
        -:     0:Data:gcov_example.gcda
        -:     0:Runs:1
5       -:     0:Programs:1
        -:     1:// \file gcov_example.cpp
        -:     2:// \author Sandeep Koranne, (C) 2010
        -:     3:// \description Example of using GNU gcov program
        -:     4:#include <iostream>        // for program IO
10      -:     5:#include <cassert>         // for assertion checking
        -:     6:#include <cstdlib>         // for exit
        -:     7:
        -:     8:static const unsigned int N = 16;
        -:     9:
15      -:    10:static int gMatrix[N][N];
```

```
        -:    11:
    #####:    12:static void InitMatrix(void) {
    #####:    13:   for( unsigned int i=0; i < N; ++i )
    +++++:    13-block  0
    +++++:    13-block  1
    +++++:    13-block  2
    #####:    14:       for( unsigned int j=0; j < N; ++j )
    +++++:    14-block  0
    +++++:    14-block  1
    +++++:    14-block  2
    #####:    15:          gMatrix[i][j] = i + j;
    #####:    16:}
    +++++:    16-block  0
        -:    17:
        1:    18:int main( int argc, char *argv[] ) {
        -:    19:
        1:    20:   std::cout << "Example of using GNU gcov.\n";
        1:    20-block  0
        -:    21:
        1:    22:   return (0);
        3:    23:}
        1:    23-block  0
        1:    23-block  1
        1:    23-block  2
        1:    23-block  3
```

Listing 3.4 gcov generated annotated source code

Since we did not call the function InitMatrix at all in the file, execution count for the section of code in that function is 0, and the whole section is marked as unused code. Calling the function, recompiling and re-running the code fixes this problem.

Branch probabilities as calculated by -fprofile-arcs are also used in *feedback directed optimization* in GNU gcc. With the deep pipeline present in current generation CPUs, such optimization is very effective and can easily reduce runtime by 15 to 20 percent.

When we run gcov -a to get information for all the blocks we see in the generated .gcov file the following type of information:

```
    -:    11:
#####:    12:static void InitMatrix(void) {
#####:    13:   for( unsigned int i=0; i < N; ++i )
$$$$$:    13-block  0
$$$$$:    13-block  1
$$$$$:    13-block  2
#####:    14:       for( unsigned int j=0; j < N; ++j )
$$$$$:    14-block  0
```

On another code block which has heavy branching, running gcov -b produced:

```
 1117:   586: if(A_set & B_set) assert(A.empty()==false);
branch  0 taken 28% (fallthrough)
branch  1 taken 72%
call    2 returned 100%
branch  3 taken 0% (fallthrough)
branch  4 taken 100%
call    5 never executed
 1117:   587: if( A.empty()) assert((A_set&B_set)==0);
call    0 returned 100%
```

```
branch  1 taken 72% (fallthrough)
branch  2 taken 28%
branch  3 taken 0% (fallthrough)
branch  4 taken 100%
call    5 never executed
```

We can use the branch counts to optimize the code by introducing GCC's intrinsic functions for branch hints; see Section 3.1.7.1. We can add

```
#if defined (__GNUC__) && ( __GNUC__ > 2 ) && defined(__OPTIMIZE__)
#define PROB1(expr) (__builtin_expect((expr),1))
#define PROB0(expr) (__builtin_expect((expr),0))
#else
#define PROB1(expr) (expr)
#define PROB0(expr) (expr)
#endif
```

Then, whenever we have a condition for which we know the branch probability with high certainty, we can code it as:

```
if( PROB0( error_case ) ) {
    std::cerr << ''Unlikely, but has to be handled...\n'';
}
```

We performed a simple experiment using this technique on a compute intensive workload of discrete geometry. We ran the unmodified program first, then we added PROB0 and PROB1 to a total of 5 conditional branch points (all branch points were within loops), and we recompiled and reran the program on same input. The results are shown below:

```
Before optimization

45.97user 1.96system 0:53.49elapsed 89%CPU
0inputs+7304outputs (0major+20828minor)pagefaults 0swaps


After optimization
41.82user 1.84system 0:46.24elapsed 94%CPU
0inputs+7304outputs (0major+20841minor)pagefaults 0swaps
```

Ofcourse, this is a very small runtime example, but over longer running programs, even a 10-15 percentage improvement is substantial.

It should be noted that the execution counts present in the '.gcda' files are cumulative over multiple runs of the program (this is useful in test coverage analysis where a collection of tests is expected to test the whole program).

3.9 Debug Tools

3.9.1 GDB

In this section we describe the GNU debugger, gdb. Almost all non-trivial programs go through a process of finding bugs (see Section 3.8 on code coverage), and this procedure is called *debugging*. A good software debugger can reduce the time to find problems in the code by an order of magnitude. The GNU debugger is called gdb, and it has the following command-line options:

```
gdb [options] [exe [core-file or pid] ]
```

- –cd = dir: change current directory to dir,
- –core=COREFILE: analyze the core dump file,
- –directory=dir: search dir for source code,
- –exec=EXECFILE: use EXECFILE as the executable binary,
- –command=FILE, -x: execute gdb commands from file,
- –version: print version and exit.

The main commands of gdb can be found using help command, we list the important and often used commands below, their usage is described later in this section.

aliases aliases of other commands
breakpoints making program stop at certain points
data examining data
files specifying files
running Running the program
stack Examining the stack

Consider the program listing shown in Listing 3.5.

```
   // \file number_test.cpp
   // \author Sandeep Koranne, (C) 2010
   // \description C++ example for GDB
   // Fibonacci = (0,1,1,2,3,5,8,13,21,34,55,89)
5  // Sum = prefix-sum of F
   // (0,1,2,4,7,12,20,
   #include <stdio.h>
   #include <string.h>
   #include <stdlib.h>
10 /**
      \function SumFibonacci
      @param s0, s1 contain initial fibonacci numbers
      @param N represents how many numbers to sum
      @return sum of N fibo numbers from (s0,s1)
15    \description Algorithm is non-recursive
   */
   unsigned int SumFibonacci( int s0, int s1, int N ) {
     int i,j;
     unsigned int next=s0+s1;
20   unsigned int sum=0;
     for(i=0; i < N; ++i ) {
       next = s0 + s1;
```

```
        s1 = next;
        s0 = s1;
25      sum += next;
      }
      return sum ;
    }

30  int main( int argc, char * argv[] ) {
      int s0=0, s1=1;
      int i,j,rc=0;
      unsigned int sum = 0;
      for( i=3; i < 8; ++i ) {
35      sum = SumFibonacci( 0, 1, i );
        printf("Sum = %d\n", sum );
      }
      return ( EXIT_SUCCESS );
    }
```

Listing 3.5 Example for debugging with GDB

Listing 3.5 computes the sub of the first N Fibonacci numbers (Fibonacci numbers
are defined by the recurrence $F_n = F_{n-1} + F_{n-2}$. We compile this program and run
it. We expect the sequence to be $(1, 2, 4, 7, 12, 20,,$ but there is atleast one error in
the program and we turn to gdb to find it and fix it.

We compile the program using -ggdb option to GCC to produce a binary with
debug symbols. We also turn off optimization using the -O0 option. We then run
gdb number_test to run GDB:

```
(gdb) b main
Breakpoint 1 at 0x80484d3: file number_test.cpp, line 31.
(gdb) list
32    int i,j,rc=0;
33    unsigned int sum = 0;
34    for( i=3; i < 8; ++i ) {
```

We also put a *breakpoint* on the function SumFibonacci. We can use TAB-
completion to query GDB's loaded symbol table from the binary to complete the
function name.

```
(gdb) b SumFibonacci(int, int, int)
Breakpoint 2 at 0x804847a: file number_test.cpp, line 19.
```

We then start execution of the program by issuing the run command to GDB;
any command line arguments to our program can be added to the run command.
The program starts to execute, and hits our first break point main. We then proceed to *step* through the program using the gdb step command, which is aliased
to 's'; the command next is aliased to 'n'. Since we know the problem is inside the
SumFibonacci function (as the inputs to the function are trivial), and we have a break-
point inside the function we can *continue* using the continue command, aliased to
conti. By debugging the function we can see that the for-loop is incorrect, and that
the order of the assignments of s0, s1 is incorrect.

The current stack (sequence of functions in last in first out) order is displayed
using where.

```
(gdb) where
#0   SumFibonacci (s0=0, s1=1, N=3) at number_test.cpp:23
#1   0x08048519 in main (argc=1, argv=0xbfffec54) at
     number_test.cpp:35
(gdb)
```

To display the value of local and global variables we can use the `display` and `print` functions. We can also dump the content of the binary, a memory location, and variables to a file. We can inspect the value of CPU registers, and if the program under debug is multi-threaded we can switch between threads. If the program was compiled with debug symbols the current code at the program counter of the currently debugged thread can be displayed using the `list` command:

```
(gdb) list
18     int i,j;
19     unsigned int next=s0+s1;
20     unsigned int sum=0;
21     for(i=0; i < N; ++i ) {
22
```

We can quit GDB using the `quit` command, aliased to 'q'. GNU gdb also has remote debugging capabilities where it can connect to a remote system running the binary, while the source code resides on the host system. It can also connect to a running process, and analyze crash dump files.

GNU `gdb` is a versatile tool and recent versions have added significant features, and more are expected in the future (such as running the program backwards to debug an error condition). Although GNU `gdb` itself is a command-line driven text tool, several frontends and GUIs have been developed around it, and it has been integrated in IDEs such as Eclipse (see Section 3.6.2) and Kdevelop (see Section 3.6.3. Emacs has a frontend for GDB called `gud`, see more details on Emacs as an IDE in Section 3.6.1.

A frontend for GNU `gdb` is Insight, and we describe it in the next section.

3.9.2 Insight

As mentioned above, Insight is a frontend of GNU `gdb`. We have already discussed the command-line usage in Section 3.9.1, and in this section we discuss a graphical frontend to `gdb`. Launching Insight on a binary which has debug symbol, we get the source window with the `main` function as shown in Figure 3.12.

In addition to the GUI controls, the existing `gdb` style commands also work; we can enter them in the console window as shown in Figure 3.13.

In GNU `gdb` we can display the content of registers and memory locations but in Insight there is a graphical display which shows data-structures as unions and classes, as shown in Figure 3.14(a) and (b); registers with their contents are shown in Figure 3.15.

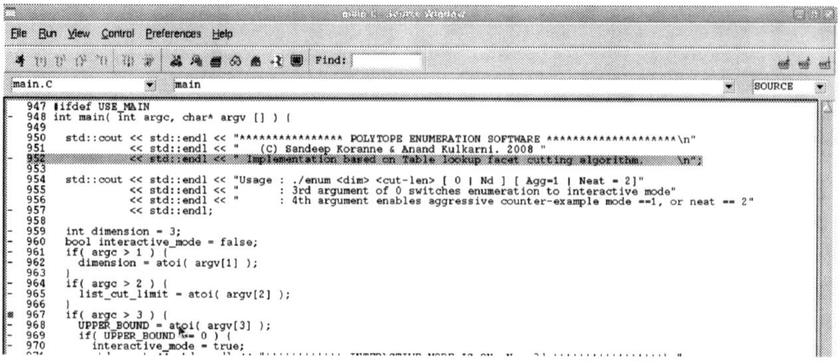

Fig. 3.12 Insight : gdb front end as debugger

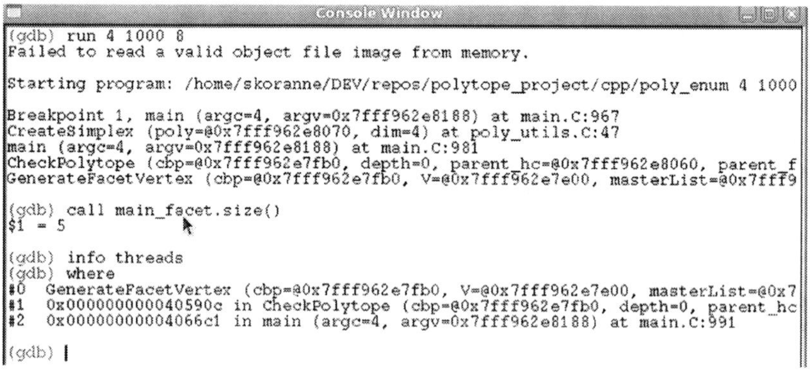

Fig. 3.13 Insight console front end for gdb

3.10 Doxygen

Doxygen is an automated document generator for source code written in C/C++, Java and many other programming languages. It is a document processor and can generate output in many formats including HTML, RTF, and LaTeX(which can then be converted to PS, PDF). Doxygen uses a control file which specifies the location of the source code and options for output generation. By default it generates HTML from source code, but the source code itself can be used to *markup* aspects of the program which are of particular interest, such as main algorithms, invariants, parameters passed to functions. For object oriented languages, Doxygen can generate class diagrams using the `dot` graph drawing tool (see Section 19.4). Doxygen is used for generating, maintaining and writing reference documentation for software. It is a valuable tool in refactoring, or understanding a complex project.

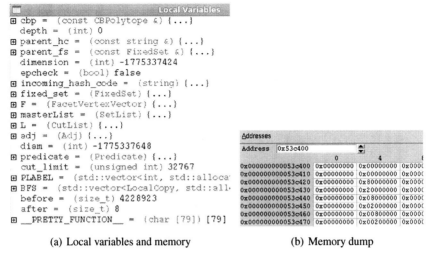

(a) Local variables and memory   (b) Memory dump

Fig. 3.14 More examples of debugging using Insight

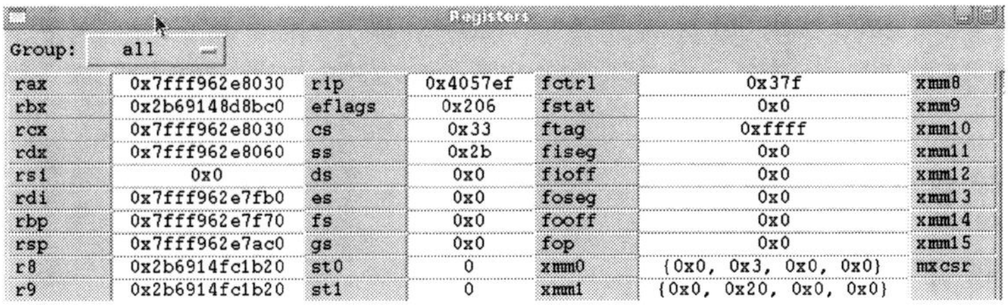

Fig. 3.15 Insight: register view and memory watch

3.10.1 Using Doxygen

Doxygen itself can be used to generate a template configuration file using the `doxygen -g` command-line option. An example config file is shown below.

3.10.1.1 Writing .doxy files

```
DOXYFILE_ENCODING    = UTF-8
PROJECT_NAME         = polytope
PROJECT_NUMBER       = 1.0
OUTPUT_DIRECTORY     = Doutput
INPUT                = cpp lisp python
FILE_PATTERNS        = *.cpp *.h *.c *.py *.lisp
```

```
CLASS_GRAPH              = YES
COLLABORATION_GRAPH      = YES
GROUP_GRAPHS             = YES
INCLUDE_GRAPH            = YES
INCLUDED_BY_GRAPH        = YES
CALL_GRAPH               = YES
DIRECTORY_GRAPH          = YES
DOT_IMAGE_FORMAT         = png
DOT_CLEANUP              = YES
```

3.10.1.2 Generating output

Output can now be generated using `doxygen Doxyfile` command. Doxygen generates HTML which can be opened using any Web browser, we show an example in Figure 3.16.

3.10.1.3 Doxygen markup in source code

The quality of the generated document from Doxygen can be enhanced if the author of the source code adds additional *markup* to the source code (using special comments and tags). We describe some of the source code markup for Doxygen below:

1. @file: name of the file,
2. @author: name of the author,
3. @version: version of the file,
4. @section: part of the source code (or algorithm),
5. @class: name of the class,
6. @param: parameter passed to function,
7. @return: return value from function.

Special comments using `///`, and `//!` can also be used to generate additional information in the Doxygen output.

3.11 Source Navigation

SourceNavigator is a source code analysis tool. It can be used to edit source, manage source code projects, display dependencies between classes and functions, and display call trees. An example of a project managed by SourceNavigator is shown in Figure 3.17(a) and Figure 3.17(b).

When importing an existing project into SourceNavigator, the project can be automatically converted to its native format. To create a new project, individual source code directories can also be added. The SourceNavigator code analysis engine then

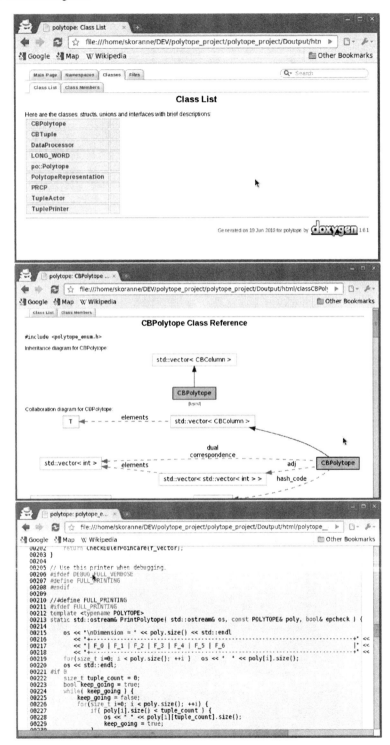

Fig. 3.16 Doxygen : document generator

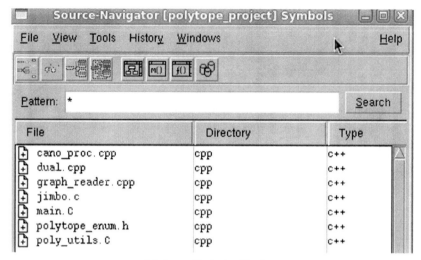

(a) SourceNavigator file view

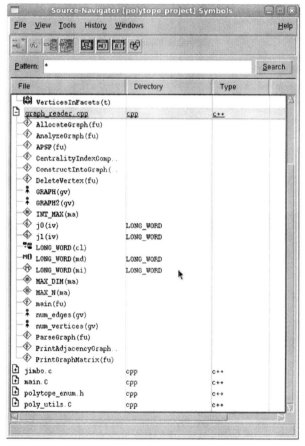

(b) SourceNavigator data view

Fig. 3.17 SourceNavigator : project management for code

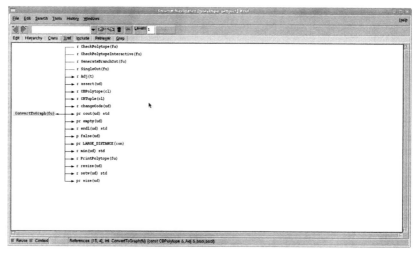

(a) SourceNavigator call graph

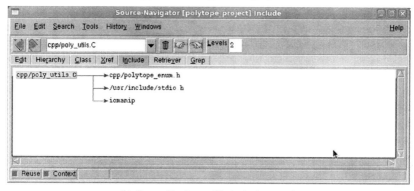

(b) SourceNavigator file inclusion graph

Fig. 3.18 SourceNavigator : project management for code

parses the code, infers the dependencies and creates the project database. As shown
in Figure 3.18, source code can be analyzed for dependencies within SourceNavigator. Even though SourceNavigator has a code editor, it is also possible to invoke
an external editor such as Emacs (see Section 3.6.1) to edit the source code. As and
when the source code is modified, SourceNavigator updates the project database to
keep the dependencies current.

3.12 Profilers

In this section we discuss code profiling. It is widely believed that "premature optimization is evil", and even when the code is mature and stable, programmers have a tendency to make mistakes when figuring out which portion of the program is the performance bottleneck. Towards this end, automatic code profiling tools have been developed. The work flow for profiling, is to compile the source program with instrumentation in order to calculate execution counts, as well as measuring the compute time spent in individual functions. Then, the program should be executed on a *representative workload*. During this execution, the program may execute several times slower than normal (due to the measurement taking place). Thus, this technique of instrumented code profiling may not be suitable for all programs, such as those requiring real-time interaction, as the measurement itself would cause the program to execute a different code pattern than the normal workload.

But for many other programs, where the program is compute bound, automatic profiling can quickly identify code sections which should be optimized; either using better algorithms and data-structures, or by compiler settings such as loop unrolling, inlining and feedback directed optimization. The two common open-source profilers are (i) GNU profiler `gprof`, and (ii) Valgrind. The former is a code instrumentation based profiler which calculates the amount of time the program counter is located within the stack frame of a function and calculates cumulative program time using sampling. Valgrind, on the other hand is not only a code profiler, but also contains tools for memory error checking, cache trace analysis, and thread error checking. Valgrind can also analyze programs without code instrumentation. Valgrind actually simulates an instruction set and can thus perform much greater levels of introspection on the program. These two profilers are discussed in the following sections.

3.12.1 GNU profiler : *gprof*

The process of code profiling using GNU `gprof` can be divided into two parts: (i) source code instrumentation and execution, and (ii) call graph display using the `gprof` program. To instrument the source code for profiling the `-pg` command-line argument to GNU `gcc` has to be used both during compilation as well as linking. While optimization settings can be left on, it is recommended to include debug symbols (using the `-ggdb` command-line argument). During the initial phase of the profiling, optimizations can be switched off also, as algorithmic optimizations are not usually affected by compiler optimizations, and in the first phase, we should strive to detect algorithmic outliers in the implementation. As stated above, the instrumented program should now be run on representative workloads. The runtime of this instrumented binary is higher than the non-instrumented code. It should also be noted that multi-threaded sampling is still (as of version 4.5.0 in the GNU compilers) not reliable, and multi-threaded execution should be switched off during profiling. But, we expect this problem to be addressed in the short term. On completion of the program, the program writes out a file (usually called `gmon.out`) which contains the summary of the execution data. This file is subsequently analyzed using the `gprof` command.

3.12.1.1 `gprof` - display call graph profile data

As we stated above, after execution of the instrumented binary we get a file, 'gmon.out' which is analyzed using `gprof`. The GNU profiler has a number of command-line options as listed below. The general command-line syntax for `gprof` is:

```
$gprof <options> <executable> <gmon.out>
```

-A=symspec: print annotated source listing, example:

```
        template <>
1468 -> inline int ChangeCode< CBTuple >( const CBTuple& v ) {
            return v.changeCode;
        }
...
Top 10 Lines:

      Line        Count

       145         1468
        89          404
        91           43
       151            3
       173            3
       447            3
       320            2
       454            2
        95            1
       103            1

Execution Summary:

        22    Executable lines in this file
        22    Lines executed
    100.00    Percent of the file executed

      1934    Total number of line executions
     87.91    Average executions per line
```

-b: be less verbose (the `brief` mode), see example shown below.
-C: print tally of functions and the number of time each function was called; for example,

```
stl_tree.h:249: \
(_ZNKSt23_Rb_tree_iteratorIiEdeEv:0x8051a46)
                    57004 executions
```

sorting the produced report file on the 3rd column gives us the function which
was invoked the maximum number of times:

```
$sort -r -n -k 3 profile.txt | more
```

-i: display information about the profile data:

```
File 'gmon.out' (version 1) contains:
1 histogram record
1984 call-graph records
0 basic-block count records
```

-p: print the *flat profile* (see discussion on profile types below),
-P: no flat profile,
--graph: print the call graph analysis; for example:

```
                        Call graph (explanation follows)


granularity: each sample covers 4 byte(s)
                               for 33.33% of 0.03 seconds

index % time    self  children    called     name
                                              <spontaneous>
[1]     100.0   0.00    0.03                      main [1]
                0.00    0.03 1/1               DP::Run() [2]
...
0.00     0.01   2290/4566 set<int, less<int> ::set() [8]
0.00     0.00   2290/4825 set<int, less<int> ::~set() [12]
```

-r: print suggested function ordering for this program; for example:

```
_ZNSt15insert_iteratorISt3setIiSt4lessIiESaIiEEEppEv
_ZSt16set_intersectionISt23_Rb_tree_const_iteratorIiES1
_ZNKSt23_Rb_tree_const_iteratorIiEneERKS0_
```

These C++ names can be demangled using the c++filt program as discussed
in Section 13.5.1.3. Using this program we can convert the above names to C++
identifiers:

```
$c++file < file.txt |more
std::insert_iterator<set<int, less<int> >::operator++()
std::insert_iterator<set<int, less<int>> > > std:: \
         __copy_move<false, false, std::bidirectional_iterat
...
```

-R mapfile: print a suggested '.o' (object file) link line ordering based on
profile data. The mapfile passed to the command-line can be generated using
the nm (see Section 13.5.1.5) GNU binutils tool. The map file has the following
syntax:

```
CP:08049690 T _init
CP:08049cc0 T _start
```

```
CP:08049cf0 T __gmon_start__
CP:0804a1cb T _Z1sRSoRK22PolyRep
CP:0804a97c T _ZN22PolyRep13AddToGraphicsEPS
CP:0804b0e4 T _ZN22PolyRep11ComputeRankEv
CP:0804b12a T _ZN22PolyRep19CalculatePropertiesEv
CP:0804dccc T _ZN13DPsor14parse_verticesEj
CP:0804eb1c T _ZN13DPsor22parse_vertices_reducedEj
CP:0804f406 T _ZN13DPsor12new_polytopeEv
CP:0804f424 T _ZN13DPsor3RunEv
CP:0804fe64 T main
```

The suggested ordering optimizes paging, TLB (translation lookahead buffer) and instruction cache behavior for the program. To use the -R option, the -a option is recommended.

gprof calculates the amount of time spent in each function and propagates this on the edges of the call graph to arrive at cumulative program times. Using gprof we can generate the following styles of reports:

1. Flat profile: displays how much time was spent in each function, and how many times that function was called. For each function it prints a tuple of information:

   a. percentage of total: running time used by the function,
   b. cumulative seconds: running sum of number of seconds,
   c. self seconds: number of seconds accounted by this function alone,
   d. number of calls: number of times this function was invoked,
   e. average number of milliseconds: spent in this function and descendants,
   f. name of the function: example is shown below.

   ```
   Each sample counts as 0.01 seconds.
   %    cumulative   self              self     total
   time sec   sec    calls  Ts/call Ts/call  name
    0.0 0.0   0.0 22159596   0.0     0.0    ConvertToPolytope
   ```

2. Call graph profile: for each function, displays which other functions called the named function, which functions did the named function call, and also an estimate of how much time was spent in each sub-routine call.
3. Annotated source listing: is a copy of the source code, labeled with the execution count of each line (similar to GNU gcov, see Section 3.8).

To get a concise overview of the performance of the program, use the *flat profile*. To eliminate spurious function calls, use *call graph profile*, and for a systematic listing of execution counts alongwith the code, use the *annotated source listing*.

```
Each sample counts as 0.01 seconds.
 %    cumulativef           self     total
time   seconds   calls  s/call  s/call  name
 8.96  0.12       1632   0.00    0.00   CBPolytope::init(bool)
 7.84  0.23    8426811   0.00    0.00   std::vector<std::vector
 6.72  0.32                              refine1
 5.97  0.40    8674249   0.00    0.00   std::vector<int>
 3.73  0.45    1960465   0.00    0.00   int const& std::min<int>
```

GNU profiler `gprof` is a versatile tool and can result in significant performance improvement. Consider the following sample of a *brief* profile output (part of the output have been edited to make the data fit in a reduced number of pages):

```
Each sample counts as 0.01 seconds.
  %   cumulative   self              self     total
 time   seconds   seconds    calls  s/call  s/call  name
 17.0   21.43     21.43      29879   0.0    0.0     CBPolytope::init(bool)
 11.0   35.35     13.93  1016658200  0.0    0.0     vector<A>::operator[](uint)
  9.0   46.75     11.39  1034828918  0.0    0.0     vector<B>::operator[](uint)
  4.3   63.24      5.46   244592208  0.0    0.0     int const& std::min<int>
```

The `CBPolytope::init()` turned out to be an algorithmic issue, where the data-structure representing the polytope could simply be copied from a basic initialized polytope (and need not be initialized per polytope). The remaining are classic C++ member function and STL optimizations. In loops dealing with STL vectors it is common to write:

```
for(size_t i=0; i < vec.size(); ++i ) {
  // process vec[i]
}
```

Due to C++ semantics, the `vector<T>::size()` function is called per trip of the `for` loop. It is indeed much better to write the code as:

```
for( size_t i=0, e=vec.size(); i < e; ++i ) {
  // process vec[i]
}
```

In this version of the loop, `vector<T>::size()` is evaluated once. The programmer must make sure that this is the intended semantics of the loop. Similarly, writing the `min` function for references as an inline function will reduce function call overhead in that example.

3.12.2 Valgrind

Valgrind is a instrumentation framework for building dynamic analysis tools. Valgrind includes: (i) cache and branch-prediction profiler, (ii) memory error detector, (iii) two thread error detectors, and (iv) a heap profiler.

3.12.2.1 Cachegrind, a cache and branch-prediction profiler

is the performance analyzer tool of Valgrind. It is invoked with the

```
Cachegrind, a cache and branch-prediction profiler
Copyright (C) 2002-2009, and GNU GPL'd, by Nicholas Nethercote et al.
Using Valgrind-3.5.0 and LibVEX; rerun with -h for copyright info
Command: ./poly_enum_i686 3 1000 9


I   refs:      306,463,674
I1  misses:        346,980
L2i misses:          2,563
I1  miss rate:        0.11%
L2i miss rate:        0.00%
```

```
D   refs:       165,173,691  (106,648,498 rd   + 58,525,193 wr)
D1  misses:         503,574  (    400,345 rd   +    103,229 wr)
L2d misses:          15,407  (      4,319 rd   +     11,088 wr)
D1  miss rate:         0.3%  (        0.3%     +        0.1%  )
L2d miss rate:         0.0%  (        0.0%     +        0.0%  )

L2  refs:           850,554  (    747,325 rd   +    103,229 wr)
L2  misses:          17,970  (      6,882 rd   +     11,088 wr)
L2  miss rate:         0.0%  (        0.0%     +        0.0%  )
```

The cachegrind tool has the following options:

--I1=<size>,<associativity>,<line size>: specify size, associativity and line size of level 1 instruction cache,

--D1=<size>,<associativity>,<line size>: specify size, associativity, and line size of level 1 data cache,

--L2=<size>,<associativity>,<line size>: specify size, associativity, and line size of level 2 cache,

--branch-sim=no|yes: enables or disables collection of cache access trace.

3.12.2.2 Callgrind, a call-graph generating cache profiler

A sample of the output from callgrind is shown below:

```
==23126==
==23126== Events    : Ir
==23126== Collected : 604392288
==23126==
==23126== I   refs:        604,392,288
```

3.12.2.3 Memcheck, memory error detector

This is the default tool of Valgrind, and performs memory leak detection as well as memory errors. A sample output is shown below:

```
HEAP SUMMARY:
in use at exit: 11,852,753 bytes in 36,614 blocks
 total heap usage: 388,258 allocs, 351,644 frees,
                   22,409,570 bytes allocated

LEAK SUMMARY:
   definitely lost: 13,834 bytes in 172 blocks
   indirectly lost: 280,784 bytes in 3,455 blocks
   possibly lost: 7,431 bytes in 195 blocks
still reachable: 11,550,704 bytes in 32,792 blocks
        suppressed: 0 bytes in 0 blocks
Rerun with --leak-check=full to see details of
leaked memory

For counts of detected and suppressed errors,
rerun with: -v
ERROR SUMMARY: 0 errors from 0 contexts
(suppressed: 15 from 8)
```

The `memcheck` tool has the following options:

`--leack-check=<no|summary|yes|full>`: search for memory leaks when the program finishes. This argument specifies the details to be computed for the leak,
`--leak-resolution=<low|med|high>`: whether to use multiple backtraces to the same memory,
`--show-reachable=<yes|no>`: when disabled the checker will only state "definitely lost",
`--malloc-fill=<hexnumber>`: can aid in detecting uninitialized memory usage by filling allocated memory with prescribed bit-pattern,
`--free-fill=<hexnumber>`: check against use-after-free type errors.

3.12.2.4 Helgrind, a thread error detector

Consider the thread-pool example of APR code (see Section 5.4) shown in Listing 5.3. We can analyze this code for data races, critical section errors, and other multi-threading problems by running the `helgrind` thread checker tool from Valgrind. We invoke the binary as:

```
valgrind -tool=helgrind ./a.out 2>thread_errors.txt
```

The output from the checker is shown below:

```
Helgrind, a thread error detector
Copyright (C) 2007-2009, and GNU GPL'd, by OpenWorks LLP et al.
Using Valgrind-3.5.0 and LibVEX; rerun with -h for copyright info
Command: ./thpool

Thread #1 is the program's root thread

Thread #2 was created
   at 0xA12DB8: clone (in /lib/libc-2.11.1.so)
   by 0xAE7FF2: pthread_create@@GLIBC_2.1
                       (in /lib/libpthread-2.11.1.so)
   by 0x400A192: pthread_create_WRK (hg_intercepts.c:229)
   by 0x400A225: pthread_create@* (hg_intercepts.c:256)
   by 0x4048871: apr_thread_create (thread.c:179)
   by 0x403125D: apr_thread_pool_create (apr_thread_pool.c:380)
   by 0x8048CB0: main (thpool.cpp:129)

Possible data race during read of size 4 at 0x41fb130 by thread #1
   at 0x403C2CE: apr_palloc (apr_pools.c:649)
   by 0x40487AC: apr_thread_create (thread.c:154)
   by 0x403125D: apr_thread_pool_create (apr_thread_pool.c:380)
   by 0x8048CB0: main (thpool.cpp:129)
 This conflicts with a previous write of size 4 by thread #2
   at 0x403C37B: apr_palloc (apr_pools.c:651)
   by 0x4030E2C: thread_pool_func (apr_thread_pool.c:222)
   by 0x40484B5: dummy_worker (thread.c:142)
   by 0x400A2A4: mythread_wrapper (hg_intercepts.c:201)
```

```
by 0xAE7AB4: start_thread (in /lib/libpthread-2.11.1.so)
by 0xA12DCD: clone (in /lib/libc-2.11.1.so)
```

A detailed analysis of this log can be helpful in uncovering race conditions, and other errors in multi-threaded applications. Valgrind can also check for *self modifying code*, which can be present in malicious programs. To enable this check we use the `--smc-check=<stack|all>` command-line option. Valgrind can be used to optimize the program's performance and check for memory leaks. There are a number of GUI frontends for Valgrind, one of which is `alleyoop`, as shown in Figure 3.19.

Fig. 3.19 Alleyoop: GUI frontend for Valgrind

3.13 Conclusions

In this chapter we have discussed software construction tools. We discussed the venerable GNU Compiler Collection (GCC) compilers, source code configuration systems (CVS, SVN and git), as well as the GNU Build Tools. Automatic build management tools GNU `gmake` and SCons are described with examples. The Bugzilla defect tracking system is described; incidentally, this book had a Bugzilla page during its development for issue tracking. Source code editors and IDEs (including Emacs, Kdevelop, and Eclipse) are shown as well as debugging, documentation, and profiling tools.

In many ways, this chapter contains foundational material for starting an open-source project, and for contributing to an existing project. Reading the documentation for each of the tools mentioned in this chapter, reading their source code, and experimenting with it will provide valuable experience to the reader in developing and using open-source software in the subsequent chapters.

Chapter 4
Standard Libraries

Abstract In this chapter we discuss the GNU C Library and the GNU C++ Standard library. Almost all open-source applications make use of the standard libraries, and indeed it is considered good programming practice to use functions wherever possible from the standard library (unless performance or other criteria clearly dictate otherwise). The important and salient functions of the C standard library are presented, in particular the use of error return code, regular expressions, and system configuration functions are presented. In the sequel of the chapter we present some of the main features of the C++ library including Standard Template Library.

Contents

In the coming chapters we discuss the various open source software libraries which are available for common computing tasks. These include the Boost C++ Project, Google's Perftools, ZLIB and bzip2 for data compression, HDF (Hierarchical Data Format), Berkeley db, MD5, Boehm garbage-collector, simplified Wrapper and Interface Generator (SWIG), and GNU Scheme.

4.1 GNU C Library

The GNU Standard C library (glibc) is the GNU implementation of the C standard library. Most C programs (except embedded platforms) use some function of the C library. C library functions are different from the system calls provided by the kernel, and execute in user-space. C library functions are provided for many of the common application processing tasks such as sorting, searching, command-line processing, FILE I/O, memory allocation, and string processing.

When using the GNU C library we should keep in mind that although it is ANSI C compliant, it also has extensions and other functions. To restrict only the ANSI

S. Koranne, *Handbook of Open Source Tools*,
DOI 10.1007/978-1-4419-7719-9_4, © Springer Science+Business Media, LLC 2011

subset we have to use the `-ansi` command-line argument to the compiler. The C library comprises of the header files, and the object code which implements the functions. The object codes are archived in the C library which we link against. To use a specific C library function the corresponding header file should be included in the application code. Any function name or macro defined in the ANSI standard should be treated by the application programmer as a reserved keyword, and should not be used in the application.

We discuss some of the common functions below.

1. Error conditions: C library functions return error codes in the global variable `errno` (while it is often described as a variable its implementation is defined as a modifiable lvalue). Once the application detects that the system function has encountered an error condition, only then should the error number be inspected to find out more details about the error. The various error codes are described in `errno.h`, and the application can use the `perror` (print error) function to print a message describing the error condition,

2. Memory allocation functions: memory can be allocated using the `malloc` family of functions which include besides `malloc`, `calloc`, `realloc`, and `alloca`. Memory thus allocated has to be freed using the `free` function. Alignment of memory can be assured using the `memalign` function (it takes the alignment as an additional argument alongwith size). To avoid locking a large memory region between small ones, very large memory allocations are performed using `mmmap` (memory mapping). A GNU extension called `memcheck` can check the consistency of the malloc heap (this checks for write after end, for example). Statistics about allocated memory can be obtained by using the `mallinfo` function which populates a structure of the same name. This structure contains the total number of blocks allocated. To reduce paging, a critical page of memory can be locked using the `mlock` function,

3. Character handling: these include the classic predicates of `isalpha`, `islower`, etc. One key addition is the support for *wide characters*, thus names as the characters can be wider than 1 byte in length,

4. String processing: another classic set of `strlen`, `strcmp`, etc. The memory copy functions are included in `string.h`, thus, to use `memcpy`, the application has to include `string.h`. The memory movement functions include, `memcpy`, `memmove`, and `memset`. String search functions such as `strchr` are also included. The string tokenizer function `strtok` finds use in lexical processing (see Section 13.3),

5. Searching and sorting: the GNU C library implements a binary search function and a sorting function. The search functions include *linear search*, names `lfind` and `lsearch` (`lsearch` is similar to `lfind` except that if the element is not present in the collection, it is added to it). For ordered collections a binary search function `bsearch` is implemented. Sorting is implemented using the `qsort` function which takes as input a comparison function which imposes a partial order on the elements in the universe,

6. Pattern matching: the function `fnmatch` implements pattern matching based on filenames and other text patterns. For *regular expression* or *regex* matching it uses a special data type `regex_t` defined in `regex.h` which has to be used to define

the pattern as a regular expression (see Section 13.3 for more details on regular expressions). Once a regular expression is defined it has to be *compiled* once before it can be used in multiple searches; compilation is done using the `regexcomp` function, while matching is done using the `regexec` function which returns 0 if the pattern matches,

7. FILE I/O functions: the GNU C library implements a plethora of functions dealing with IO. These include directory handling, file name query and resolution, and actual IO using streams (`FILE*`), which include functions such as `fopen`, `fclose`, `fread` and `fwrite`. The function `getline` reads an entire line from the stream. In addition to reading and writing raw bytes, the library also has functions for formatted input (`scanf`) and output (`printf`), which are well known, still an example is shown in Listing 4.1.

```
static int ReadCustomer( FILE* fp,
                              Customer* customer ) {
    static const int MAX_LEN = 1024;
    char temp[MAX_LEN];
5   int rc = 0;
    rc = fscanf( fp, "Customer = [ %d %s %f %d ]\n",
                   &customer->number, temp,
                   &customer->amount, &customer->status );
    if( rc < 4 ) return 0;
10  customer->len = strlen( temp );
    customer->name = malloc( ( customer->len+1 )*sizeof( char ) );
    strcpy( customer->name, temp );
    return 1;
}
```

Listing 4.1 Example of `scanf`

The lower level functions which operate not on streams but on actual files include `open`, `close` `read`, and `write`. The function `pread` and `pwrite` are similar to `read`, and `write`, but they accept a fourth argument which represents the file offset at which to perform the action, and these functions do not update the current file offset. The function `lseek` updates the current file offset for subsequent operations on that file descriptor. Consider the example shown in Listing 4.2.

```
    current_offset = lseek( fd, 0, SEEK_SET ); /* rewind */
    current_offset = lseek( fd, customer->number * RECORD_SIZE, SEEK_SET );
    if( current_offset == (off_t) -1 ) error_exit("lseek failed..\n");
    rc =  write( fd, &customer->number, sizeof( int ) );
5   rc += write( fd, &customer->len, sizeof( int ) );
    rc += write( fd, customer->name, customer->len ) ;
    rc += write( fd, &customer->amount, sizeof( float ) );
    rc += write( fd, &customer->status, sizeof( int ) );
    current_offset = lseek( fd, ( RECORD_SIZE - rc -1 ), SEEK_CUR );
10  write( fd, &customer->number, 1 ); /* marker */
    if( current_offset == (off_t) -1 ) error_exit("lseek failed..\n");
```

Listing 4.2 Example of `lseek`

GNU C library implements fast *scatter-gather* of data which is spread in memory but contiguous on the file. Memory mapped file is implemented using `mmap` function (and has to be unmapped using the `unmap` function). A recent addition to POSIX (POSIX 1b) defines asynchronous IO using the `aio_read` and `aio_write` functions. File control and status modes are also available using functions such as `fstat`. An example of using `fstat` is shown in Listing 4.3.

```
    static void CalculateFileInformation( int fd ) {
      struct stat sb;
      int rc;
      rc = fstat( fd, &sb );
5     if( ( sb.st_mode & S_IFMT ) != S_IFREG )
        error_exit("db Index should be a regular file.\n" );
      fprintf( stdout, "Preferred block size = %ld\n",
               (long) sb.st_blksize );
      fprintf( stdout, "File size = %lld\n",(long long) sb.st_size );
10    fprintf( stdout, "Blocks allocated = %lld\n",
               (long long) sb.st_blocks );
      fprintf( stdout, "Last file access = %s", ctime( &sb.st_atime ) );
      fprintf( stdout, "Last file modification = %s",
```

Listing 4.3 Example of `stat`

Directory information beyond the file level is obtained using file-system information functions defined in `unistd.h`. Files can be renamed (`rename`), or deleted (`unlink`). File size can be obtained using `stat`, and changed using `truncate` functions. Temporary files (which is guaranteed to be unique) can be opened using `tmpfile` (this function is reentrant). Interprocess communication channels can be created using `pipe`, and `popen` functions. For unrelated processes (which do not share file descriptors) a file-system file operating as a FIFO can be opened (using `mkfifo`). Thus, from shared memory, to pipes (for related processes), to FIFOs (on common file system) we come to the problem of communication channels between remote computers. This is implemented using *sockets*, which provide networking channels with similar interface as regular file descriptors.

8. Mathematical functions: including trigonometric functions are available upon including `math.h` and linking with `libm`,

9. System information :this category includes process resource usage, system information, job control, user, and groups. We have discussed resource usage functions in Section 1.1.1.1. The functions for querying the system for user, group, and other service information is done using the Name Service Switch (NSS) module. The GNU C library has functions to query group id and user id (called the persona of the process) using similarly named functions (`getuid` for get user id). An example is shown in Listing 4.4.

```
    /* \file user_info.c
       \author Sandeep Koranne (C) 2010
       \description Example of using passwd structure
    */
5   #include <stdio.h>          // for program IO
    #include <unistd.h>         // system functions
    #include <string.h>         // memory allocation
    #include <stdlib.h>         // library functions
    #include <grp.h>            // 'group' functions
10  #include <pwd.h>            // 'passwd' functions
    #include <sys/types.h>      // predefined types

    /**
       \description print information about the process' user
15  */
    static void PrintSelfUserInformation(void) {
      uid_t self;
```

```
       struct passwd *self_pwd;
       struct group  *self_grp;
20     char **member_of = NULL;

       self = getuid();
       printf("Self UID = %d\n", self );
       self_pwd = getpwuid( self );
25     if( self_pwd == 0 ) {
         perror( "pwd retrieval failed...\n");
         exit( 1 );
       }
       printf(" Self LOGIN = %s", self_pwd->pw_name );
30     printf(" Name = %s\n", self_pwd->pw_gecos );
       printf(" HOME = %s\n", self_pwd->pw_dir );
       self_grp = getgrgid( self_pwd->pw_gid );
       if( self_grp == 0 ) {
         perror( "group retrieval failed..\n");
35       exit( 1 );
       }
       printf(" GROUP = %s\n", self_grp->gr_name );
       member_of = self_grp->gr_mem;
       while( *member_of ) {
40       printf( "\tmember of %s\n", *( member_of ) );
       }
     }

     int main( int argc, char *argv[] ) {
45
       PrintSelfUserInformation();

       return (0);
     }
```

Listing 4.4 GNU libc example of passwd structure

Compiling and running this program produces the following output on my GNU/Linux system:

```
Self UID = 500
 Self LOGIN = skoranne Name = Sandeep Koranne
 HOME = /home/skoranne
 GROUP = skoranne
```

System information can be gathered using the sysconf function as shown in Listing 4.5.

```
     /* \file page_size.c
        \author Sandeep Koranne, (C) 2010
        \description Utility to print page size
     */
5    #include <unistd.h>
     #include <stdio.h>
     int main() {
       long sz = sysconf(_SC_PAGESIZE);
       long num_phys_pages = sysconf( _SC_PHYS_PAGES );
10     long num_avphys_pages = sysconf( _SC_AVPHYS_PAGES );
       printf("\n PAGE_SIZE=%ld NUM_PHYS_PAGES=%ld "
              "NUM_AV_PHYS_PAGES=%ld\n",
              sz, num_phys_pages, num_avphys_pages );
       return 0;
15   }
```

Listing 4.5 GNU libc example of sysconf

In addition to the functions described above, the GNU C library also supports *internationalization* of programs.

4.2 C++ Library

The standard C++ library is called the Standard Template Library (STL). It is a library of containers, algorithms, iterators, and associated runtime support functions. STL was designed as a generic library; the data-structures of the containers are decoupled from the algorithm which operate on them. The various containers in STL are:

1. `vector`: template class of resizable array,
2. `list`: non-intrusive doubly linked list,
3. `deque`: double ended queue,
4. `set`: template class of items having partial order (implemented using height balanced trees),
5. `multisets`: same as above (sets), except allows for more than one item with the same value,
6. `map`: dictionary class with key-value semantics, where there is a partial order in the key,
7. `multimap`: same as above (map), except allows for more than one key to have same value,
8. `string`: venerable character array.

A simple example of using `std::vector` class is shown in Listing 4.6.

```cpp
// \file stl_example.cpp
// \author Sandeep Koranne (C) 2010
// \description Standard Template Library (STL)
#include <iostream>          // for program IO
#include <vector>            // Vector class STL
#include <cassert>           // assertion checking
#include <cstdlib>           // exit

int main( int argc, char *argv[] ) {

   std::vector<int> A(5,1); // initialize contents to 1
   std::cout << "sizeof(A) = " << sizeof(A)
             << "\n A.size() = " << A.size()
          << "\nA[0] = " << A[0];
   std::cout << std::endl;
   return (0);
}
```

Listing 4.6 Using STL std::vector

The design of STL is based on the model of:

1. Concepts: a *concept* in C++ is a generalization of types to include semantic information,

2. Containers: abstract data types which are template based, and use the memory allocator of STL to provide non-intrusive container like data-structures, including, list, vectors, sets and maps. Containers in STL have strict requirements on the amortized runtime for insertion, deletion, and lookup. These requirements are part of the STL definition of the type, and a compliant implementation will ensure that the runtime expectations of the containers are met, e.g., `std::set` has $\log(n)$ requirement for `insert` and `find`,

3. Algorithms: algorithms in STL have two flavors. They can be predefined member functions on the containers (such as `std::map<K,V>::find`), or they can be generic functions such as `std::reverse( Range A, Range B)`. Algorithms, like containers, have strict requirement on runtime and memory, as part of their specification,

4. Iterators: iterators connect data-structures to algorithms (and vice-versa). Iterators also provide a level of indirection in the implementation of STL algorithms, which is necessary to prevent the combinatorial explosion which would ensue in its absence. For example, the `std::reverse` algorithm operates on iterator ranges. These iterators could be vector iterators, or list iterators. As long as an iterator meets the requirement (the requirement is presented as a *concept*) of the algorithm, that iterator can be passed to the algorithm.

On GNU/Linux, there are atleast two portable implementations of STL, (i) the GNU C++ library and (ii) STLport.

4.3 Conclusion

In this chapter we discussed the GNU C Library and the GNU C++ Standard library. The most important functions of the C standard library were presented, in particular the use of error return code, regular expressions, and system configuration functions. We also discussed the use of STL and the C++ library.

Chapter 5
Apache Portable Runtime (apr)

Abstract In this chapter we discuss the Apache Portable Runtime (APR) application development framework library API. In particular we discuss the APR memory pool, process, thread, and thread pool. We present example which use these functions in a real-life setting. APR file information functions are used with memory mapped IO. APR hash tables are used to develop a word frequency counting applications. We use the Memcache library with APR, and also present an example which uses APR shared memory.

Contents

The Apache Portable Runtime was written to support the portability of the Apache HTTP web server, and as such provides an operating system abstraction. The Subversion (SVN) version control system also uses APR. Another application development framework is Nokia/Trolltech Qt Framework (see Section 19.1.3 for more details). APR can be used to provide the application with basic functionalities in a consistent API on multiple platforms. The main functionalities of APR are shown below:

1. Memory pool,
2. Atomic operations: interface to *atomics* (see Section 3.1.7.1),
3. Dynamic Object handling,
4. Environment functions,
5. Signal handling,
6. File information and IO,

S. Koranne, *Handbook of Open Source Tools*,
DOI 10.1007/978-1-4419-7719-9_5, © Springer Science+Business Media, LLC 2011

7. Hash tables,
8. Memory map and allocation, shared memory, network library,
9. Thread and process library,
10. Mutex and condition variables, read/writer locks.

These features are described in detail with the help of examples in this chapter.

5.1 APR Memory Pool

The distinct advantages of using a memory pool over standard `malloc`, **new** are the following:

1. Efficiency: a custom memory pool can be more efficient than standard library implementation,
2. Resource tracking: a memory pool can be used as a dynamic heap where the life time of memory is tracked. Once the pool is destroyed all the memory is automatically reclaimed,
3. Constructor/Destructor: APR memory pools provide raw memory bits, so cost of constructor is not added (caveat, neither is its convenience of initializing complex data structures).

The key functions for APR memory pools are:

```
  apr_pool_create( apr_pool_t **pool, apr_pool_t *parent);
  apr_allocator_t * apr_pool_allocator_get( apr_pool_t*);
  apr_allocator_max_free_set( allocator, 64 );
  void* apr_palloc( apr_pool_t *pool, apr_size_t size);
5 apr_pool_clear( apr_pool_t *p);
  apr_pool_destroy( apr_pool_t *p );
```

Using a parent pool we can build *hierarchy* of pools, such that when the parent pool is destroyed all child pools are destroyed as well. It is also possible to register a *callback* function to be called when a pool is destroyed. Examples of using APR memory pools are presented in the next section.

5.2 APR Processes

Consider the program listing as shown in Listing 5.1.

```
  // \file calendar_proc.pp
  // \author Sandeep Koranne (C) 2010
  // \description Example of APR process type
  #include <iostream>
5 #include <stdlib.h>
  #include <assert.h>
  #include <apr_general.h>
  #include <apr_thread_proc.h>

10 static const char *PROGRAM_ARG[ 16 ];
```

```
     int main( int argc, char *argv[] ) {
       apr_status_t retval;
       apr_pool_t    *pool;
15     apr_initialize();
       apr_pool_create( &pool, NULL );
       apr_procattr_t *attribute;
       /* first create the attribute */
       retval = apr_procattr_create( &attribute, pool );
20     retval = apr_procattr_io_set( attribute, APR_NO_PIPE,
             APR_FULL_BLOCK, APR_NO_PIPE );
       //retval = apr_procattr_cmdtype_set( attribute, APR_PROGRAM_ENV );
       retval = apr_procattr_cmdtype_set( attribute, APR_PROGRAM_PATH );
       apr_proc_t calendar_process;
25     PROGRAM_ARG[0] = "cal";
       PROGRAM_ARG[1] = "1";
       PROGRAM_ARG[2] = "2010";
       PROGRAM_ARG[3] = NULL;
       retval = apr_proc_create( &calendar_process, PROGRAM_ARG[0],
30           PROGRAM_ARG, NULL, attribute, pool );
       if( retval != APR_SUCCESS ) exit( 1 ); //
       while( true ) { // read data from child process
         char buf[ 1024 ];
         retval = apr_file_gets( buf, sizeof( buf ), calendar_process.out );
35       if( APR_STATUS_IS_EOF( retval ) ) break;
         std::cout << buf;
       }
       apr_file_close( calendar_process.out );
       int status;
40     apr_exit_why_e why;
       retval = apr_proc_wait( &calendar_process, &status, &
             why, APR_WAIT );
       if( APR_STATUS_IS_CHILD_DONE( retval ) ) {
         std::cout << "WHY = ";
45       switch( why ) {
         case APR_PROC_EXIT:
           { std::cout << "APR_PROC_EXIT"; break; }
         case APR_PROC_SIGNAL:
           { std::cout << "APR_PROC_SIGNAL"; break; }
50       case APR_PROC_SIGNAL_CORE:
           { std::cout << "APR_PROC_SIGNAL_CORE"; break; }
         }
         std::cout << " Status = " << status;
       } else {
55       std::cout << "still processing...";
       }
       std::cout << std::endl;
       apr_terminate();
       return (0);
60   }
```

Listing 5.1 APR process running 'cal' program

In the above listing we see the use of several APR functions and data structures. Before using APR functionality, APR has to be initialized using: `apr_initialize()`. APR has a memory pool functionality which is used by most of APR's functions, as well as application programs. We create a memory pool using `apr_pool_create` function. The next several lines are used to create process attributes which control the process's behavior, detach state, and IO. In this case we want to read from the child process output, thus we use `APR_FULL_BLOCK` for its output.

We *wait* for the process to terminate (otherwise we would create a *zombie* process), and print its exit code and reason for exiting.

5.3 APR Threads

In addition to supporting process based multi-tasking, APR also has support for multi-threading as shown in Listing 5.2. In this program we create a thread which prints a message. The thread creation function in APR is shown below:

```
/**
 * Create a new thread of execution
 * @param new_thread The newly created thread handle.
 * @param attr The threadattr to use to determine how to create the thread
 * @param func The function to start the new thread in
 * @param data Any data to be passed to the starting function
 * @param cont The pool to use
 */
apr_thread_create(apr_thread_t **new_thread,
                  apr_threadattr_t *attr,
                  apr_thread_start_t func,
                  void *data, apr_pool_t *cont);
```

```cpp
// \file tapr.cpp
// \author Sandeep Koranne, (C) 2010
// APR test and thread support

#include <stdio.h>
#include <stdlib.h>
#include <assert.h>
#include <apr_general.h>
#include <apr_thread_proc.h>

#define NUM_THREADS 4

static void* APR_THREAD_FUNC comp( apr_thread_t *T, void* arg);

int main( int argc, char *argv [] ) {
  apr_status_t retval;
  apr_pool_t   *pool;
  apr_thread_t *THREADS[ NUM_THREADS ];
  apr_threadattr_t *thread_attribute;
  int i, j;

  apr_initialize();
  apr_pool_create( &pool, NULL );
  apr_threadattr_create( &thread_attribute, pool );
  for(i=0; i < NUM_THREADS; ++i ) {
    retval = apr_thread_create( &THREADS[i], thread_attribute,
        comp, NULL, pool );
  }
  for(i=0; i < NUM_THREADS; ++i ) {
    retval = apr_thread_join( &retval, THREADS[i] );
  }
  apr_terminate();
  return 0;
}

static void* APR_THREAD_FUNC comp( apr_thread_t *T, void* arg ) {
  printf("\n Inside apr_thread function: comp");
  apr_thread_exit( T, APR_SUCCESS );
  return NULL;
}
```

Listing 5.2 APR thread example

5.4 APR Thread Pool

Scheduling jobs such that CPUs are efficiently utilized is a non-trivial problem when the problem is not uniform, and has data dependent runtime. In such situation it is often easier to design a pool of threads which can be dynamically assigned to perform computation. Towards this end, APR has a *thread pool* facility. An example of a thread pool to perform compute intensive calculation is shown in Listing 5.3.

```cpp
/* \file thpool.cpp
   \author Sandeep Koranne, (C) 2010
   Fibonacci = (0,1,1,2,3,5,8,13,21,34,55,89)
   Sum = prefix-sum of F
   (0,1,2,4,4,12,20,
*/
#include <stdio.h>
#include <pthread.h>
#include <string.h>
#include <stdlib.h>
#include <apr_general.h>
#include <apu.h>
#include <apr_reslist.h>
#include <apr_thread_pool.h>
#include <cassert>
#include <iostream>

struct Data {
  int s0, s1, N;
  unsigned int sum;
};

unsigned int CountCollatzReturn( unsigned int N ) {
  unsigned int count = 0;
  while( 1 ) {
    count++;
    if( N <= 1 ) return count;
    if( N % 2 ) N = 3*N+1;
    else N = N/2;
  }
  return count;
}

#if 0
/**
   \function SumFibonacci
   @param s0, s1 contain initial fibonacci numbers
   @param N represents how many numbers to sum
   @return sum of N fibo numbers from (s0,s1)
   \description Algorithm is non-recursive
*/
unsigned int SumFibonacci( int s0, int s1, int N ) {
  int i,j;
  unsigned int next=s0+s1;
  unsigned int sum=0;
  for(i=0; i < N-2; ++i ) {
    next = s0 + s1;
    s0 = s1;
    s1 = next;
    sum += next;
  }
  return sum ;
}
#else
unsigned int SumFibonacci( int s0, int s1, int N ) {
  std::cout << "Running function " << __FUNCTION__ << std::endl;
```

```
     int i,j;
     unsigned int next=s0+s1;
     unsigned int sum=0;
60   for(j=0; j < 100000000; ++j ) {
       sum = 0;
       for(i=0; i < N-2; ++i ) {
         next = s0 + s1;
         s0 = s1;
65       s1 = next;
         sum += next;
       }
     }
     return sum ;
70   }
     #endif

     void* APR_THREAD_FUNC FUNCTION( apr_thread_t *thd, void* arg ) {
       struct Data* data = (struct Data*) arg;
75     data->sum = SumFibonacci(data->s0, data->s1, data->N );
       data->sum = CountCollatzReturn( data->sum );
       return APR_SUCCESS;
     }

80   //#define SERIAL_VERSION
     #ifdef SERIAL_VERSION
     int main( int argc, char * argv[] ) {
       apr_status_t retval;
       apr_pool_t *pool;
85
       const int LOOP_COUNT  = 8;
       const int Z = 7;
       int s0=0, s1=1;
       int i,j,rc=0;
90     unsigned int sum = 0;
       struct Data data[LOOP_COUNT];
       for( i=0; i < LOOP_COUNT; ++i ) {
         data[i].s0 = i; data[i].s1 = i+1; data[i].N = (i*Z);
       }
95
       apr_initialize();
       apr_pool_create( &pool, NULL );
       for(j=0; j < LOOP_COUNT; ++j ) {
         FUNCTION( NULL, (void*) &data[j] );
100    }
       // all work has been done
       for( i=0; i < LOOP_COUNT; ++i ) {
         printf("\n sum (%d,%d,%d) = %d", i,(i+1),(i*Z), data[i].sum );
       }
105    printf("\n");
       return ( EXIT_SUCCESS );
     }
     #endif

110  #define THREADED_VERSION
     #ifdef THREADED_VERSION
     int main( int argc, char * argv[] ) {
       apr_status_t retval;
       apr_pool_t *pool;
115
       const int LOOP_COUNT  = 8;
       const int Z = 7;
       int s0=0, s1=1;
       int i,j,rc=0;
120    unsigned int sum = 0;
       struct Data data[LOOP_COUNT];
       for( i=0; i < LOOP_COUNT; ++i ) {
         data[i].s0 = i; data[i].s1 = i+1; data[i].N = (i*Z);
```

```
125    }

       apr_thread_pool_t *thrp;
       apr_initialize();
       apr_pool_create( &pool, NULL );
       retval = apr_thread_pool_create( &thrp, 2, 4, pool );
130    if( retval != APR_SUCCESS ) assert( 0 && "apr_thread_pool_create");


       for( j=0; j < LOOP_COUNT; ++j ) {
         retval = apr_thread_pool_push( thrp, FUNCTION, &data[j], 0, NULL );
135      if( retval != APR_SUCCESS )
           assert( 0 && "apr_thread_pool_push");
       }

       apr_size_t scount = apr_thread_pool_scheduled_tasks_count( thrp );
140    std::cout << std::endl << scount
                   << " tasks have been scheduled..." << std::endl;
       scount = apr_thread_pool_tasks_count( thrp );
       std::cout << std::endl << scount
                   << " tasks are waiting..." << std::endl;
145    scount = apr_thread_pool_tasks_run_count( thrp );
       std::cout << std::endl << scount
                   << " tasks have completed..." << std::endl;

       while( scount != LOOP_COUNT ) {
150      scount = apr_thread_pool_tasks_run_count( thrp );
         if( scount == LOOP_COUNT ) break;
         std::cout << std::endl << scount
                     << " tasks have completed..." << std::endl;
         sleep(5);
155    }
       // all work has been done
       retval = apr_thread_pool_destroy( thrp );
       for( i=0; i < LOOP_COUNT; ++i ) {
         printf("\n sum (%d,%d,%d) = %d", i,(i+1),(i*Z), data[i].sum );
160    }
       printf("\n");
       return ( EXIT_SUCCESS );
     }
     #endif
```

Listing 5.3 APR thread pool example

5.5 File information, IO, and Memory mapped files

APR supports an abstract file type, `apr_file_t` as well as a *file info* type, `apr_finfo_t`.
To open a file for reading we can use the following APR function:

```
     apr_status_t retval;
     apr_pool_t *pool;
     apr_pool_create( &pool, NULL );
     apr_finfo_t file_info;
5    apr_file_t  *fp;
     retval = apr_file_open( &fp, fileName,
             APR_READ|APR_BINARY,
             APR_OS_DEFAULT, pool );
     retval = apr_file_info_get( &file_info, APR_FINFO_SIZE, fp );
```

Moreover, memory mapped files can be created using the `apr_mmap_create` function, e.g.:

```
retval = apr_mmap_create( &mmap, fp, 0, file_info.size,
        APR_MMAP_READ, pool );
// .. use the mmap->mm data as const char *
apr_mmap_delete( mmap ); // unmap the file
```

5.6 Hash tables

Hash tables in APR are implemented using APR *pools*. The following functions are useful when developing applications using APR hash tables:

```
apr_hash_t* apr_hash_make( apr_pool_t* );
void* apr_hash_get( apr_hash_t*, void* key, apr_ssize_t keylen, void* val);
```

A simple word frequency counting application using APR hash tables is presented in Listing 5.4.

```
     // \file apr_hash_example.cpp
     // \author Sandeep Koranne (C) 2010
     // \description Use of hash table in APR
     #include <iostream>        // Program IO
5    #include <fstream>         // File IO
     #include <cstdlib>         // exit
     #include <string>          // STL string
     #include <sstream>         // string iterator
     #include <cassert>         // assertion checking
10   #include <cstring>         // memory functions
     #include <apr_general.h>   // APR basic functions
     #include <apr_hash.h>      // APR hash tables

     static int PrintHashFunction( void* dummy, const void* key,
15                                 apr_ssize_t klen, const void* val ) {
       std::cout << (const char*)key << "\t" << (int)val << "\n";
     }

     static void CountWordFrequency( const char* filename ) {
20     std::ifstream ifs( filename );
       if( !ifs ) {
         std::cerr << "Unable to open file: " << filename;
         exit(1);
       }
25     apr_status_t retval;
       apr_pool_t *pool;
       apr_pool_create( &pool, NULL );
       apr_hash_t *word_hash = apr_hash_make( pool );
       assert( word_hash && "Word hash not created." );
30     char line[1024];
       while( ifs ) {
         ifs.getline( line, 1024 );
         std::istringstream sstr( line );
         while( sstr ) {
35         std::string word; sstr >> word;
           if( word == "" ) break;
           std::cout << "Processing " << word << " " << word.size() << "\n";
           char *next_word = new char[ word.size()+1];
           strcpy( next_word, word.c_str() );
40         void* val = apr_hash_get( word_hash, (void*) next_word, word.size() );
```

```
          if( val == NULL ) { // first time
            apr_hash_set( word_hash, (void*) next_word, word.size(),
              (const void*)(1) );
          } else {
45          int count = int(val) + 1;
            apr_hash_set( word_hash, (void*) next_word, word.size(),
              (const void*)(count) );
          }
        }
50    }
      // now the hash table has been constructed.
      std::cout << "Hash has " << apr_hash_count( word_hash ) << " elements.\n";
      {
        apr_hash_index_t *hi;
55      for (hi = apr_hash_first(pool, word_hash); hi; hi = apr_hash_next(hi)) {
          void *val;
          const void *key_word;
          apr_ssize_t key_len;
          apr_hash_this(hi, &key_word, &key_len, &val);
60        std::cout << (const char*)(key_word) << "\t" << (int)val << "\n";
        }
      }
      //(void)apr_hash_do( PrintHashFunction, NULL, word_hash );
      apr_pool_destroy( pool );
65  }

    int main( int argc, char *argv [] ) {
      if( argc != 2 ) {
        std::cerr << "Usage: ./apr_hash_example <file>..\n";
70      exit(1);
      }
      apr_initialize();
      CountWordFrequency( argv[1] );
      std::cout << std::endl;
75    apr_terminate();
      return (0);
    }
```

Listing 5.4 Example using APR hash tables for word frequency counting

In Listing 5.4 it should be noted that the key to the hash table is the address of
an allocated memory containing the word. Although, the hash function correctly
calculates the hash for different words, reusing the same address as the key for
different words will not work, as the hash table does not store keys inside the hash
table (it only stores the value for that key, indexed by the computed hash function).
Compiling the program shown in Listing 5.4 and running it on a small data set gives
us:

```
./apr_hash_example small_words.txt
Processing jill 4
Processing the 3
Processing jack 4
Processing jack 4
Processing jill 4
Hash has 3 elements.
the 1
jack 2
jill 2
```

The program (in Listing 5.4) also shows the two hash table iteration mechanisms available in APR; (i) the use of the hash index type and (ii) using the callback function.

5.7 Using Memcache with APR

As we will discuss in Section 9.6, *memcached* is a high-performance distributed memory object caching system. It is designed for reducing the database load in web applications and speed up dynamic web content generation. It comprises of a server component which manages the lifetime of the cached objects which can be accessed on the network. Using APR, we can integrate Memcache functionality in an application. Ofcourse, a valid running memcached process should be executing on the server with the port accessible to the client program. An example of using APR with Memcache is shown in Listing 5.5.

```cpp
// \file aprcache.cpp
// \author Sandeep Koranne (C) 2010
// \description Use of memcache server with APR
#include <iostream>           // Program IO
#include <cassert>            // assertion checking
#include <cstdlib>            // exit
#include <fstream>            // C++ STL for file IO
#include <apr_general.h>      // APR functions
#include <apr_memcache.h>     // memcache client in APR
#if 0
echo "stats settings" | nc localhost 11211
#endif
int main( int argc, char *argv []) {
  apr_initialize();
  apr_status_t retval;
  apr_pool_t *pool;
  apr_pool_create( &pool, NULL );
  // create a server object
  apr_memcache_server_t *myserver;
  retval = apr_memcache_server_create( pool, "localhost", 11211,
                                       1, 10, 10, 1000, &myserver );
  if( retval != 0 )
    { std::cerr << "Unable to create server object..\n"; exit(1); }

  apr_memcache_t *mycache;
  retval = apr_memcache_create( pool, 2, 0x0, &mycache );
  if( retval != 0 ) exit(1);

  retval = apr_memcache_add_server( mycache, myserver );
  if( retval != 0 )
    { std::cerr << "Unable to add server object..\n"; exit(1); }

  retval = apr_memcache_enable_server( mycache, myserver );
  if( retval != 0 )
    { std::cerr << "Unable to enable server object..\n"; exit(1); }

#if 0
  retval = apr_memcache_set( mycache, "ABCDE", "KEY", 3, 100, 0x0 );
  if( retval != 0 )
    { std::cerr << "Unable to store data on server object..\n"; exit(1); }
#endif
  // now check to see if you can retrieve this information
  char *baton = NULL;
```

```
         apr_size_t baton_len = 0;
45       retval = apr_memcache_getp( mycache, pool, "ABCDE",
                     &baton, &baton_len, 0x0 );
         // An application which uses memcache must be prepared
         // to assume data is not present on the server.
         if( ( retval != 0 ) || ( baton == NULL ) || ( baton_len == 0 ) ) {
50         std::cout << "Data for key ABCDE does not exist in server.\n";
         } else {
           std::cout << "Memcache server returned : " << baton << std::endl;
         }

55       apr_memcache_server_t *server =
           apr_memcache_find_server( mycache, "localhost", 11211 );
         if( server ) {
           std::cout << "Found memcached server running...\n";
         } else {
60         std::cout << "Unable to find memcached server running...\n";
           exit(1);
         }

         // get server statistics
65       apr_memcache_stats_t *stats;
         retval = apr_memcache_stats( myserver, pool, &stats );
         if( retval != 0 ) {
             std::cerr << "Unable to collecte memcache server stats..\n";
             exit( 1 );
70       }
         std::cout << "\nMemcache server stats...." << std::endl
                   << "\nVersion information     : " << stats->version
                   << "\nPID                     : " << stats->pid
                   << "\nUptime                  : " << stats->uptime
75                 << "\nCurrent items stored    : " << stats->curr_items
                   << "\nBytes used              : " << stats->bytes
                   << "\nCurrent connections     : " << stats->curr_connections
                   << "\nBytes written           : " << stats->bytes_written;

80       std::cout << std::endl;
         apr_terminate();
         return (0);
       }
```

Listing 5.5 Using memcache with APR

Compiling and running this program gives us:

```
./aprcache
Memcache server returned : KEY
Found memcached server running...
Memcache server stats....
Version information     : 1.4.5
PID                     : 30661
Uptime                  : 1920
Current items stored    : 1
Bytes used              : 57
Current connections     : 10
Bytes written           : 9251

$ ./aprcache
Data for key ABCDE does not exist in server.
Found memcached server running...
Memcache server stats....
Version information     : 1.4.5
```

```
PID                     : 30661
Uptime                  : 1991
Current items stored    : 0
Bytes used              : 0
Current connections     : 10
Bytes written           : 10050
```

It can be seen on line 38 of Listing 5.5 that we install the data on the server with a lifetime of 100 seconds. This means, that after 100 seconds, the data is ejected from the cache. By compiling the program with and without the data insertion we can see the effect of cache ejection as shown in the above transcript.

5.8 Shared memory with APR

As we have seen above APR has functions for invoking sub-processes and multi-processing. Inter-process communication involving large shared data segments is often accomplished using *shared memory*. Shared memory segments are memory locations which can be accessed by multiple processes. All the processes must agree (or be informed of) the *segment name* through which the shared memory will be access. As with any shared resource, critical sections need to be protected against inadvertent use by another process. The APR functions apr_shm_create creates a shared segment of the given name and size. Function, apr_shm_destroy destroys the shared segment. During the time the shared segment is active, its memory address can be calculated using the apr_shm_baseaddr_get function which returns a handle to the underlying memory. An example of creating a shared memory segment is shown in Listing 5.6.

```cpp
   // \file aprshmA.cpp
   // \author Sandeep Koranne, (C) 2010
   // \description Example of shared memory using APR
   #include <iostream>          // for program IO
5  #include <cstdlib>           // for exit
   #include <apr_general.h>     // APR functions
   #include <apr_shm.h>         // APR shared memory

   static const char SHM_FILE_NAME[] = "SHM1234MHS"; // should be common
10
   int main( int argc, char *argv [] ) {
     apr_initialize();
     apr_status_t retval;
     apr_pool_t *pool;
15   apr_pool_create( &pool, NULL );
     apr_shm_t *shm_segment;
     retval = apr_shm_create( &shm_segment, 1024, SHM_FILE_NAME, pool );
     if( retval != 0 ) {
       std::cerr << "Unable to create shared memory segment..\n";
20     exit( 1 );
     }
     int *data = (int*) apr_shm_baseaddr_get( shm_segment );
     for( int i=0; i < 10; ++i ) data[i] = i;

25   int x;
     std::cout << "Enter a number when done with programB:";
     std::cin >> x;
```

```
      std::cout << std::endl;

30    apr_shm_destroy( shm_segment );
      apr_pool_destroy( pool );
      apr_terminate();
      return (0);
   }
```

Listing 5.6 Example of using shared memory with APR

The other process requests access to the shared segment using the `apr_shm_attach` function using the same filename as the process which created the shared segment. Once the segment has been attached the baseaddress and segment size can be inspected. An example of creating a shared memory segment is shown in Listing 5.7.

```
   // \file aprshmA.cpp
   // \author Sandeep Koranne, (C) 2010
   // \description Example of shared memory using APR
   #include <iostream>          // for program IO
5  #include <cstdlib>           // for exit
   #include <apr_general.h>     // APR functions
   #include <apr_shm.h>         // APR shared memory

   static const char SHM_FILE_NAME[] = "SHM1234MHS"; // should be common
10
   int main( int argc, char *argv [] ) {
      apr_initialize();
      apr_status_t retval;
      apr_pool_t *pool;
15    apr_pool_create( &pool, NULL );
      apr_shm_t *shm_segment;
      retval = apr_shm_attach( &shm_segment, SHM_FILE_NAME, pool );
      if( retval != 0 ) {
        std::cerr << "Unable to attach shared memory segment..\n";
20      exit( 1 );
      }
      apr_size_t data_size = apr_shm_size_get( shm_segment );
      std::cout << "size of shared memory segment = " << data_size << "\n";
      int *data = (int*) apr_shm_baseaddr_get( shm_segment );
25    for( int i=0; i < 10; ++i ) {
        if( data[i] != i ) {
          std::cerr << "Data corrupted in shared segment..\n";
          exit(1);
        }
30    }
      std::cout << std::endl << "Data in shared segment is valid.\n";
      apr_shm_detach( shm_segment );
      apr_pool_destroy( pool );
      apr_terminate();
35    return (0);
   }
```

Listing 5.7 Using shared memory segment with APR

The Qt library also has shared memory segment functionality available as part of its application programming API, see Section 19.1.4.1 for more details. For POSIX shared memory functionality, run `man shm_overview` for an overview of the shared memory functions in POSIX.

5.9 Conclusion

Apache Portable Runtime (APR) is indeed a versatile application development library. It not only provides an operating system independent set of routines with predetermined semantics on a wide variety of machine, OS combinations, but also provides useful application development framework utilities such as hash tables, memory pools, multi-threading and multi-process supports, memory mapping, and network access.

Chapter 6
Boost C++ Libraries

Abstract In this chapter we discuss the Boost C++ API. Boost is a peer-reviewed C++ class library which implements many interesting and useful data structures and algorithms. In particular we discuss the use of Boost smart pointers, Boost asynchronous IO, and IO Streams. Boost also implements many data structures which are not present in the C++ standard library (e.g. bimap). Boost Graph Library (BGL) is presented with the help of real-life example. We compare Boost multi-threading and memory pool performance to APR. We discuss the integration of Python with C++ using Boost. We conclude the chapter with a discussion of Boost Generic Image Processing Library.

Contents

Boost is a collection of C++ libraries and header files which are peer-reviewed, portable, and work well with standard C++ libraries. Indeed, some of the Boost libraries have even become part of the new C++ standard. The main libraries present in Boost (as of version 1.4.2) are given below:

1. Template meta programming and C++ enhancements: these include (i) `any`, a generic container of single values of different types, (ii) `array`, STL compliant wrapper for fixed sized arrays, (iii) `bimap`, bidirectional map for C++, (iv) concept checking tools, (v) `foreach` in C++, (vi) `functional/hash` for TR1 C++,

2. Data structures: disjoint sets, date/time, dynamic bitset, property maps, unordered associative containers, universally unique identifier (UUID),

3. Asynchronous IO (asio): see Section 6.2 below,

S. Koranne, *Handbook of Open Source Tools*, 127
DOI 10.1007/978-1-4419-7719-9_6, © Springer Science+Business Media, LLC 2011

4. Memory Pool and Smart Pointer: flyweight pattern, memory pool, see Section 6.1 below, pointer container, serialization (see also XDR in Section 9.1.1), smart pointers,
5. Mathematics: linear algebra, quaternions, octonions, interval arithmetic, special functions (see also GNU scientific library, rational number class, uBLAS, in Section 16.5), generic image library,
6. Boost Graph Library (BGL): library for manipulating graphs (vertices and edges),
7. Lexical analysis and parsing: regular expression parsing, Spirit LL parser framework, tokenizer. An example of using Boost Spirit framework is given in Section 13.4.1.
8. Multi-threading: interprocess, memory mapped files, shared memory, system interface, threading library interface, timer functions, interface to MPI (see Section 12.3.1 for more details on Boost MPI),
9. CRC checksum: see Section 9.2 for a detailed example,
10. Python/C++ integration: see Section 9.7 for an example.

We discuss some of the important Boost C++ libraries below:

6.1 Boost smart pointer and memory pool

Memory pools are an important memory optimization technique. The `malloc` provided with the system library has been optimized for best average case performance; in many situations where the programmer is aware of the number, lifetime, usage patterns of objects, it is certainly possible to come up with allocation schemes which either reduce memory consumption, or optimize runtime (or both). We had previously seen the Apache Portable Runtime library (APR) which has a memory pool (see Section 5.4). It is instructive to compare performance of Boost memory pool with that of APR. See Figure 6.1(a) for a runtime comparison. Figure 6.1(b) denotes the number of minor page faults associated with the memory allocation. We will discuss performance optimization tools (perftools) in Section 7.1, and by linking against the provided `libtcmalloc` we reran the experiments. The comparison with and without `libtcmalloc` are shown in Figure 6.2(a), for Boost and Figure 6.2(b) for APR.

The internal memory allocation scheme for Boost and APR is readily apparent from the graphs.

The source code for this example, which can also be used as a short model of using Boost pools and APR is given in Listing 6.1.

```
   // \file compare_pool.cpp
   // \author Sandeep Koranne, (C) 2010
   // \description Example of Boost memory pool class
   #include <fstream>
 5 #include <boost/pool/pool.hpp>
   #include <boost/pool/object_pool.hpp>
   #include <apr_general.h>
   #include <iostream>
   #include <cassert>
```

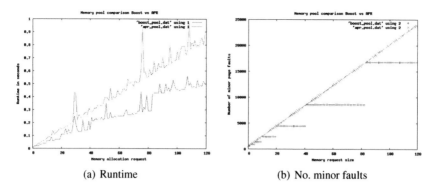

Fig. 6.1 Comparison of Boost vs APR memory pool

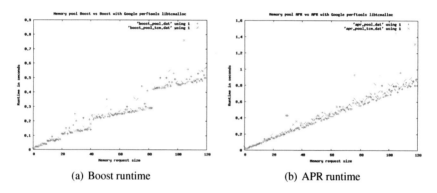

Fig. 6.2 Effects of linking with Google's perftools

```
10   struct MyObject {
       int X;
       char *p;
       MyObject( ) { std::cout << __PRETTY_FUNCTION__ << "\n"; }
15     ~MyObject( ) { std::cout << __PRETTY_FUNCTION__ << "\n"; }
     };

     int normal_pool() {
       boost::pool<> pool( sizeof( double ) );
20
       double *allocated = (double*)pool.malloc();
     }

     int object_pool() {
25     boost::object_pool<MyObject> pool;

       MyObject *A = pool.malloc();
     }

30   unsigned long BoostPoolTest( unsigned long N ) {
       boost::pool<> pool( sizeof( int ) );
```

```
        unsigned long sum = 0;
        for( unsigned long i=0; i < N; ++i ) {
          int * const A = (int*) pool.malloc();
35        *A = (int)A;
          sum += *A;
        }
        return sum;
      }
40
      unsigned long APRPoolTest( apr_pool_t *pool,
            unsigned long N ) {
        unsigned long sum = 0;
        for( unsigned long i=0; i < N; ++i ) {
45        int * const A = (int*) apr_palloc( pool, sizeof( int ) );
          *A = (int)A;
          sum += *A;
        }
        return sum;
50    }
      static size_t GetMemorySize(void)
      {
          std::ifstream ifs;
          ifs.open("/proc/self/statm",std::ios::in);
55        size_t memoryUsed = 0;
          ifs >> memoryUsed;
          return memoryUsed;
      }

60    static void DisplayMemoryUsed(void)
      {
          size_t memUsed = GetMemorySize();
          std::cout << std::endl << "Program consumed "<< memUsed << " kb.\n";
      }
65
      int main(int argc, char *argv[] ) {
        static const unsigned long COUNT = atol( argv[1] );
        int use_boost = atoi( argv[2] );
        apr_status_t retval;
70      apr_pool_t *apr_pool;
        normal_pool();
        object_pool();
        return 0;
        if( use_boost ) {
75        BoostPoolTest( COUNT );
        } else {
          apr_initialize();
          apr_pool_create( &apr_pool, NULL );
          APRPoolTest( apr_pool, COUNT );
80      }
        //DisplayMemoryUsed();
        return 0;
      }
```

Listing 6.1 Example of Boost memory pool

One important thing to keep in mind when using Boost `object_pool` is that the
destructor of the object in the pool is called automatically when the pool is destroyed. In the above listing we see:

```
MyObject::~MyObject()
```

when the pool goes out of scope.

6.2 Boost asio framework

A fundamental component of the Boost `asio` framework for communication is the *socket* which provides an abstraction for the usual networking socket available in POSIX. The socket as defined in `basic_datagram_socket` is a template type which can be instantiated with different protocols. The socket type can be configured to specify the endpoint-type and other parameters. It supports functions such as `bind`, `connect`, `close`, `receive`, `receive_from` and `shutdown`.

Consider a UDP server returning the server's date and time on a specified port. The code using Boost asio is shown in Listing 6.2. The corresponding client code is shown in Listing 6.3. We can trivially change the server code to instead return the load average as shown.

```
   // \file udp_server.cpp
   // \author Sandeep Koranne, (C) 2010
   // \description Example of Boost ASIO UDP connection
   #include <ctime>              // for the date/time function
 5 #include <iostream>           // program IO
   #include <fstream>            // for ifstream
   #include <string>             // string to store date
   #include <boost/array.hpp>    //
   #include <boost/asio.hpp>

10
   static const int PORT_NUMBER = 8954; // port of server
   using boost::asio::ip::udp;


   #ifdef PRINT_TIME
15 std::string make_info_string() {
     using namespace std; // For time_t, time and ctime;
     time_t now = time(0);
     return ctime(&now);
   }
20 #endif
   std::string make_info_string() {
     char temp[1024];
     std::ifstream ifs("/proc/loadavg");
     ifs.getline( temp, 1024 );
25   return std::string( temp );
   }


   int main() {
     boost::asio::io_service io_service;

30
     udp::socket socket(io_service, udp::endpoint(udp::v4(), PORT_NUMBER));
     while( true ) {
       boost::array<char, 1> recv_buf; // this is just the trigger
       udp::endpoint remote_endpoint;
35     boost::system::error_code error;
       socket.receive_from(boost::asio::buffer(recv_buf),
                           remote_endpoint, 0, error);
       std::string message = make_info_string();
       boost::system::error_code ignored_error;
40     socket.send_to(boost::asio::buffer(message),
                      remote_endpoint, 0, ignored_error);
     }
     return 0;
   }
```

Listing 6.2 Example of Boost asio UDP server

```
// \file udp_client.cpp
// \author Sandeep Koranne, (C) 2010
// \description Example of Boost ASIO UDP connection
#include <iostream>          // for program IO
5   #include <stdlib.h>          // exit
#include <string>            // hostname
#include <boost/array.hpp>   // boost::array
#include <boost/asio.hpp>    // Boost ASIO library

10  using boost::asio::ip::udp;

    int main(int argc, char* argv[]) {
      std::string hostName;
      if (argc != 2) {
15      std::cerr << "Usage: udp_client <host>" << std::endl;
        exit( 1 );
      } else {
        hostName = argv[1];
      }
20    boost::asio::io_service io_service;
      udp::resolver resolver(io_service);
      // the port number is specified as a string argument
      udp::resolver::query query(udp::v4(), hostName, "8954");
      udp::endpoint receiver_endpoint = *resolver.resolve(query);
25
      udp::socket socket(io_service);
      socket.open(udp::v4());              // Use UDP socket

      boost::array<char, 1> send_buf  = { 0 };
30    socket.send_to(boost::asio::buffer(send_buf), receiver_endpoint);

      boost::array<char, 128> recv_buf;
      udp::endpoint sender_endpoint;
      size_t len = socket.receive_from(
35           boost::asio::buffer(recv_buf),
             sender_endpoint);
      if( len > 128 ) len=128;
      std::cout.write(recv_buf.data(), len);
      return 0;
40  }
```

Listing 6.3 Example of Boost asio UDP client

Compiling this requires the `boost_system` and `pthread` library to be linked:

```
$g++ -o udp_server udp_server.cpp  -lboost_system -lpthread
```

6.2.1 Boost IOStreams framework

The architecture of Boost IOStreams is based on a pipeline of components in which data is passed from *source* to *sink* and undergoes filtering and conversion at each stage of the pipeline. The source for the stream can be a disk file, memory mapped file, or array of bytes. The sink can be a disk file, memory-mapped file, or array of bytes as well. The filters included in the IOStreams library include compression filters based on `zlib`. We present an example of a writer and reader based on Boost IOStreams which store the data on disk in a compressed manner.

The writer is shown in Listing 6.4, while the reader is shown in Listing 6.5.

```cpp
// \file file_io_write.cpp
// \author Sandeep Koranne, (C) 2010
// \description Using Boost IO Streams
#include <boost/iostreams/device/file.hpp>
5   #include <boost/iostreams/filtering_stream.hpp>
#include <boost/iostreams/filter/gzip.hpp>
using namespace boost::iostreams;
int main() {
  filtering_ostream out; // STL compliant io stream
10    out.push(gzip_compressor()); // step 1 : compress
  out.push(file_sink("am.txt"));// step 2: write to disk
  out << 1 << " " << 2 << " " << 3 << " " << "Johnson";
  return 0;
}
```

Listing 6.4 Example of Boost IOStreams writer

```cpp
// \file file_io_read.cpp
// \author Sandeep Koranne, (C) 2010
// \description Using Boost IO Streams
#include <string>
5   #include <iostream>
#include <boost/iostreams/device/file.hpp>
#include <boost/iostreams/filtering_stream.hpp>
#include <boost/iostreams/filter/gzip.hpp>
using namespace boost::iostreams;
10  int main() {
  filtering_istream data_in; // STL iostream compliant
  data_in.push(gzip_decompressor()); // step 2 : decompress
  data_in.push(file_source("am.txt")); // step 1: read from source
  int x,y,z;  data_in >> x >> y >> z;
15    std::string name;  data_in >> name;
  std::cout << "Read " << x << " " << y << " " << z
            << " Name = " << name << std::endl;
  return 0;
}
```

Listing 6.5 Example of Boost IOStreams reader

Compiling and running these programs gives:

```
$g++ file_io_write.cpp -o file_io_write -lboost_iostreams -lz
./file_io_write
$file am.txt
am.txt: gzip compressed data
$g++ file_io_read.cpp -o file_io_read -lboost_iostreams -lz
$./file_io_read
Read 1 2 3 Name = Johnson
```

Later in Section 8.1 we shall see more examples of using zlib for compression and decompression.

6.3 Boost data structures

6.3.0.1 Dynamic bitset

Boost dynamic bitset provides a class to represent a set of bits which can be accessed using the `[]` operator. It is identical to the `std:bitset`, except that the size of underlying container is not fixed at compile time (for `std::bitset` it has to set at compile time).

6.3.0.2 Bimap: bidirectional map

Boost Bimap is a map which represents bidirectional relations between elements. This container is designed to work as two independent maps, from X to Y, and Y to X can be written as:

```
typedef boost::bimap<X,Y> XY_BM;
```

For a given bimap, the left map view represents the X to Y map, while the right map view represents the Y to X map. For example a representation for a dictionary can be:

```
typedef boost::bimap< std::string, int > DICT;
DICT d;
d.insert( DICT::value_type("Jack", 1 ) );
```

A full example is given in Listing 6.6.

```cpp
// \file bimap_example.cpp
// \author Sandeep Koranne, (C) 2010
// \description Example of Boost bimap
#include <boost/bimap.hpp>
#include <string>
#include <iostream>

typedef boost::bimap< std::string, int > DICT;

int main( int argc, char * argv[] ) {
  DICT d;
  d.insert( DICT::value_type("Jack", 1 ) );
  d.insert( DICT::value_type("Jane", 2 ) );
  d.insert( DICT::value_type("Jill", 3 ) );
  for( DICT::const_iterator it = d.begin();
       it != d.end(); ++it ) {
    std::cout << it->left << "\t" << it->right << std::endl;
  }
  std::cout << "-------\n";
  for( DICT::left_map::const_iterator it=d.left.begin();
       it != d.left.end(); ++it ) {
    std::cout << it->first << "\t" << it->second << std::endl;
  }
  std::cout << "-------\n";
  for( DICT::right_map::const_iterator it=d.right.begin();
       it != d.right.end(); ++it ) {
    std::cout << it->first << "\t" << it->second << std::endl;
  }
  return 0;
```

```
30  }
```

Listing 6.6 Example of Boost bimap container

Compiling and running this program gives us:

```
Jack  1
Jane  2
Jill  3
-------
Jack  1
Jane  2
Jill  3
-------
1  Jack
2  Jane
3  Jill
```

6.3.0.3 Array: STL compliant container for fixed size array

```
boost::array< int, 1024 > integer_buffer;
```

By providing the type and the length of the array as the template parameter Boost array provides an abstraction which is STL compliant.

6.4 Boost Graph Library

Fig. 6.3 Graph of dependencies using BGL

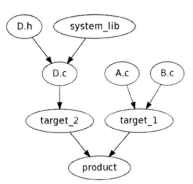

We present an example of using Boost Graph Library to solve a real world problem. Consider the problem of dependency analysis of programs. Given the input *make* type file:

```
begin
D.c        : D.h system_lib ;
target_1 : A.c B.c ;
product  : target_1 target_2 ;
target_2 : D.c ;
end
```

the system has to compute the correct order in which to process the files. The graph of dependencies is shown in Figure 6.3. The C++ program using BGL is shown in Listing 6.7.

```
// \file bgl_make.cpp
// \author Sandeep Koranne, (C) 2010
// \description Example of BGL for dependency analysis

#include <iostream>           // messaging
#include <fstream>            // reading Make file
#include <cassert>            // assertions
#include <vector>             // vertices
#include <list>               // edges
#include <deque>              // topological sort
#include <string>             // string
#include <sstream>            // string parsing
#include <map>                // string->id
#include <boost/graph/vector_as_graph.hpp> // BGL
#include <boost/graph/topological_sort.hpp>// topological_sort

typedef std::map< std::string, int > FileTable;
FileTable gFileTable;

std::vector< std::string > gFileNames;

static int LookupOrAdd( const std::string& Name ) {
  int retval = gFileTable.size();
  if( gFileTable.find( Name ) == gFileTable.end() ) {
    gFileTable[ Name ] = retval;
    gFileNames.push_back( Name );
    return retval++;
  } else {
    return gFileTable[ Name ];
  }
}

static const size_t MAX_N = 100;
typedef std::vector< std::list< int > > Graph;
Graph gDependency( MAX_N );
// the input format is simple
// target : <dep1> <dep2> <dep3>

static void ParseFileAndCreateGraph( const std::string& fileName ) {
  std::ifstream ifs( fileName.c_str() );
  if( !ifs ) { std::cerr << "Unable to open file: " << fileName << std::endl;
    exit( 1 );
  }
  char a_line[ 1024 ];
  ifs.getline( a_line, 1024 );
  std::string line( a_line );
  if( line != "begin" ) {
    std::cerr << "File: " << fileName << " is not a .mak file\n";
    exit( 1 );
  }
  while( true ) { // parse the remaining file
    ifs.getline( a_line, 1024 );
    std::string line( a_line );
    if( line == "end" ) break;
```

```
55       // actual parsing
         std::istringstream sis;
         sis.str( line );
         std::string target; sis >> target;
         std::string token; sis >> token;
60       if( token != ":" ) {
           std::cerr << "Parsing error: " << token << " expecting : " << std::endl;
           exit( 1 );
         }
         int target_id = LookupOrAdd( target );
65
         while( sis ) {
           sis >> token;
           if( token == ";") break;
           int dep_id = LookupOrAdd( token );
70         gDependency[ target_id ].push_back( dep_id );
         }
       }
     }

75  std::deque<int> t_order;
     static void ComputeDependency(void) {
       boost::topological_sort(
         gDependency, std::front_inserter( t_order ),
         boost::vertex_index_map( boost::identity_property_map() ) );
80   }

     static void PrintDependency(void) {
       for( std::deque<int>::reverse_iterator it=t_order.rbegin();
            it != t_order.rend(); ++it ) {
85       if( *it >= gFileNames.size() ) continue;
         std::cout << "Action " << gFileNames[*it] << std::endl;
       }
       std::cout << std::endl;
     }
90
     int main( int argc, char * argv[] ) {
       std::string MakeFileName( argv[1] );
       ParseFileAndCreateGraph( MakeFileName );
       ComputeDependency();
95     PrintDependency();
       return 0;
     }
```

Listing 6.7 Example of BGL

The main features of Boost Graph Library are:

- Graph construction: in BGL graphs can be *directed* or *undirected*. Once a graph has been constructed, vertices and edges can be added into it. The underlying data-structure for the graph depends on the application, and determines the run-time complexity of the algorithms using the graph. Graphs can be dense (with a high ratio of edges (M to N^2), where M denotes the number of edges, and N denotes the number of vertices. For sparse graphs (such as those found in large networks), a list of nodes per vertex (called an adjacency list) implementation is efficient. The functions `add_vertex` adds and returns an opaque handle to the newly added vertex. Similarly, the `add_edge` function returns a 2-tuple, a handle to the edge, and a Boolean to specify whether the edge was actually inserted or was already present.
- Vertex and Edge descriptors: access to the graph's nodes and edges is made through opaque handles called *descriptors*. Graphs may use arbitrary types for

handles, although integers and pointers are common. The descriptor of a graph are accessible using the `graph_traits` class. Vertex descriptors can only be constructed, copied and compared for equality. Edge descriptors can return the constituent vertex descriptors.

- Property maps: property maps are a generic method to attach auxiliary information to vertices and edges. In most examples the vertices represent and model some parameter in the system. Without property maps, the programmer must ensure that the graph, and the model remain in sync, but by using property maps (which can be as simple as a 1-1 index into the model), this problem is removed. The 1-1 map is called the *identity map*, and is a common property map for integer typed vertices. BGL has a number of predefined property types, such as `vertex_name_t`, to install an integer as name for each vertex, `property< vertex_name_t, int>` Multiple property types can be added to vertices and edges using the third parameter of `property` which allows for chaining.
- Algorithms on graphs: Boost graph library has implemented a number of useful graph algorithms which are immediately available to all users of BGL. The algorithms are based on (i) graph traversal, (ii) graph search, and (iii) graph manipulation. Graph traversal is performed using *iterators*, BGL has iterators for vertices and edges. An example of using graph iterators:

```
typedef typename graph_traits<G>::vertex_iterator VIT;
std::pair< VIT, VIT > v_of_g = vertices(g);
for( ; v_of_g.first != v_of_g.second; ++v_of_g.first )
```

The `vertices(Graph)` function returns a 2-tuple (called a *tie*) as a pair of vertex iterators. The first points to a valid vertex, while the second points to the *end* of the vertex iterator sequence. The algorithms implemented in BGL include:

1. Breadth first search (BFS) and Depth first search (DFS),
2. Shortest path,
3. Minimum spanning tree (MST),
4. Connected components,
5. Maximum flow,
6. Graph search,

- Adapters: graph adapters in BGL are filters on the input graph which produce another BGL graph (through a transformation), or produce a graph in an external format such as DOT (using Graphviz .dot, see Section 19.4), or LEDA graphs, or Stanford GraphBase format.

6.5 Boost Spirit Framework

The Spirit Parser Framework is an object oriented recursive descent parser generator framework implemented using template meta-programming techniques. Expression templates allow users to approximate the syntax of Extended Backus Naur Form

(EBNF) completely in C++. Parser objects are composed through operator overloading and the result is a backtracking LL(∞) parser that is capable of parsing rather ambiguous grammars. Spirit can be used for both lexing and parsing, together or separately.

Consider a parser modeled after the classic recursive descent method for parsing mathematical expressions of the form:

```
start <- assignment;
assignment <- lhs '=' rhs;
lhs <- literal;
rhs <- expression;
expression <- '(' expression ')' | term op expression;
op <- XOR | NEG;
term <- term AND factor;
factor <- literal | literal OR factor;
```

The SPIRIT implementation of this parser is shown below:

```
struct EqnParser : public grammar<EqnParser> {
  template <typename ScannerT>
  struct definition  {
    definition(EqnParser const& /*self*/)  {
      expression
        = str_p("INPUT()") [&do_input]
        | str_p("OUTPUT()")[&do_output]
        | term
        >> *(   ('#' >> term)    [&do_or]
              |  ('$' >> term)[&do_xor]
              |  ('-' >> term)[&do_subt]
              )
        ;

      term
        =   factor
        >> *(   ('&' >> factor)[&do_and]
              |   ('/' >> factor)[&do_div]
              )
        ;

      factor
        =   lexeme_d[(+digit_p)[&do_int]]
        |   lexeme_d[(+(alnum_p|'_'|'['|']'))[&do_literal]]
        |   '(' >> expression[&do_expr] >> ')'
        |   ('!' >> factor)[&do_neg]
        |   ('#' >> factor)
        ;

      assign
        =   lexeme_d[(+(alnum_p|'_'|'['|']'))[&do_lhs]]
        >> '=' >> expression[&do_final] >> ';'
        ;
  }
```

```
   rule<ScannerT> expression, term, factor, assign;
   rule<ScannerT> const&
   start() const { return assign; }
  };
};
```

Using Boost SPIRIT framework, the formal grammar is translated to C++ code
using the following conventions: (i) production choices are delineated using the |
operator, (ii) Kleene star closures are defined using the * operator, (iii) the stream
redirection operator >> is overloaded to define production rules, and (iv) the array
index operator [] is overloaded to define the function to be called when the production is matched. In the above example, we see that when the operator # is matched
for example, the do_or function is called.

The function to be called is defined to accept the *left* and *right* operand, e.g.:

```
void do_xor(char const* l, char const* r) {
  // take the 2 top elements from the stack
  // and make their tree and add to sub tree
  Tree* pT = new Tree( XOR, GetName() );
5 pT->left = treeStack.top();   treeStack.pop();
  pT->right = treeStack.top();  treeStack.pop();
  treeStack.push(pT);
}
```

For more details on parsing see Section 13.4.1.

6.6 Boost multi-threading

The boost::thread class is responsible for multi-threading in Boost C++ programs.
The Boost threading interface provides an operating system independent threading
API which is compatible with C++ programming idioms. Threads can be created
with user defined functions and class member functions. The usual POSIX threading
capabilities of critical section protection, using *mutex*, thread local variables, and
pthread once functions are also supported. An example of using Boost threads with
member functions representing the work function is shown in Listing 6.8.

```
  // \file boost_thread_example.cpp
  // \author Sandeep Koranne (C) 2010
  // \description Example of using Boost threading API
  #include <iostream>          // for Program IO
5 #include <cassert>           // assertion checking
  #include <boost/thread/thread.hpp> // Boost threads

  struct MyComputation {
    int how_many;
10  void operator()(void) {
      std::cout << "Thread function called : " << how_many << "\n";
    }
    MyComputation( int x ): how_many( x ) {}
  };
15
  int main( int argc, char* argv[] ) {
    MyComputation C1(10), C2(20);
```

```
      boost::thread A( C1 ), B( C2 );

20    A.join();
      B.join();

      std::cout << std::endl;
      return (0);
25 }
```

Listing 6.8 Example of using Boost threads

6.7 Boost Python integration

Boost Python library is a C++ framework for interfacing C++ functions and objects
with the Python language. Using the Boost Python module arbitrary C++ functions
and objects can be interfaced with Python. Consider the following example:

```
   // C++ code
   char* PrintHello(void) {
     return ''Hello, World!'';
   }
5  #include <boost/python.hpp>
   BOOST_PYTHON_MODULE( hello_ext )
   {
     using namespace boost::python;
     def(''hello_world'', PrintHello);
10 }
```

Now, we can import this module in Python and issue this function call from Python.

6.8 Boost Generic Image Processing Library (GIL)

Boost also has a generic image processing library (GIL) which implements image
processing functions such as gradient calculation, color space conversion, image
transforms, as well as reading and writing common image file formats such as JPEG.
An example of using Boost GIL to compute the histogram of a given JPEG file is
shown in Listing 6.9. The output histogram is shown in Figure 6.4.

```
   // \file gil_hist_example.cpp
   // \author Sandeep Koranne (C) 2010
   // \description Example of using Boost GIL (generic image library)
   #include <iostream>           // for program IO
5  #include <iterator>           // ostream iterator
   #include <fstream>            // reading image files.
   #include <cstdlib>            // for exit
   #include <boost/gil/image.hpp> // GIL image
   #include <boost/gil/typedefs.hpp> // GIL typedefs
10 #include <boost/gil/extension/io/jpeg_io.hpp> // reading JPEG files.

   using namespace boost::gil;

   template <typename GrayView, typename R>
15 void gray_image_hist(const GrayView& img_view, R& hist) {
```

```
        for (typename GrayView::iterator it=img_view.begin(),
          en=img_view.end(); it != en; ++it)
          ++hist[ *it ];
      }
20
      template <typename V, typename R>
      void get_hist(const V& img_view, R& hist) {
          gray_image_hist(color_converted_view<gray8_pixel_t>(img_view), hist);
      }
25
      int main( int argc, char *argv[] ) {
        if( argc != 2 ) {
          std::cerr << "Usage: gil_hist_example <file.jpg>...\n";
          exit(1);
30      }
        rgb8_image_t img;
        jpeg_read_image( argv[1], img);
        std::vector<int> histogram(256, 0);

35      get_hist( const_view(img),histogram );
        std::fstream output_file( "output.txt", std::ios::out );
        std::copy( histogram.begin(), histogram.end(),
            std::ostream_iterator<int>( output_file, "\n" ));
        output_file.close();
40      std::cout << std::endl;
        return (0);
      }
```

Listing 6.9 Example of using Boost GIL

In addition to reading JPEG files, GIL also supports PNG (portable network graphics) and TIFF input and output. Using these IO functions pixel data from JPEG, PNG files can be read into an Image data-structure in memory. Thereafter, calculations and transforms can be applied to the in-memory representation of the image. The transformed image can be written back to disk in JPEG, TIFF or PNG format.

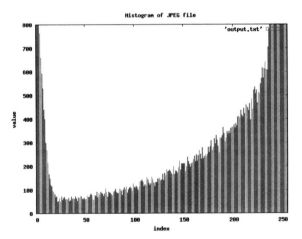

Fig. 6.4 Boost Generic Image Library (GIL) computing histograms of JPEG files

Boost GIL supports many different types of image representations, including: (i) planar and interleaved, (ii) color space and alpha channel, (iii) multi-depth channels, of 8-bit, 16-bit and 24-bit colors, (iv) multi-ordered channels, RGB vs BGR, and (v) row alignment policy.

6.9 Conclusion

Boost C++ libraries provide many well designed data structures and application framework tools, which are useful in a broad range of applications. Using Boost libraries, not only can the developer reuse high-quality code, but also adopt an operating system independent approach. Boost APIs have been effectively used in many scientific and engineering applications, and with the continued development of Boost C++ library, this trend is only expected to increase.

Chapter 7
Performance Libraries

Abstract In this chapter we discuss performance optimization libraries which concentrate on memory issues. We first describe the Google Perftool set of tools which include the `tcmalloc` memory allocation and thread-key based memory allocation library. Perftools also contains a heap-checker and heap-profiler, their usage alongwith examples which are optimized by their usage is shown in this chapter. We compare the performance of perftools memory allocation with APR and the results are presented in this chapter. Another technique for memory optimization is the use of garbage collection. This is included in Java and Common Lisp family of languages, but using the Boehm garbage collector, we also use garbage collection (GC) in C and C++ programs. Examples using Boehm GC are presented in this chapter, and we investigate the impact of GC on performance and memory consumption.

Contents

7.1 Google perftools

Google's *perftools* libraries are an open-source performance enabling API for memory allocation, leak checking, and profiling. Version 1.5 of Google perftools include:

1. `tcmalloc`: thread caching memory allocator,
2. `heap_checker`: heap checking and leak detection API,
3. `heap_profiler`: performance monitoring of heap allocation,
4. `cpu_profiler`: performance monitoring for functions.

We discuss these below.

S. Koranne, *Handbook of Open Source Tools*,
DOI 10.1007/978-1-4419-7719-9_7, © Springer Science+Business Media, LLC 2011

7.1.1 perftools : `tcmalloc`

Google Perftool `tcmalloc` is designed as a fast, optimized thread aware replacement for C lib standard `malloc`. On multi-threaded systems, `tcmalloc` reduces thread contention by reducing threading overhead for small objects, and by using fine grained thread aware locking. While standard `malloc` also uses thread specific arenas for memory allocation, migration of free space from one thread arena to another is not implemented in the standard C library, while perftools `tcmalloc` reduces this wastage of memory space. Another advantage of `tcmalloc` is the optimized representation of small objects. To make use of `tcmalloc` in their own code, the application can either link to `-ltcmalloc` or use LD_PRELOAD to load the `tcmalloc` shared library (this is the way to use `tcmalloc` in applications which cannot be recompiled to take advantage of `tcmalloc`, which should not be a problem for the software considered in this book).

7.1.1.1 Implementation

Perftool `tcmalloc` assigns each thread a local cache from which small allocations are directly serviced. A page is divided into a sequence of small objects, while a large object is allocated directly from the main heap. `tcmalloc` uses a class of 60 allocatable size-classes. Object size requests outside this range are rounded up to the nearest available size. Since thread local arenas are available for small objects, if the *free list* for an arena is not empty, an object of the appropriate size is returned directly (no locking is required). If the free list for the requested size class is empty, a collection of objects from a central free list is requested and placed on the thread local free list, subsequently an object from this list is returned to the application. If the central list is also empty, then a run of pages is carved up into objects of the requested class and placed on the lists. Since the API is identical to `malloc` for use, it can be quickly integrated with existing code.

On an existing discrete geometry application using STL in C++, standard GNU C `malloc` had a runtime of : 266.12 seconds. Linking the same application with `tcmalloc` reduced the runtime to 208 seconds. Another example is shown in Figure 7.1 where the runtime performance of GNU C `malloc` is compared to perftools `tcmalloc`.

7.1.2 perftools : `heap_checker`

The heap checker `heap_checker` is part of `tcmalloc` so its use is already possible by linking the `libtcmalloc` library with the application. However, linking to `libtcmalloc` does not turn on heap checking. The heap checker tracks usage of memory before `main()` and confirms that all allocated memory has been freed at `exit()`. If the heap checker detects any memory that has been allocated but not

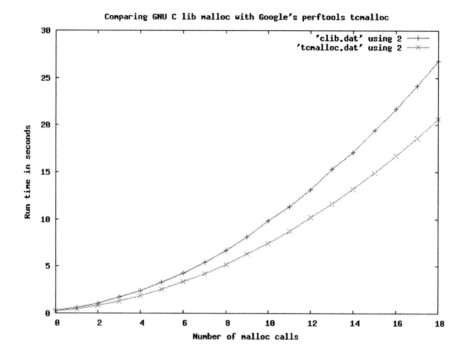

Fig. 7.1 Performance measurement of GNU C `malloc` with perftools `tcmalloc`.

freed at `exit()` it aborts the program and prints a message. It should be noted that `heap_checker` records the stack trace for each allocation (thus increasing the memory usage and runtime of the application significantly).

To switch on the `heap_checker`, the environment variable HEAPCHECK should be set one of the following: (i) `minimal`, (ii) normal, (iii) strict, and (iv) draconian. Example of `minimal` output is shown below:

```
sizeof(Object) = 72

Leak check _main_ detected leaks of 3120 bytes in 101 objects
The 2 largest leaks:
Leak of 2400 bytes in 100 objects allocated from:
@ 8048979
@ 94ebb6
@ 8048801
Leak of 720 bytes in 1 objects allocated from:
@ 80488e8
@ 94ebb6
@ 8048801
```

It should be noted that since the heap-checker uses the heap-profiling framework (see Section 7.1.3) internally, it is not possible to use both tools at the same time. To disable the heap-checker for a block of code we can use the following method:

```
   // code block has know memory allocation
   {
     HeapLeakChecker::Disabler myDisabler;
     int *n = new int[1024];
5    if( wanted_elsewhere) globalArray = n; //
     // Or we can use
     IgnoreObject( n ); // which also waives the heap error
   }
```

In addition to basic heap-checking, it is possible to check for memory alignment problems (on modern processors, memory latency can be heavily influenced by memory alignment). The environment variable HEAP_CHECK_TEST_POINTER_ALIGNMENT can be set to enable this check.

7.1.3 perftools : heap_profiler

The primary uses of heap_profiler are to detect places that perform a lot of memory allocation in C++ program code. It can also be used to detect memory leaks. Similar to the HEAP checker, the heap profiler does not switch on by default. To switch on the heap profiler, the environment variable HEAPPROFILE needs to set to a file location. It is possible to analyze portions of code for heap-profiling as:

```
   // performance sensitive code block
   void RegenrateMatrix( double* data, unsigned int N ) {
     // do computation
     #ifdef _ENABLE_PROFILING_
5    if( IsHeapProfilerRunning() ) HeapProfilerDump();
     #endif
   }
   void CheckStrata(double* data, unsigned int N) {
     #ifdef _ENABLE_PROFILING_
10   HeapProfilerStart(''/tmp/strata_profile'');
     #endif
     // algorithm phase I, setup FFT vectors
     #ifdef _ENABLE_PROFILING_
     HeapProfilerDump();
15   #endif
     // call actual complex function
     RegenrateMatrix( data, N );
     #ifdef _ENABLE_PROFILING_
     HeapProfilerStop();
20   #endif
   }
```

The function HeapProfilerStart expects the filename prefix as the argument. The function IsHeapProfilerRunning can be used to check if the function is already being profiled. Once the profile data has been generated into a file the pprof tool can be used to analyze the data and generate reports. The reports can be generated using TEXT formar, or PostScript output (as shown in Figure 7.2.

The TEXT format report example is shown below:

```
$ ~/OSS/bin/pprof --text poly_enum_i686.tcmalloc
   /tmp/enum_profile.0001.heap
Total: 6.8 MB
```

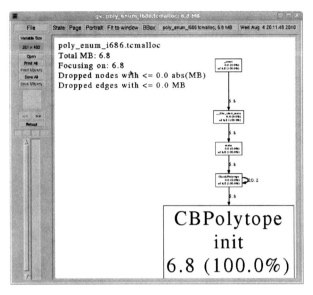

Fig. 7.2 Google Perftools `pprof` output graph

```
6.8 100.0% 100.0%       6.8 100.0% CBPolytope::init
0.0   0.0% 100.0%       0.0   0.0% nauty
0.0   0.0% 100.0%       0.0   0.0% refine1
0.0   0.0% 100.0%       0.0   0.0% firstpathnode0
0.0   0.0% 100.0%       0.0   0.0% doref
0.0   0.0% 100.0%       0.0   0.0% bestcell
0.0   0.0% 100.0%       0.0   0.0% CBPolytope::compute_orbits
0.0   0.0% 100.0%       6.8 100.0% CheckPolytope
```

At least in this example, the polytope `init()` function does all the memory allocation (which is expected). We can also ask `pprof` to focus on specific functions using the `--focus=<name>` command-line option:

```
$ ~/OSS/bin/pprof --focus=nauty --text
      poly_enum_i686.tcmalloc /tmp/enum_profile.0001.heap
Total: 6.8 MB
      0.0   0.0%   0.0%       0.0   0.0% nauty
      0.0   0.0%   0.0%       0.0   0.0% refine1
      0.0   0.0%   0.0%       0.0   0.0% firstpathnode0
      0.0   0.0%   0.0%       0.0   0.0% doref
      0.0   0.0%   0.0%       0.0   0.0% bestcell
```

7.1.4 perftools : cpu_profiler

To add CPU profiling to the application, we have to link with the `lprofiler` library of perftools. Similar to the heap profiling, this does not switch on the profiling code, it only links the code. To enable CPU profiling, the environment variable `CPUPROFILE` needs to set to the filename where the cpu profile results will be generated. The environment variable `CPUPROFILE_FREQUENCY` and `CPUPROFILE_REALTIME` control the behavior of this profiling. Once the data has been generated the `pprof` tool can be used to analyze the trace, as shown below:

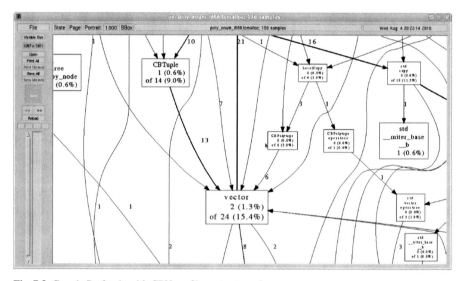

Fig. 7.3 Google Perftools with CPU profiler output graph

```
$ ~/OSS/bin/pprof  --text poly_enum_i686.tcmalloc
        /tmp/enum_profile
Removing _init from all stack traces.
Total: 156 samples
   48   30.8%   30.8%    48   30.8% __mcount_internal
   25   16.0%   46.8%    71   45.5% mcount
   12    7.7%   54.5%    12    7.7% _init
    8    5.1%   59.6%    49   31.4% CBPolytope::init
    5    3.2%   62.8%     5    3.2% __i686.get_pc_thunk.bx
    5    3.2%   66.0%     5    3.2% std::vector::operator[]
    3    1.9%   67.9%     3    1.9% std::_Rb_tree_increment
    3    1.9%   69.9%     6    3.8% std::__copy_move::__copy_m
    3    1.9%   71.8%    17   10.9% std::__copy_move_a2
    2    1.3%   73.1%     2    1.3% refine1
    2    1.3%   74.4%     4    2.6% std::_Rb_tree::_M_destroy_node
    2    1.3%   75.6%    11    7.1% std::_Rb_tree::_M_insert_unique
    2    1.3%   76.9%     2    1.3% std::_Rb_tree::_S_value
```

```
2    1.3%   78.2%    8    5.1% std::set_difference
```

As with the heap-profiler, it is possible to restrict the report to match a regular expression string using the `--focus` command-line option. We can generate the graphical report for the CPU profiler as well, as shown in Figure 7.3.

7.2 Boehm GC : garbage collection

Boehm Garbage Collector (GC) is a conservative garbage collection for C and C++. Garbage collection can be defined as the automatic reclaiming of dynamic memory which is no longer in use (referenced) by any object in the program. Use of garbage collection was pioneered in the Common Lisp community, which also contributed large advances in the field of garbage collection, mostly in performance and capabilities.

GNU Compiler Collection uses the Boehm GC internally. The Java language also uses garbage collection. The use of garbage collection not only frees the developer from the responsibility of keeping track of dynamic variable lifetime, as well as prevent leaks and improves system performance (when used properly). When the developer understands that dynamic memory can be reclaimed automatically, better algorithms can be chosen (which without GC would be too complicated to implement correctly). This can counteract the slight performance and capacity hit that GC entails. Since the system has to recognize all dynamic memory as alive or not, extra information (meta-data) has to be kept for dynamic memory. Boehm GC is implemented to reduce this impact. Moreover, the Boehm GC can also be used as a leak detector.

7.3 Using Boehm GC

Boehm GC's API have been designed as a replacement for the C library `malloc`. The basic idea behind garbage collection is that all valid dynamic memory must be pointed to by some object in the program. Such an object can be (i) other dynamic memory, (ii) stack variables, (iii) data, (iv) statically allocated BSS segments, or (v) registers. Pointer recognition is a fundamental problem, and Boehm GC is a conservative GC (which means that it errs on the side of caution) when deciding which memory to reclaim. Memory allocated by system `malloc` are not seen by the GC so mixing memory allocations is not a good idea.

The C interface of the allocator is listed below:

- GC_malloc:
- GC_malloc_atomic:
- GC_realloc:
- GC_free:

A simple way to use Boehm GC is to define:

```
#define malloc(n) GC_malloc(n)
#deine calloc(m,n) GC_malloc((m)*(n))
```

There is also a C++ interface to the garbage collector. A garbage collector is most useful when the lifetime of the object cannot be easily tracked. Maintaining reference count within objects makes the code complicated and error prone. Using automatic garbage collection can solve this problem. Consider a small simulation kernel as shown in Listing 7.1.

```cpp
     // \file unknown_lifetime.cpp
     // \author Sandeep Koranne, (C) 2010
     // \description Example of using garbage-collection as object
     // life time is not known apriori
5    #ifdef USE_GC
     #include <gc.h>                  // Boehm garbage collector
     #endif
     #include <fstream>               // reading /proc/self/statm
     #include <iostream>              // program IO
10   #include <cassert>               // assertion checking
     #include <cstdlib>               // exit
     #include <cstdio>                // perror
     #include <sys/resource.h>        // setrlim

15   static unsigned long num_mallocs = 0;
     #ifdef USE_GC
       #define MALLOC GC_malloc
     #else
       #define MALLOC malloc
20   #endif

     struct Node {
       struct Node *left, *right;
       int data;
25     int mark;
       char BUF[10240];
       Node() : left( NULL ), right( NULL ), data( 0 ), mark( 0 ) {}
       void reset() { left = right = NULL; data = mark = 0; }
     };
30
     std::ostream& operator<<( std::ostream& os, const Node& N ) {
       return os << N.data << "\t" << N.mark;
     }

35   Node ** ConstructNodes( unsigned int N ) {
       Node ** retval = (Node**)MALLOC( N * sizeof( Node * ) );
       assert( ( retval != NULL ) && "retval allocated" );
       for( int i=0; i < N; ++i ) {
         retval[i] = (Node*)MALLOC( sizeof( Node ) );
40       retval[i]->reset();
         retval[i]->data = i;
       }
       return retval;
     }
45
     static void PrintNodeData( Node **data, unsigned int N ) {
       for( int i=0; i < N; ++i )
         std::cout << (*data[i]) << std::endl;
     }
50
     static void RunSimulation( Node **data,
                                unsigned int N,
                                unsigned int LOOP_COUNT ) {
       for( int i=0; i < LOOP_COUNT; ++i ) {
```

```
55      unsigned int source_id = rand() % N;
        unsigned int dest_id   = rand() % N;
        assert( ( data[ source_id ] != NULL ) && "source is invalid" );
        assert( ( data[ dest_id ] != NULL ) && "dest is invalid" );
        Node *source = data[source_id], *dest = data[dest_id];
60      Node *temp = (Node*) MALLOC( sizeof( Node ) );
        num_mallocs++;
        temp->reset();
        temp->data = source->data + dest->data;
        if( source->data  < dest->data ) {
65        source->mark |= 0x01;
          source->right = temp;
          data[ dest_id ] = temp;
        } else if( source->data > dest->data ) {
          dest->mark |= 0x01;
70        dest->left = temp;
          data[ source_id ] = temp;
        } else {
          data[ source_id ] = data[ dest_id ] = temp;
        }
75    }
    }
    static size_t GetMemorySize(void) {
        std::ifstream ifs;
        ifs.open("/proc/self/statm",std::ios::in);
80      size_t memoryUsed = 0;
        ifs >> memoryUsed;
        return memoryUsed;
    }

85  static void DisplayMemoryUsed(void) {
        size_t memUsed = GetMemorySize();
        std::cout << std::endl << "Program consumed "<< memUsed << " kb.\n";
    }

90  int main( int argc, char *argv [] ) {
      struct rlimit mem;
      mem.rlim_cur = mem.rlim_max = 102400;
      int rc = setrlimit( RLIMIT_DATA, &mem );
      if( rc ) { std::perror( "setrlimit"); exit( 1 ); }
95    unsigned int N = 10;
      Node **data = ConstructNodes( N );
      RunSimulation( data, N, 100000 );
      std::cout << "number mallocs = " << num_mallocs << std::endl;
      DisplayMemoryUsed();
100   std::cout << std::endl;
      return 0;
    }
```

Listing 7.1 Example of using Boehm garbage collection

We compile this program with and without the garbage collector:

```
g++ -O3  unknown_lifetime.cpp
$ /usr/bin/time ./a.out
number mallocs = 100000
Program consumed 251876 kb.

0.08user 0.94system 0:01.10elapsed 93%CPU
  (0avgtext+0avgdata 0maxresident)k
0inputs+0outputs (0major+102255minor)pagefaults 0swaps
# Now we compile with garbage collection
$ g++ -O3 -DUSE_GC unknown_lifetime.cpp   -lgc
$ /usr/bin/time ./a.out
```

```
number mallocs = 100000

Program consumed 1491 kb.

3.65user 0.01system 0:03.77elapsed 97%CPU
  (0avgtext+0avgdata 0maxresident)k
0inputs+0outputs (0major+625minor)pagefaults 0swaps
```

The garbage collected binary is several times slower (this is a very artificial example where the runtime cost of the memory allocator dominate any processing),
but the important aspect is the significant reduction in memory consumption from
251 MB (for the non-gc binary), to 1.5 MB for the garbage collected binary. Although, the example is simplistic (the dangling pointers can be fixed with some
extra code and reference counts), having the flexibility of automatic garbage collection allows the programmer to concentrate on the actual simulation kernel in this
example, without running out of memory.

7.4 Conclusion

In this chapter we discussed performance optimization libraries which concentrate
on memory issues. We first described the Google Perftool set of tools which include
the tcmalloc memory allocation and thread-key based memory allocation library.
We compared the performance of perftools memory allocation with APR and the results are presented in this chapter. Another technique for memory optimization is the
use of garbage collection. Examples using Boehm GC are presented in this chapter,
and we investigate the impact of GC on performance and memory consumption.

Chapter 8
Compression Engines

Abstract In this chapter we present libraries and APIs for lossless compression. Even with the geometric rise in memory capacity, the information theoretic compression gains provided at relatively cheap computing power can result in impressive performance and capacity gains. In this chapter we present with the help of examples, the use of the ZLIB and BZIP2 libraries. Examples are presented using C/C++ API as well as Python modules. More recently, LZMA and XZ Utilities also provide even better compression, albeit at slightly slower speed. Their use is also described in this chapter.

Contents

Although network speeds and memory capacities have expanded at a rapid rate in recent years, application demands for ever more features and capabilities have continued to place a constraint on the total memory space consumed by the program, or data which is communicated between cooperating processes. One solution to alleviating this concern is the use of compression, both lossless (for applications which need bit-equivalent data on both ends of the communication) or lossy (such as image compression). In the next section we discuss techniques for lossless compression including the use of *deflate*, and its implementation in ZLIB. As the performance gap between CPU and memories continue to expand, spending CPU cycles for compression and decompression, which reduces the memory requirement is indeed beneficial.

S. Koranne, *Handbook of Open Source Tools*,
DOI 10.1007/978-1-4419-7719-9_8, © Springer Science+Business Media, LLC 2011

8.1 ZLIB Compression Library

zlib is a loss-less compression and decompression library. It has the advantage
of input independent memory requirements, as well as the compression algorithm
produces excellent compression of data. The zlib library is an implementation of
the deflate algorithm which in turn uses Huffman coding to perform compression. The zlib library can be used as a buffer compression library (which in turn
can be used to compress files), or it can be used to read files compressed using
the gzip tool. The zlib functions gzread allows reading compressed files. See
Section 8.1.2 for more details on the file functions of zlib.

The compressed data format used by the in-memory functions is the zlib format
(by default), which is a zlib wrapper (RFC 1950), around a deflate stream (RFC
1951). A Deflate stream consists of a series of blocks. Each block is preceded by a
3-bit header:

1. Bit 0 :marker,

   - 1: last block in stream,
   - 0: more blocks.

2. Bit 1 and 2: represent the encoding method:

   - 00: a stored/raw/literal section follows, between 0 and 65,535 bytes in length,
   - 01: static Huffman compressed block (pre-agreed tree),
   - 10: dynamic Huffman block alongwith tree,
   - 11: reserved.

Most data blocks are compressed using the dynamic Huffman coding, where the
tree is contained after the header. The two techniques used for data compression
in DEFLATE are (i) duplicate string referencing and (ii) bit-reduction by symbol
encoding. Duplicate strings are *back referenced* (upto a maximum distance of 32K
bytes), and symbol encoding is performed using the Huffman encoding method.
The lookback for the string matching contributes significantly to the quality and
performance of the compression library. In this section we discuss the reference
zlib implementation since it is widely used in many open-source software, as well
as being part of Internet standard (RFC 1950) and the Java JAR format.

The zlib functionality is accessed using an opaque ZSTREAM object as shown
in Listing 8.1. The declaration of this stream structure can be found in zlib.h and
is shown below:

```
    typedef struct z_stream_s {
        Bytef    *next_in;  /* next input byte */
        uInt     avail_in;  /* no. bytes available next_in */
        uLong    total_in;  /* input bytes read so far */
5
        Bytef    *next_out; /* location of next output */
        uInt     avail_out; /* free space at next_out */
        uLong    total_out; /* no, bytes output so far */

10      char     *msg; /* error message, NULL if no error */
        struct internal_state FAR *state; /* opaque */
```

```
      alloc_func zalloc;    /* internal allocator */
      free_func  zfree;     /* internal deallocator */
15    voidpf     opaque;    /* pvt. data to allocator */

      int        data_type; /* binary or text */
      uLong      adler;     /* adler32 value */
      uLong      reserved;  /* reserved  */
20  } z_stream;
```

During compression or decompression, `next_in` and `avail_in` must be updated when `avail_in` becomes zero. Similarly, the members `next_out` and `avail_out` have to be updated. The memory allocation and `opaque` member must be initialized. When compressing data-buffers, compression can be performed in a single step if the input/output buffers have appropriate memory allocated for them. A comprehensive example of using ZLIB with XDR is given in Listing 8.1, which uses the single-shot method of compression. Otherwise, the stream object's `avail_out` member has to be monitored to ensure that space is available for output. The fields `total_in` and `total_out` are used for statistics or progress reports and contain the number of input, and output bytes, read and written, respectively. The main functions provided by the `zlib` library are shown in Table 8.1.

Table 8.1 Major functions of the zlib library

`zlib` function	Description
zlib_version	return version number
deflateInit	initialize the stream state for compression
deflateInit2	same as above, but has more options
deflate	compress as much data as possible
deflateEnd	free dynamically allocated structures
inflateInit	initialize stream for decompression
inflateInit2	same as above, but has more options
inflate	decompress as much data as possible
inflateEnd	free dynamically allocated structures
compress	utility function to compress buffer
compress2	same as above, but has `level`
compressBound	function to return upper bound on length of output
uncompress	utility function to perform decompression
adler32	update running Adler-32 checksum
crc32	update running CRC32

The functions `deflateInit2` has customizable options including (i) compression method, (ii) windowBits, (iii) memory level, and (iv) compression strategy. Similarly, the `inflateInit2` has only one extra parameter, the number of window bits (specified as a base two logarithm), which controls the maximum window size.

```
// \file rdb.cpp
// \author Sandeep Koranne, (C) 2010
```

```
     // \description Use of zlib, bzip2, LZMA with XDR
     #include <iostream>
5    #include <stdlib.h>
     #include <stdio.h>
     #include <string.h>
     #include <rpc/types.h>
     #include <rpc/xdr.h>
10   #include <boost/crc.hpp>        // actual CRC
     #include <zlib.h>

     static size_t MAX_LEN = 0;
     static char *BUFFER = NULL;
15   static char *ZLIB_BUFFER = NULL;
     static unsigned long CHECKSUM = 0;
     // Note: use of CRC32 is not endian-proof as we always
     // across machines, but is acceptable as a test case
     // against data corruption.
20
     int WriteNumberStream( int N, char** buf ) {
       boost::crc_32_type boost_crc;
       XDR xdrs;
       if( MAX_LEN <= ( N * sizeof( int ) ) ) {
25       *buf = new char[ ( N * sizeof( int ) ) ];
         MAX_LEN = N * sizeof( int );
       }
       xdrmem_create( &xdrs, *buf, MAX_LEN, XDR_ENCODE );
       for( int i=0; i < N; ++i ) {
30       int x = ( rand() % 100 ) ^ (i) ;
         boost_crc.process_byte( x & 0x0FF0 ); // middle order bits
         if (!xdr_int( &xdrs, &x ) ) return -1;
       }
       int written_len = xdr_getpos( &xdrs );
35     //std::cout<<written_len<<"\t"<<boost_crc.checksum()<<"\t";
       //std::cout << "Wrote " << written_len << " bytes.\n";
       //std::cout << "CRC32 checksum = " << boost_crc.checksum();
       CHECKSUM = boost_crc.checksum();
       return written_len;
40   }

     int ReadNumberStream( int N, char* buf ) {
       boost::crc_32_type boost_crc;
       XDR xdrs;
45     xdrmem_create( &xdrs, buf, N, XDR_DECODE );
       int x;
       int count=0;
       while( true ) {
         if( !xdr_int( &xdrs, &x ) ) break;
50       boost_crc.process_byte( x & 0x0FF0 );
         count++;
       }
       if( CHECKSUM != boost_crc.checksum() ) {
         std::cerr << "CHECKSUM failed.\n";
55       exit( 1 );
       }
       //std::cout << "Read " << count << " integers.\n";
       //std::cout << "CRC32 checksum = " << boost_crc.checksum();
       return count;
60   }

     int CompressXDRStream( char *input, int len, char* output, int MAX_LEN ) {
       z_stream zstrm;
       int flush = Z_NO_FLUSH;
65     int compressed_till_now = 0;
       zstrm.zalloc = Z_NULL;
       zstrm.zfree = Z_NULL;
       zstrm.opaque = Z_NULL;
       int ret = deflateInit( &zstrm, 3 ); // level = 3
```

```
70    if( ret != Z_OK ) {
        std::cerr << "ZLIB stream construction failed.\n";
        exit( 1 );
      }
      zstrm.avail_in = len;
75    zstrm.next_in  = (unsigned char*)input;
      zstrm.avail_out = MAX_LEN;
      zstrm.next_out = (unsigned char*)output;
      do {
        // we expect the whole buffer in single shot
80      ret = deflate( &zstrm, flush );
        if( ret != Z_OK ) {
          std::cerr << "ZLIB failure\n";
          exit( 1 );
        }
85    } while( zstrm.avail_in > 0 );
      ret = deflate( &zstrm, Z_FINISH );
      compressed_till_now = MAX_LEN - zstrm.avail_out;
      //std::cout<<"Wrote "<<compressed_till_now<<" bytes during deflate.\n";
      deflateEnd( &zstrm );
90    return compressed_till_now;
    }

    int DecompressXDRStream( char *input, int len, char* output, int MAX_LEN ) {
      z_stream zstrm;
95    int flush = Z_NO_FLUSH;
      int decompressed_till_now = 0;
      zstrm.zalloc = Z_NULL;
      zstrm.zfree = Z_NULL;
      zstrm.opaque = Z_NULL;
100   int ret = inflateInit( &zstrm );
      if( ret != Z_OK ) {
        std::cerr << "ZLIB stream construction failed.\n";
        exit( 1 );
      }
105   zstrm.avail_in = len;
      zstrm.next_in  = (unsigned char*)input;
      zstrm.avail_out = MAX_LEN;
      zstrm.next_out = (unsigned char*)output;
      do {
110     // we expect the whole buffer in single shot
        ret = inflate( &zstrm, flush );
        if( ret == Z_STREAM_ERROR ) {
          std::cerr << "ZLIB failure\n";
          exit( 1 );
115     }
      } while( zstrm.avail_in > 0 );
      ret = deflate( &zstrm, Z_FINISH );
      decompressed_till_now = MAX_LEN - zstrm.avail_out;
      //std::cout<<"Read "<<decompressed_till_now<<" bytes during deflate.\n";
120   inflateEnd( &zstrm );
      return decompressed_till_now;
    }

    int main( int argc, char *argv[] ) {
125   int number_to_write = 1000;
      if( argc > 1 ) number_to_write = atoi( argv[1] );
      int numWritten = WriteNumberStream( number_to_write, &BUFFER );
      if( numWritten < 0 ) {
        std::cerr << "XDR stream overflowed, allocate more memory.\n";
130     exit( 1 );
      }
      if( MAX_LEN ) ZLIB_BUFFER = new char[ MAX_LEN * sizeof( int ) ];
      int comp_len = CompressXDRStream( BUFFER, numWritten,
                                        ZLIB_BUFFER, MAX_LEN );
135   std::cout << numWritten << "\t" << comp_len << "\n";
      memset( BUFFER, 0, MAX_LEN ); // initialize to 0
```

```
      int decomp_len = DecompressXDRStream( ZLIB_BUFFER, comp_len,
                                            BUFFER, MAX_LEN );
      int numRead = ReadNumberStream( numWritten, BUFFER );

140
      return 0;
    }
```

Listing 8.1 Using ZLIB with XDR

Using the utility functions uncompress requires the transmission of the length of the uncompressed stream alongwith the compressed data as that is not stored in the compressed data. The compress function is declared as:

```
compress( Bytef *dest, uLongf *destLen,
          const Bytef *source, uLong sourceLen );
```

The size of the compressed stream is returned in destLen. Using the compressBound function an output buffer to accumulate the compressed data can be allocated prior to calling compress. The uncompress function is similarly declared:

```
uncompress( Bytef *dest, uLongf *destLen,
            const Bytef *source, uLong sourceLen );
```

The output buffer length has to be known to the application in advance. On return, destLen contains the number of bytes (which can be compared with the application data). Both these functions return z_ok if the action was performed without error.

8.1.1 Compression ratio

This section is subjective, and depending on the data present in the stream compression ratio can vary significantly. We presented an example in Listing 8.1. Running this example on several different sizes of input yielded the plot as shown in Figure 8.1.

8.1.2 gzip file access functions

The zlib library supports reading and writing gzip (RFC 1952) files (also called '.gz' files). The main functions are shown in Table 8.2.

8.1.3 Integration of zlib and gzip in Python

In addition to using the library from your own C/C++ programs as we have shown in Listing 8.1, it is also possible to use zlib and gzip from Python. Consider the example shown below:

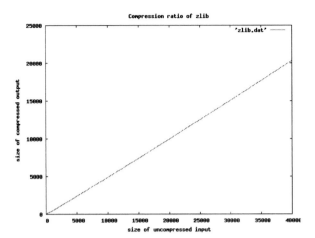

Fig. 8.1 Compression ratio for zlib

Table 8.2 Major `gzip` functions of the zlib library

`gzip` function	Description
gzopen	open filename for reading with mode,
gzdopen	open file descriptor for reading with mode
gzbuffer	set internal buffer size
gzsetparams	set level and strategy for file
gzread	read length (uncompressed) of bytes into buffer
gzwrite	write length (uncompressed) to file
gzprintf	format, convert and write to file
gzputs	write null-terminated string to file
gzgets	read null-terminated string of given length from file
gzflush	flush output to file
gzseek	sets starting position for next read/write
gztell	return position (uncompressed) of next read/write
gzoffset	return current offset of file
gzeof	return 1 if end-of-file indicator is set
gzclose	close the file

```
import sys
import gzip
def ParseFile( fileName ):
    print "Reading file %s" %fileName
    if poly_file_name.find( '.gz' ) != -1:
        file = gzip.open( fileName, 'rb' )
        lines = file.readlines()
        file.close()
    else:
```

```
      f = open( fileName, 'r' )
      lines = f.readlines()
      f.close()
   for l in lines:
      # computation on contents
```

As the above example shows using `zlib` in Python is simple, and transparent.

8.2 LIBBZ2 and BZIP2

The `bzip2` implements the Burrows-Wheeler block-sorting text compression algorithm. The underlying library of `bzip2` is `libbz2`. An advantage of using `bzip2`
is that internally it organizes data as chunks of 900k, so partially damaged files
can be recovered (using the `bzip2recover` program). The major functions of the
`libbz2` library are shown in Table 8.3.

Table 8.3 Major functions of the libbz2 library

`zlib` function	Description
BZ2_bzCompressInit	prepares for compression
BZ2_bzCompress	compress the data, change state
BZ2_bzCompressEnd	cleanup data structures
BZ2_bzDecompressInit	prepare for decompression
BZ2_bzDecompress	decompress the data
BZ2_bzDecompressEnd	cleanup data structures
BZ2_bzReadOpen	prepare to read data from FILE*
BZ2_bzRead	read data into buffer
BZ2_bzReadClose	close FILE
BZ2_bzWriteOpen	open FILE for writing
BZ2_bzWrite	write compressed data to FILE
BZ2_bzWriteClose	close FILE
BZ2_bzBuffToBuffCompress	one-shot buffer compression
BZ2_bzBuffToBuffDecompress	one-shot buffer decompression

The stream for `libbz2` is shown below:

```
   struct {
      char *next_in;              /* data to be compressed  */
      unsigned int avail_in;      /* length of input data remaining */
      unsigned int total_in_lo32;/* statistics */
 5    unsigned int total_in_hi32;
      char *next_out;             /* output buffer */
      unsigned int avail_out;     /* length of output buffer remaining */
      unsigned int total_out_lo32;/* statistics */
      unsigned int total_out_hi32;
10    void *state;                /* internal state */
      void *(*bzalloc)(void *,int,int); /* memory allocator function */
```

```
    void (*bzfree)(void *,void *);    /* deallocator function */
    void *opaque;                     /* pointer given to allocator*/
} bz_stream;
```

8.2.1 Integration of bzip2 in Python

In addition to using the library from your own C/C++ programs it is also possible to use `bzip2` from Python. Consider the example shown below:

```
import bz2
def ParseFile( fileName ):
    print "Reading file %s" %fileName
    if poly_file_name.find( '.bz2' ) != -1:
5       file = bz2.BZ2File( fileName, 'rb' )
        lines = file.readlines()
        file.close()
    else:
        f = open( fileName, 'r' )
10      lines = f.readlines()
        f.close()
    for l in lines:
        # computation on contents
```

8.3 LZMA and XZ Utils

The Lempel-Ziv-Markov chain algorithm is implemented in the LZMA toolchain and library. The underlying library for the XZ tools is the `liblzma` library. We describe the compression and decompression functions of this library in this section.

The API for decoding LZMA compressed data is shown below:

```
SRes LzmaDecode(Byte *dest,
                SizeT *destLen,
                const Byte *src,
                SizeT *srcLen,
5               const Byte *propData,
                unsigned propSize,
                ELzmaFinishMode finishMode,
                ELzmaStatus *status,
                ISzAlloc *alloc);
```

The memory allocators can be set to use the system wide `malloc` and `free` as:

```
void *SzAlloc(void *p, size_t size)
    { p = p; return malloc(size); }
void SzFree(void *p, void *address)
    { p = p; free(address); }
5   ISzAlloc alloc = { SzAlloc, SzFree };
```

Compressing data using LZMA is shown below:

```
    CLzmaEncHandle enc;
    enc = LzmaEnc_Create(&g_Alloc);
```

```
     if (enc == 0) return SZ_ERROR_MEM;
     LzmaEncProps_Init(&props);
 5   res = LzmaEnc_SetProps(enc, &props);
     HRes LzmaEncode(Byte *dest,
                     SizeT *destLen,
                     const Byte *src,
                     SizeT srcLen,
10                   CLzmaEncProps *props,
                     Byte *propsEncoded,
                     SizeT *propsSize,
                     int writeEndMark,
                     ICompressProgress *progress,
15                   ISzAlloc *alloc,
                     ISzAlloc *allocBig);
```

The distinction between the allocators is done for systems which support large page
size tables for efficient allocation of large memory buffers.

8.3.1 XZ Utils

An implementation of LZMA in the XZ utilities is available on GNU/Linux systems.
The compression engines are also available with the GNU `tar` program using the
`-J` command-line option (see Section 11 for a description of the command-line
options). We experimented with compressing a file of 13 Mb using `gzip`, `bzip2`
and `xz`, the results are shown in Table 8.4. It shows that XZ utilities can achieve
significantly better compression, albeit at a higher runtime. As shown in Table 8.4
XZ (with the `-e`, extreme option) achieves almost 50% compression, while `gzip`
and `bzip2` achieve (only) 0.84.

Table 8.4 Compression ratio of various tools

Tool name	Ratio	Comp. time (s)	Decomp. time (s)
gzip -9	0.84	02.59	0.46
bzip2 -9	0.84	11.99	3.96
xz -e	0.55	24.20	2.18

8.4 Conclusion

In this chapter we presented libraries and APIs for lossless compression. Using short
examples, we presented the use of the ZLIB and BZIP2 libraries. Examples were
presented using C/C++ API as well as Python modules. The use of LZMA and XZ
Utilities was also discussed.

Chapter 9
Application Development Libraries

Abstract In this chapter we present several useful libraries and API which we could not categorize with any of the other libraries we presented in this part of the book. We first present the RPC (remote procedure call) library for remote invocation of predefined functions. We present a real-life example motivated from physical chemistry to use RPC for optimizing runtime using remote servers. In any discussion of cluster computing, data endianess issues always crop up, and a part of the RPC library is XDR (extensible data representation). We continue to present APIs for checksum, and hash signature computation to check and prevent transmission errors. An alternative to XDR is to use XML files, and library APIs for XML processing is presented in this chapter. For persistent storage we show examples of using Berkeley DB, and present examples using C++. A network based caching library (Memcache) is presented with examples in C++ and Python. To seamlessly use Python and other interactive languages, SWIG (the Simplified Wrapper Interface Generator) is presented.

Contents

9.1 RPC (remote procedure call) library

The remote procedure call (RPC) library provides an API and abstraction to invoke pre-defined and pre-registered functions on remote servers. The library supports messaging, data-representation, client and server programming. There is a RPC protocol compiler *rpcgen* which converts a high level description of a function in RPC Language (a language similar to C), and generates C language code for client and server. The server needs to be registered with the `rpcbind` RPC server running on the server, by adding its service number into the `/etc/rpc` file.

We present a small example inspired from computational biology. Given a molecule with *N* atoms, and spectroscopic data containing *M* distances between the atoms, the goal of the *PDF Solver* is to assign x, y, z coordinates to the atoms which best matches the distances. We use a greedy minimization algorithm to solve for x, y, we assume z is fixed. The molecule data type is a simple collection of x, y, d vector of double. An example RPC Language file is shown below in Listing 9.1.

```
    /*
     * RPC file for molecule computation
     * Sandeep Koranne, (C) 2010
     */
5   const MAX_MOLECULE_SIZE = 64;
    const MAX_NUM_DISTANCES = 4096;

    struct molecule {
            int N;
10          int M;
            double coordinate_x< MAX_MOLECULE_SIZE >;
            double coordinate_y< MAX_MOLECULE_SIZE >;
            double distances< MAX_NUM_DISTANCES >;
    };
15
    typedef struct molecule* MPTR;

    program MOLECULEPROG {
      version MOLVERS {
20      MPTR  SOLVEPDF(MPTR) = 1;
      }
      = 1;
    } = 0x20000002;
```

Listing 9.1 RPC Language description of molecule

We can generate the client and server code by running *rpcgen* as follows:

```
$rpcgen molecule.x
```

This generates the files:

1. molecule_clnt.c: the client functionality of `pdfsolver`,
2. molecule_svc.c: the server functionality of `pdfsolver`,
3. molecule_xdr.c: the XDR representation of molecule. See Section 9.1.1 for more details on XDR library.

In addition to these code files, we need to write additional code to actually perform the computation, and to call the `client` code which connects to the service running on the server. These are:

```
    /* Molecular PDF Processing using RPC
     * (C) Sandeep Koranne, 2010
     */
    #include <stdlib.h>
5   #include "molecule.h"
```

```
      #include "molecule_ds.h"

      extern  MPTR * solvepdf_1(MPTR *, CLIENT *);
      extern  MPTR * solvepdf_1_svc(MPTR *, struct svc_req *);
10
      static double ComputeDistanceEnergy( MPTR M ) {
        int i;
        double energy = 0.0;
        for( i=0; i < M->M; ++i ) {
15          energy += M->distances.distances_val[i];
        }
        return energy;
      }

20    static double ComputePointEnergy( int N,
              const double* X,
              const double* Y)
      {
        int i,j;
25      double energy = 0.0;
        for( i=0; i < N; ++i )
          for( j=0; j < N; ++j ) {
            double dx = X[i] - X[j];
            double dy = Y[i] - Y[j];
30          if( ( X[i] == 100.0 ) || ( X[j] == 100.0 ) ) dx += 10000;
            if( ( Y[i] == 100.0 ) || ( Y[j] == 100.0 ) ) dy += 10000;
            energy += ( dx * dx ) + ( dy * dy );
          }

35      return energy;
      }


      static double ComputeEnergy( MPTR M ) {
40      int i;
        double energy = 0.0;
        const double *X = M->coordinate_x.coordinate_x_val;
        const double *Y = M->coordinate_y.coordinate_y_val;
        energy = ComputePointEnergy( M->N, X, Y );
45      return energy;
      }

      /*
       * \function ProcessMoleculeForPDF
50     * \description memory has been allocated for output
       */
      static void ProcessMoleculeForPDF( MPTR input, MPTR output ) {
        const static unsigned int LOOP_COUNT = 1000;
        int i,j;
55      int N, count;
        const double *iX = input->coordinate_x.coordinate_x_val;
        const double *iY = input->coordinate_y.coordinate_y_val;
        const double *iD = input->distances.distances_val;

60      double *oX = output->coordinate_x.coordinate_x_val;
        double *oY = output->coordinate_y.coordinate_y_val;
        double *oD = output->distances.distances_val;

        double *X = malloc( input->N * sizeof( double ) );
65      double *Y = malloc( input->N * sizeof( double ) );
        double d_energy = ComputeDistanceEnergy( input );
        double energy, min_energy = 1e10;
        N = input->N;

70      for( i=0; i < input->N; ++i ) {
          X[i] = oX[i] = iX[i];
          Y[i] = oY[i] = iY[i];
```

```
      }
      for( i=0; i < input->N; ++i ) {
75       oD[i] = iD[i];
      }
      /* for every X,Y which is 0, we have to solve */
      energy = ComputeEnergy( output ) - d_energy;
      for( i=0; i < LOOP_COUNT; ++i ) {
80       double cur_energy = 0.0;
         /* set unsolved to random solutions */
         for( j=0; j < N; ++j ) {
           double prev_X;
           if( iX[j] != 100.0 ) continue;
85         prev_X = X[j];       X[j] = drand48();
           /* now compute new energy */
           cur_energy = ComputePointEnergy( N, X, Y );
           if( cur_energy < energy )
      energy = cur_energy;
90         else
      X[j] = prev_X;
         }
         /* Now do the Y loop */
         for( j=0; j < N; ++j ) {
95         double prev_Y;
           if( iY[j] != 100.0 ) continue;
           prev_Y = Y[j];       Y[j] = drand48();
           /* now compute new energy */
           cur_energy = ComputePointEnergy( N, X, Y );
100        if( cur_energy < energy )
      energy = cur_energy;
           else
      Y[j] = prev_Y;
         }
105      /* printf("\n E %d  = %f ", i, energy ); */
      }
      count=0;
      for( i=0; i < N; ++i ) {
         oX[i] = X[i]; oY[i] = Y[i];
110   }
      for( i=0; i < N; ++i )
         for( j=0; j < N; ++j ) {
           double dx = oX[i] - oX[j];
           double dy = oY[i] - oY[j];
115        oD[ count++] = ( dx * dx ) + ( dy * dy );
         }
      free( X );
      free( Y );
}
120
MPTR* solvepdf_1( MPTR* input, CLIENT* client ) {
   static MPTR retval = NULL;
   MPTR actual_input;
   //if( retval ) { DestroyMolecule( retval ); retval = NULL; }
125   actual_input = *input;
   retval = ConstructMolecule( actual_input->N, actual_input->M );
   ProcessMoleculeForPDF( actual_input, retval );
   return &retval;
}
130
MPTR* solvepdf_1_svc( MPTR* input, struct svc_req* req ) {
   CLIENT *client;
   printf("\n Memory used = ");
   system("cat /proc/self/statm");
135   printf("\n");
   return( solvepdf_1( input, client ) );
}
```

Listing 9.2 Molecule processing service

The client code has to be wrapped as:

```
/* \file rmolecule.c
   \author Sandeep Koranne, (C) 2010
   \description Remote molecule processing
*/
5   #include <stdio.h>
    #include "molecule.h"
    #include "molecule_ds.h"

    int main( int argc, char *argv [] ) {
10
      CLIENT *clnt;
      MPTR *result;
      MPTR input;
      char *server, *mfile;
15    FILE *fp;
      server = argv[1];
      mfile =  argv[2];
      fp = fopen( mfile, "rt" );

20    input = ReadMolecule( fp );
      InitializeMolecule( input );
      fclose( fp );

      clnt = clnt_create( server, MOLECULEPROG,
25            MOLVERS, "udp" );
      if( clnt == NULL ) {
        clnt_pcreateerror(server);
        return 1;
      }
30    result = solvepdf_1( &input, clnt );
      if( result == NULL ) {
        clnt_perror( clnt, server );
        return 1;
      }
35    PrintMolecule( input );
      printf("\nMessage sent to server");
      clnt_destroy( clnt );
      /* now print the returned molecule */
      printf("\n Returned molcule = \n");
40    PrintMolecule( *result );
      printf("\n");
      return 0;
    }
```

Listing 9.3 Client for molecule processing service

We add the new service to /etc/rpc:

```
molecule_server 536870914
```

9.1.1 XDR : External Data Representation Library

XDR's approach to standardizing data representations is canonical. For example XDR defines a single byte order (big-endian) and uses the IEEE format for representing floating point numbers. Any program running on any machine can use XDR to create portable data by translating its local representation to the XDR standard representations. The single standard completely de-couples programs that create or send portable data from those that use or receive portable data. Consider the program below:

```
/* \file rpc_writer.c
 * \author Sandeep Koranne, (C) 2010
 * \description Using XDR for serialization
 */

#include <stdio.h>
#include <rpc/types.h>
#include <rpc/xdr.h>

int main(int argc, char *argv[] ) {
  XDR xdrs;
  long i;
  xdrstdio_create( &xdrs, stdout, XDR_ENCODE );
  for( i = 0; i < 8; i++) {
    if( !xdr_long( &xdrs, &i ) ) {
      fprintf( stderr, "ERROR: xdr_long" );
      exit( 1 );
    }
  }
  return( 0 );
}
```

Listing 9.4 XDR writer example

and

```
/* \file rpc_reader.c
 * \author Sandeep Koranne, (C) 2010
 * \description Using XDR for serialization
 */

#include <stdio.h>
#include <rpc/types.h>
#include <rpc/xdr.h>

int main(int argc, char *argv[] ) {
  XDR xdrs;
  long i,j;
  xdrstdio_create( &xdrs, stdin, XDR_DECODE );
  for( i = 0; i < 8; i++) {
    if( !xdr_long( &xdrs, &j ) ) {
      fprintf( stderr, "ERROR: xdr_long" );
      exit( 1 );
    } else {
      printf("%d = %ld", i, j);
    }
  }
  return( 0 );
}
```

Listing 9.5 XDR reader example

The major XDR primitives are shown in Table 9.1.

At this time it is instructive to inspect the XDR description generated by rpcgen for a complex structure. Above in Section 9.1 we had discussed the RPC library with the example of computational biology using the molecule structure. The generated XDR is given below:

```
struct molecule {
```

Table 9.1 XDR functions for writing primitives

Directive	Description
xdr_bool	Boolean
xdr_chars	Char
xdr_u_chars	Unsigned char
xdr_int	Integer
xdr_u_int	Unsigned integer
xdr_long	Long
xdr_u_long	Unsigned long
xdr_float	Single precision floating point
xdr_double	Double precision floating point
xdr_void	void
xdr_string	null terminated byte sequence
xdr_bytes	byte sequence with length
xdr_arrays	arrays of arbitrary elements
xdr_union	Discriminated union
xdr_reference	Pointers
xdrstdio_create	initialize XDR stream
xdrmem_create	create XDR mem stream
xdrrec_create	create TCP/IP stream

```
      int N;
      int M;
      struct {
5       u_int coordinate_x_len;
        double *coordinate_x_val;
      } coordinate_x;
      struct {
        u_int coordinate_y_len;
10      double *coordinate_y_val;
      } coordinate_y;
      struct {
        u_int distances_len;
        double *distances_val;
15    } distances;
    };
    typedef struct molecule molecule;

    typedef molecule *MPTR;
```

Listing 9.6 Generated XDR for molecule.x

It can be seen that the arrays have been augmented with the length variable.

Another example with user defined XDR functions for writing structures is shown below:

```
    /* \file system_load.h
     * \author Sandeep Koranne, (C) 2010
     * \description Using XDR for serialization
     */
5   #ifndef _system_load_h
    #define _system_load_h
```

```
     #include <rpc/types.h>
     #include <rpc/xdr.h>
10
     struct SystemLoad {
       char* system_name;
       float avg1, avg5, avg15;
     };
15
     extern bool_t SystemLoad_XDR( XDR *xdrs, struct SystemLoad* S );
     #endif
```

Listing 9.7 XDR file for system load structure

```
     /* XDR utility for own structure */
     #include "system_load.h"

     bool_t SystemLoad_XDR( XDR *xdrs, struct SystemLoad* S ) {
5      return ( xdr_string( xdrs, &S->system_name, 64 ) &&
           xdr_float ( xdrs, &S->avg1 ) &&
           xdr_float ( xdrs, &S->avg5 ) &&
           xdr_float ( xdrs, &S->avg15 ) );
     }
```

Listing 9.8 XDR file system load

```
     /* \file rpc_info_writer.c
      * \author Sandeep Koranne, (C) 2010
      * \description Using XDR for serialization
      */
5
     #include <stdio.h>
     #include "system_load.h"

     int main(int argc, char *argv[] ) {
10     XDR xdrs;
       struct SystemLoad S;
       S.system_name = "celex";
       S.avg1 = 1.38;
       S.avg5 = 1.30;
15     S.avg15 = 0.78;
       xdrstdio_create( &xdrs, stdout, XDR_ENCODE );
       SystemLoad_XDR( &xdrs, &S );
       return( 0 );
     }
```

Listing 9.9 Server RPC file for system load

```
     /* \file rpc_info_reader.c
      * \author Sandeep Koranne, (C) 2010
      * \description Using XDR for serialization
      */
5
     #include <stdio.h>
     #include "system_load.h"

     int main(int argc, char *argv[] ) {
10     XDR xdrs;
       struct SystemLoad S;
       S.system_name = NULL;
       xdrstdio_create( &xdrs, stdin, XDR_DECODE );
       SystemLoad_XDR( &xdrs, &S );
15     printf("\n Name = %s : %f %f %f\n", S.system_name,
         S.avg1, S.avg5, S.avg15 );
```

```
    return( 0 );
}
```

Listing 9.10 Client RPC file for system load

```
$./rpc_info_writer | ./rpc_info_reader

Name = celex : 1.380000 1.300000 0.780000
```

When using `xdr_string` we have to initialize the memory pointed to by the structure, to NULL, or a valid memory location (which can atleast be of size MAXLEN). If the memory is NULL, XDR will call `malloc`; this memory should be freed afterwards. Another example of XDR, this time using it in conjunction with `zlib` is given in Listing 8.1.

9.2 Checksum computation

Alongwith data transmission and storage, comes the possibility of data corruption. Data can be corrupted due to hardware or software failures, and reliable communication channels thus always include a *checksum* with each packet of data. The capability of the checksum depends on the expected rate of failure, system requirements on reliability, and performance. Cyclic redundancy check (CRC) is a common checksum function and has been optimized for implementation in hardware. Boost includes a CRC function, and we describe its use by an example as shown in Listing 9.11.

```
    // \file crc_example.cpp
    // \author Sandeep Koranne, (C) 2010
    // \description CRC example using Boost
    #include <iostream>          // for program IO
5   #include <fstream>           // for data IO
    #include <boost/crc.hpp>     // actual CRC
    #include <string>            // for filename

    int CalculateCRC32( std::ifstream& ifs ) {
10    ifs.seekg( 0, std::ios::beg ); // seek to begining
      boost::crc_32_type boost_crc;
      unsigned long file_size = 0;
      while( ifs ) {
        unsigned char uc = ifs.get();
15      boost_crc.process_byte( uc );
        file_size++;
      }
      std::cout << "File checksum = " << boost_crc.checksum()
                << " processed " << file_size << " bytes.\n";
20  }

    int main( int argc, char * argv[] ) {
      if( argc != 2 ) { std::cerr << "Usge: crc32 <filename>\n";
        exit( 1 );
25    }
      std::string fileName( argv[1] );
      std::ifstream ifs( fileName.c_str() );
      if( !ifs ) {
```

```
             std::cerr << "Unable to open file: " << fileName << " for reading.\n";
30           exit( 1 );
         }
         CalculateCRC32( ifs );
         return 0;
     }
```

Listing 9.11 Example of Boost CRC32 checksum

In addition to CRC32, boost provides the following checksum algorithms:

1. `crc_16_type`
2. `crc_ccitt_type`
3. `crc_xmodem_type`
4. `crc_32_type`

9.2.1 MD5

Simple CRC32 (as discussed above in Boost CRC) is prone to *aliasing*. Using 128-bit MD5 (RFC 1321) can alleviate this concern to a great extent. Although the GNU/Linux command md5sum can compute the MD5 checksum of a given file, we can also use MD5 checksum on arbitrary frames of data (e.g., when devising an error correcting protocol, or disk file system). Using APR MD5 functionality we can quickly integrate MD5 checksum into an application. See the small example listing shown in Listing 9.12.

```
    // \file md5example.cpp
    // \author Sandeep Koranne (C) 2010
    // \description Example of using MD5 checksum
    #include <iostream>           // for program IO
5   #include <iomanip>            // for std::hex
    #include <cstdio>             // for C FILE*
    #include <cstdlib>            // for exit
    #include <apr_general.h>      // APR general
    #include <apr_mmap.h>         // APR mmap
10  #include <apr_md5.h>          // MD5 computation

    static const int MAX_NUMBER = APR_MD5_DIGESTSIZE;
    static void ComputeMD5Checksum( const char *filename,
            unsigned char digest[MAX_NUMBER]) {
15      apr_status_t retval;
        apr_pool_t *pool;
        apr_pool_create( &pool, NULL );
        apr_mmap_t *mmap;
        apr_finfo_t file_info;
20      apr_file_t  *fp;
        retval = apr_file_open( &fp, filename,
                APR_READ|APR_BINARY,
                APR_OS_DEFAULT, pool );

25      retval = apr_file_info_get( &file_info, APR_FINFO_SIZE, fp );
        std::cout << "File size = " << file_info.size << std::endl;
        // do the actual memory mapping
        retval = apr_mmap_create( &mmap, fp, 0, file_info.size,
                APR_MMAP_READ, pool );
30      if( retval != APR_SUCCESS ) {
          std::cerr << "APR mmap failed..\n";
```

```
        exit( 1 );
    }
    apr_file_close( fp ); // file can be closed now
35  // now mmap->mm is a const char* which can be read
    apr_md5_ctx_t md5context;
    apr_md5_init( &md5context );
    apr_md5( digest, mmap->mm, file_info.size );
    apr_mmap_delete( mmap ); // return back memory
40  apr_pool_destroy( pool );
}


int main( int argc, char *argv[] ) {
45  // Step 1. initialize APR
    if( argc != 2 ) {
        std::cerr << "Usage: ./md5example <filename>\n";
        exit(1);
    }
50  const char *filename = argv[1];
    apr_initialize();
    unsigned char digest[ MAX_NUMBER ];
    ComputeMD5Checksum( filename, digest );
    std::cout << "MD5 digest for " << filename << " = \n";
55  for( int i=0; i < MAX_NUMBER; ++i )
        std::cout << std::hex << (int)digest[i];
    std::cout << std::endl;
    apr_terminate();
    return (0);
60  }
```

Listing 9.12 Example of computing MD5 checkum using APR

Compiling and running this program on the binary itself gives us:

```
$md5sum -b md5example.cpp
a7af280c4d7c627b97f080a98074ecf5 *md5example.cpp
$ ./md5example md5example
File size = 9723
MD5 digest for md5example =
a0cf97256ae47ba16df66eee1fe42c3
```

We first run the GNU/Linux program with the '-b' option (for binary files), there-after, we run our own program which uses mmap on the given file.

9.2.2 SHA1 checksum

Simple CRC32 (as discussed above in Boost CRC) is prone to *aliasing*. Using SHA1 (NIST Secure Hash Algorithm) can alleviate this concern to a great extent. Although the GNU/Linux command sha1sum can compute the SHA1 checksum of a given file, we can also use SHA1 checksum on arbitrary frames of data (e.g., when devising an error correcting protocol, or disk file system). Using APR SHA1 functionality we can quickly integrate SHA1 checksum into an application. See the small example listing shown in Listing 9.13.

```
// \file sha1example.cpp
```

```
   // \author Sandeep Koranne (C) 2010
   // \description Example of using MD5 checksum
   #include <iostream>            // for program IO
5  #include <iomanip>             // for std::hex
   #include <cstdio>              // for C FILE*
   #include <cstdlib>             // for exit
   #include <apr_general.h>       // APR general
   #include <apr_mmap.h>          // APR mmap
10 #include <apr_sha1.h>          // SHA1 computation

   static const int MAX_NUMBER = APR_SHA1_DIGESTSIZE;
   static void ComputeSHA1Checksum( const char *filename,
                                    unsigned char digest[MAX_NUMBER]) {
15   apr_status_t retval;
     apr_pool_t *pool;
     apr_pool_create( &pool, NULL );
     apr_mmap_t *mmap;
     apr_finfo_t file_info;
20   apr_file_t *fp;
     retval = apr_file_open( &fp, filename,
                             APR_READ|APR_BINARY,
                             APR_OS_DEFAULT, pool );

25   retval = apr_file_info_get( &file_info, APR_FINFO_SIZE, fp );
     std::cout << "File size = " << file_info.size << std::endl;
     // do the actual memory mapping
     retval = apr_mmap_create( &mmap, fp, 0, file_info.size,
                               APR_MMAP_READ, pool );
30   if( retval != APR_SUCCESS ) {
       std::cerr << "APR mmap failed..\n";
       exit( 1 );
     }
     apr_file_close( fp ); // file can be closed now
35   // now mmap->mm is a const char* which can be read
     apr_sha1_ctx_t sha1context;
     apr_sha1_init( &sha1context );
     apr_sha1_update_binary( &sha1context, (unsigned char*)(mmap->mm),
            file_info.size );
40   apr_sha1_final( digest, &sha1context );
     apr_mmap_delete( mmap ); // return back memory
     apr_pool_destroy( pool );
   }


45
   int main( int argc, char *argv[] ) {
     // Step 1. initialize APR
     if( argc != 2 ) {
       std::cerr << "Usage: ./sha1example <filename>\n";
50     exit(1);
     }
     const char *filename = argv[1];
     apr_initialize();
     unsigned char digest[ MAX_NUMBER ];
55   ComputeSHA1Checksum( filename, digest );
     std::cout << "SHA1 digest for " << filename << " = \n";
     for( int i=0; i < MAX_NUMBER; ++i )
       std::cout << std::hex << (int)digest[i];
     std::cout << std::endl;
60   apr_terminate();
     return (0);
   }
```

Listing 9.13 Example of computing SHA1 checkum using APR

Compiling and running this program on the binary itself gives us:

```
$ sha1sum sha1example
```

```
9ca4b2a1fb03d3038c22e313a1cebb8a39069400  sha1example
./sha1example sha1example
File size = 9872
SHA1 digest for sha1example =
9ca4b2a1fb3d338c22e313a1cebb8a396940
```

9.3 OpenSSL

The OpenSSL Project is an open-source effort to develop a robust toolkit implementing the Secure Socket Layer (SSL v2/v3) and Transport Layer Security (TLS v1) as well as a general purpose cryptographic library. The `openssl` command-line tool can be used for:

1. Creation and management of public/private keys,
2. Public key cryptography,
3. Creation of X.509 certificate,
4. Calculation of message digests,
5. Encryption and decryption with Ciphers.

`openssl` command-line tool is a general purpose utility which can invoke the processing functions listed above. For example, the MD5 digest can be computed using:

```
$openssl version
OpenSSL 1.0.0-fips-beta4 10 Nov 2009
$openssl md5 math.lisp
MD5(/home/skoranne/math.lisp) =
    d82e9c3808c986bce26df25640c80886
```

Some of the other tools in OpenSSL can be listed using `openssl -h`

```
Standard commands
asn1parse   ca          ciphers     cms
crl         crl2pkcs7   dgst        dh
dhparam     dsa         dsaparam    enc
engine      errstr      gendh       gendsa
genpkey     genrsa      nseq        ocsp
passwd      pkcs12      pkcs7       pkcs8
pkey        pkeyparam   pkeyutl     prime
rand        req         rsa         rsautl
s_client    s_server    s_time      sess_id
smime       speed       spkac       ts
verify      version     x509

Message Digest commands
md2         md4         md5         rmd160
sha         sha1

Cipher commands
```

```
aes-128-cbc aes-128-ecb   aes-192-cbc  aes-192-ecb
aes-256-cbc aes-256-ecb   base64       bf
bf-cbc      bf-cfb        bf-ecb       bf-ofb
camellia    camellia-256  cast         cast-cbc
cast5-cbc   cast5-cfb     cast5-ecb    cast5-ofb
des         des-cbc       des-cfb      des-ecb
des-ede     des-ede-cbc   des-ede-cfb  des-ede-ofb
des-ede3    des-ede3-cbc  des-ede3-cfb des-ede3-ofb
des-ofb     des3          desx         rc2
rc2-40-cbc  rc2-64-cbc    rc2-cbc      rc2-cfb
rc2-ecb     rc2-ofb       rc4          rc4-40
seed        seed-cbc      seed-cfb     seed-ecb
seed-ofb    zlib
```

9.4 XML Processing

XML (Extended Markup Language) is a standard for representing markup languages. SVG files, configuration files for GNOME, and many other systems now use XML to communicate and store data. For application programs to read and write XML files, they can use software API. Some libraries for XML handling are:

1. SAX: SAX (Simple API for XML) is a standard interface for event-based XML parsing,
2. libXML: a SAX (Simple API for XML) like parser,
3. Expat: another event-based parsing framework,
4. APR XML API: Apache Portable Runtime (APR) has XML processing capabilities,
5. Boost XML: Boost C++ API also has XML handling.

Consider an example XML file:

```
<settings>
  <profiles>
    <profile>
      <id>JPP</id>
        <repositories>
          <repository>
            <id>internal</id>
            <layout>jpp</layout>
            <url>file:///builddir/build/BUILD/\
                 jetty-6.1.20/.m2/repository</url>
          </repository>
          <repository>
            <id>external</id>
            <layout>jpp</layout>
            <url>file:///builddir/build/BUILD/\
                 jetty-6.1.20/external_repo</url>
  ...
```

9.4.1 Expat : XML processing

An example of using Expat for XML parsing is shown in Listing 9.14.

```
    static void XMLDeclaration( void *ud, // user data
            const XML_Char *version,
            const XML_Char *encoding,
            int standalone ) {
5     std::cout << "XML Declaration : Version = " << version << "\n"
            << "Encoding = " << encoding << "\n";
    }


10  static void ParseElementDecl( void *ud,
            const XML_Char *name,
            XML_Content *model ) {
      std::cout << "Read ElementDecl : " << name << "\n";
    }
15
    static void EntityDecl( void *ud, // user data
          const XML_Char *name,
          int is_parameter,
          const XML_Char *value,
20        int value_length,
          const XML_Char *base,
          const XML_Char *systemId,
          const XML_Char *publicId,
          const XML_Char *notationName ) {
25    std::cout << "Entity : " << name << "\n";
    }

    static void StartDocHandler( void *ud,
            const XML_Char *doctypeName,
30          const XML_Char *sysid,
            const XML_Char *pubid,
            int has_internal_subset ) {
      std::cout << "Doc type name = " << doctypeName << "\n";
    }
35
    static void DefaultHandler( void *ud,
            const XML_Char *s,
            int len ) {
      std::cout << "Default : " << s << "\n";
40  }


    static void ProcessXMLStream( const char* data, size_t len ) {
      XML_Parser parser = XML_ParserCreate( NULL );
45    XML_SetElementDeclHandler( parser, ParseElementDecl );
      XML_SetXmlDeclHandler( parser, XMLDeclaration );
      XML_SetEntityDeclHandler( parser, EntityDecl );
      XML_SetStartDoctypeDeclHandler( parser, StartDocHandler );
      XML_SetDefaultHandler( parser, DefaultHandler );
50    XML_Parse( parser, data, len, 1 ); // is final
    }
```

Listing 9.14 Example of using Expat for XML parsing

9.4.2 libXML : XML processing library

Another API for XML processing is the `libxml` library, which can be configured
to read/write XML files using event-processing, or data-driven mode. An example
of the data-driven mode is shown in Listing 9.15.

```cpp
// \file libxml_example.cpp
// \author Sandeep Koranne (C) 2010
// \description XML parsing using libxml
#include <cstdio>            // for FILE
#include <iostream>          // for program IO
#include <libxml/parser.h>   // libxml

static void ProcessFile( const char *filename ) {
  xmlDocPtr doc;
  xmlNodePtr cur;
  doc = xmlParseFile( filename );
  if( doc == NULL ) {
    std::cerr << "Unable to parse XML file : " << filename << "\n";
    exit(1);
  }
  cur = xmlDocGetRootElement( doc );
  std::cout << "Root = " << cur->name << "\n";
  while( cur ) {
    std::cout << cur->name << "\n";
    xmlNode *children = cur->children;
    while( children ) {
      std::cout << "\t Child = " << children->name << "\n";
      children = children->next;
    }
    cur = cur->next;
  }
  xmlFreeDoc( doc );
}

int main( int argc, char *argv [] ) {
  if( argc != 2 ) {
    std::cerr << "Usage: ./libxml_example <file>.xml...\n";
    exit(1);
  }
  ProcessFile( argv[1] );

  std::cout << std::endl;
  return (0);
}
```

Listing 9.15 Example of using libxml for XML processing

The code in Listing 9.15 should be contrasted with Listing 9.14; in Listing 9.14 we
defined *callbacks* which the API called for us on specific events during the parsing
of the XML file. In Listing 9.15 we explore the XML data-structure tree using our
own control loops. Compiling the code in Listing 9.15 and running it on the example
XML file shown above gives:

```
Root = settings
settings
 Child = text
 Child = profiles
 Child = text
 Child = activeProfiles
```

```
Child = text
. . . . .
```

XML handling is also present in other languages such as Perl and Python, which often provide *bindings* to one or more of the above libraries, thus, the syntax and method of using XML libraries is as described above.

9.5 Berkeley DB

Berkeley DB is an open-source embedded database library that provides scalable, high-performance, transaction protected data management services to applications. Berkeley DB provides a simple API for data-access and management. Bindings for Berkeley DB are provided for C, C++, Perl, Java, Tcl, Python, Lisp and many other languages. Berkeley DB runs in the same memory space as the application, and as such no inter-process or network communication is required to use Berkeley DB.

Berkeley DB supports many different storage structures including: (i) hash tables, (ii) BTrees, (iii) simple record-number based storage,dow and (iv) persistent queues. Programmers can create tables using these structure. Hash tables are best suited for large applications with a need for predictable search time, while BTrees are better for range-based searches.

Berkeley DB also provides data management services such as concurrency, transactions and recovery. Records in Berkeley DB are represented as *(key,value)* pair, and it provides functions for:

1. Insertion: insert a record in a table,
2. Search: find a record in a table,
3. Deletion: delete a record from a table,
4. Update: update a record in a table.

Berkeley DB is not a relational database and does not support SQL. It is possible to build a relational database using the transactional library and data storage facilities of Berkeley DB (infact, MySQL used Berkeley DB for its storage).For SQL oriented databases refer to Chapter 20 for a discussion on open-source relational databases. Keys and values can be arbitrary bit strings (fixed-length or variable length, depending on the application).

9.5.1 DB open function

The `DB->open()` function opens the database for writing as well as reading. The form of this function is as follows:

```
#include <db.h>
int DB->open( DB *db, DB_TXN *txnid, const char *file,
              const char *database, DBTYPE type,
```

```
                u_int32_t flags, int mode);
```

The DB is the *handle* to the database, and `txnid` is the transaction id if transactions are used, else NULL. The file name of the database is passed in `file`.
The `database` argument is optional; it allows for the application to have several
databases in the same logical file. If there is a single database, this argument
can be left NULL. The `DBTYPE` type specifies the type of the database, whether
DB_BTREE, DB_HASH, DB_QUEUE, DB_RECNO, or DB_UNKNOWN. The
`flags` specify whether the database should be created (DB_CREATE), or opened
for read-only access (DB_RDONLY), or whether the database should be multi-
versioned (DB_MULTIVERSION). The `mode` parameter is used to set the file permission on the created file.

A database handle is created using `db_create` and subsequently a database can
be opened using `DB->open()` function. When the processing is complete the database
should be closed using the `DB->close()` function. Consider an example application
written using the Berkeley DB as shown in Listing 9.16.

```cpp
   // \file bdb_example.cpp
   // \author Sandeep Koranne, (C) 2010
   // \description Example of using Berkeley DB

5  #include <cassert>         // for assertion checking
   #include <cstdio>          // C stdio
   #include <cstdlib>         // C stdlib for exit, atoi
   #include <cstring>         // memset
   #include <iostream>        // program IO
10 #include <string>          // std::string
   #include <sys/types.h>     // C standard types
   #include <db.h>            // Berkeley db

15 int main( int argc, char *argv[] ) {
     if( argc != 3 ) {
       std::cerr << "Usage: bdb_example <db-name> <op-code>\n";
       exit(1);
     }
20   std::string dbName( argv[1] );
     int db_operation = atoi( argv[2] );
     DB *dbp = NULL;
     int rc = db_create( &dbp, NULL, 0 );
     if( rc != 0 ) {
25     std::cerr << "db_create " << db_strerror( rc );
       std::exit(1);
     }

     if( db_operation == 1 ) { // database create
30     rc = dbp->open( dbp, NULL, dbName.c_str(),
                       NULL, DB_BTREE, DB_CREATE, 0664 );
       if( rc != 0 ) {
         perror("Berkeley db open error");
         exit(1);
35     }
       // successfully opened the db
       DBT key, data;
       memset( &key, 0, sizeof( DBT ) );
       memset( &data, 0, sizeof( DBT ) );
40     key.data = (void*)"subject_A";
       key.size = sizeof( "subject_A" );
       data.data = (void*)"Math";
       data.size = sizeof( "Math" );
```

```
       rc = dbp->put( dbp, NULL, &key, &data, 0 );
45     switch( rc ) {
       case 0 : { break; /* key added */ }
       case DB_KEYEXIST:{
         std::cerr << "already stored "<<key.data<< "\n"; break;
         }
50     }
       key.data = (void*) "subject_B";
       data.data = (void*) "Physics";
       data.size = sizeof( "Physics" );
       rc = dbp->put( dbp, NULL, &key, &data, 0 );
55     dbp->close( dbp, 0 );
     } else if( db_operation == 2 ) { // database open and print
       rc = dbp->open( dbp, NULL, dbName.c_str(),
                       NULL, DB_BTREE, DB_RDONLY, 0664 );
       if( rc != 0 ) {
60       perror("Berkeley db open error");
         exit(1);
       }
       DBT key, data;
       memset( &key, 0, sizeof( DBT ) );
65     memset( &data, 0, sizeof( DBT ) );
       key.data = (void*)"subject_A";
       key.size = sizeof( "subject_A" );
       rc = dbp->get( dbp, NULL, &key, &data, 0 );
       if( rc == 0 ) {
70       std::cout << (char*)key.data << "\t" << (char*)data.data << "\n";
       } else {
         perror(" db->get error...\n");
         exit(1);
       }
75     dbp->close( dbp, 0 );
     } else {
       std::cerr << "Unknown db-operation, " << db_operation << std::endl;
     }

80     std::cout << "Example of Berkeley db\n";

       return (0);
     }
```

Listing 9.16 Example of using Berkeley DB

The functions DB->put() and DB->get() are described below:

```
#include <db.h>
int
DB->put( DB *db, DB_TXN *txnid, DBT *key, DBT *data,
         u_int32_t flags);
5  int
DB->get( DB *db, DB_TXN *txnid, DBT *key, DBT *data,
         u_int32_t flags);
```

The DB->put() function stores key/data value pairs into the database. It can either replace an existing key/value pair, or add duplicate (if allowed).

9.5.2 Other Berkeley DB functions

In addition to these above functions, Berkeley DB has a number of useful function as listed in Table 9.2.

Table 9.2 Berkeley DB functions

Database functions	Description
`DB->compact()`	compact a database
`DB->del()`	delete items from database
`DB->err()`	retrieve error message
`DB->exists()`	return if an item exists in the database
`DB->fd()`	return underlying file descriptor
`DB->get()`	get key/value pair (see above)
`DB->get_byteswapped()`	check if db is in host order
`DB->get_dbname()`	return file and db name
`DB->get_type()`	return database type
`DB->join()`	perform db join on cursors
`DB->key_range()`	return estimate on key location
`DB->open()`	open database (see above)
`DB->put()`	put key/value pair in db (see above)
`DB->remove()`	remove a database
`DB->rename()`	rename a database
`DB->stat()`	return database statistics
`DB->sync()`	flush database to disk
`DB->verify()`	verify database integrity
`DB->cursor()`	create a database cursor
BTree / Recno config Description	
`DB->set_bt_compare()`	set BTree comparison function
`DB->set_bt_compress()`	set BTree compression function

Berkeley DB also has a simple Python interface, an example of which is shown in Listing 9.17.

```
#!/usr/bin/python
#example file for using Berkeley DB with Python
import bsddb
db = bsddb.btopen('example.db', 'c')
5 print db.keys()
for k,v in db.iteritems():
    print k,v
```

Listing 9.17 Using Berkeley DB with Python

Running the C++ binary, followed by the Python program gives:

```
./bdb_example example.db 1
Example of Berkeley db
[skoranne@celex BDB]$ python bsdb.py
['subject_A\x00', 'subject_B\x00']
subject_A Math
subject_B Physics
```

Berkeley DB has a number of other features such as *cursors* and database *join* operations which can be used by the application software, but even by the above functions we can see that Berkeley DB provides an elegant and effective data storage mechanism. It is thus no surprise that a number of open-source applications (such as Subversion and MySQL) use Berkeley DB for their data storage and database management needs.

9.6 Memcached Library

Memcached is a high-performance distributed memory object caching system. It is designed to reduce database load in web applications and speed up dynamic web content generation. `libmemcached` is a client library for the *memcached* protocol. Memcache implements an in-memory key-value store for small chunks of arbitrary data. This data can be the result of applications running on the server, and the results are *cached*, so that subsequent calls to the same procedure by any application running on the server can simply return the pre-computed result, thus increasing performance.

Memcached comprises of the following components:

1. Server software: stores arbitrary key/value pairs in an internal hash table,
2. Server algorithms: for cache management and purging,
3. Client software: given a list of memcached servers,
4. Client hashing algorithm: chooses server based on key, for load balancing.

Running `memcached` on a server is a simple matter of invoking the `memcached` binary. The command accepts a number of command line options:

```
-p : port number
-s : UNIX socket path (disables network support)
-m : maximum memory (default is 64 MB)
-d : run memcache as a daemon
-M : disable automatic removal of objects from cache
-t : specify number of threads (default 4)
```

The status of a running `memcached` server can be checked using the `nc` tool. By sending the message "stats settings" to the specified port of the `memcached` server we can query its disposition, as shown below:

```
$echo "stats settings" | nc localhost 11211
STAT maxbytes 67108864
STAT maxconns 1024
STAT tcpport 11211
STAT udpport 11211
STAT inter NULL
STAT verbosity 0
STAT oldest 0
STAT evictions on
STAT domain_socket NULL
STAT umask 700
STAT growth_factor 1.25
```

```
STAT chunk_size 48
STAT num_threads 4
STAT stat_key_prefix :
STAT detail_enabled no
STAT reqs_per_event 20
STAT cas_enabled yes
STAT tcp_backlog 1024
STAT binding_protocol auto-negotiate
STAT auth_enabled_sasl no
STAT item_size_max 1048576
END
```

To use the `memcached` server as a cache for key/value pair, the client software must implement the `memcached` protocol. The protocol specifies a key, flag value, an expiration time, and arbitrary data. The `memcached` protocol can be implemented in the client, an example with Python client is shown in Listing 9.18.

```python
#!/usr/bin/python
# example of using memcached server
import pylibmc
mc = pylibmc.Client(["localhost"], binary = True )
5   mc.behaviors = {"tcp_nodelay" : True }

    mc.set("Subject_A", "Math")
    mc.set("Subject_B", "Physics")
    mc.set("Subject_C", "Chemistry")
10
    mc.get("Subject_A")
```

Listing 9.18 Example of memcached protocol in Python

`memcached` can help in improving the performance of an application when the application has the characteristics of being compute bound in calculating results which have also been previously computed at some point in the past. The cache replacement logic, as well as the distributed aspects of the data management can be handled by `memcached`.

9.7 SWIG interface generator

Simplified Wrapper and Interface Generator (SWIG) is an open source tool to generate bindings for programs written in C/C++ to other languages (usually scripting languages) such as Python, Tcl, Perl, and Lua.

Using SWIG is deliberately simple; by adding a small number of *directives* the C/C++ header file a complete module of code can be converted into appropriate form by SWIG. The tool generates source code which enables the *calling* of the C/C++ code in the target language. SWIG has enabled the conversion of many utility libraries written in C/C++ for use in Lisp (for example).

Consider the example of generating bindings for Python as shown below:

```c
int factorial( int n ) {
  if ( n < 2 ) return 1;
```

```
        else return n * factorial(n-1);
    }
```

Listing 9.19 Example of Python binding generation

We write a SWIG wrapper `factorial.i` as:

```
    /* SWIG file for math */
    %module math
    %{
    #define SWIG_FILE_WITH_INIT
5   #include "factorial.h"
    %}
    %include "factorial.h"
```

Listing 9.20 SWIG interface file for Python binding

The we run SWIG as:

```
$swig -python factorial.i
$gcc -O2 -c factorial.c
$gcc -O2 -c -I/usr/include/python2.6 factorial_wrap.c
$gcc -shared factorial.o factorial_wrap.o -o _math.so
$python
Python 2.6.2 (r262:71600, Aug 21 2009, 12:22:21)
[GCC 4.4.1 20090818 (Red Hat 4.4.1-6)] on linux2
Type "help", "copyright", "credits" ...
>>> import math
>>> math.factorial(10)
3628800
```

A more detailed example using C++ is shown below:

```
    #ifndef _layer_db_h_
    #define _layer_db_h_
    struct Box {
      int x1,y1,x2,y2;
5   };

    struct LayeredBox {
      int layer;
      int n;
10    Box *D;
    };

    LayeredBox ReadData( const char* );
    void PrintLayerData( const LayeredBox& B );
15  #endif
```

Listing 9.21 SWIG Python integration with C++ header file

```
    // simple db format library
    #include <stdio.h>
    #include <stdlib.h>

5   #include "db.h"

    LayeredBox ReadData( const char* fileName ) {
      int i;
```

```
        LayeredBox retval;
10      FILE *fp = fopen( fileName, "rb" );
        if( !fp ) exit(1);
        fscanf( fp, "%d %d", &retval.layer, &retval.n );
        retval.D = (Box*) malloc( retval.n * sizeof( Box ) );
        for(i=0; i < retval.n; ++i) {
15        fscanf( fp, "%d %d %d %d",
            &retval.D[i].x1, &retval.D[i].y1,
            &retval.D[i].x2, &retval.D[i].y2 );
        }
        return retval;
20    }

      void PrintLayerData( const LayeredBox& B ) {
        fprintf( stdout, "Layer = %d", B.layer );
        for( int i=0; i < B.n; ++i ) {
25        fprintf( stdout, "\n %d %d %d %d",
            B.D[i].x1, B.D[i].y1,
            B.D[i].x2, B.D[i].y2 );
        }
      }
```

Listing 9.22 SWIG Python integration with C++ file

```
      /* SWIG file for math */
      %module layerdb
      %{
      #define SWIG_FILE_WITH_INIT
5     #include "db.h"
      %}
      %include "db.h"
```

Listing 9.23 SWIG interface file for Python binding

```
$swig -python -c++ db.i
$g++ -shared db.o db_wrap.o -o _layerdb.so
$python
Python 2.6.2 (r262:71600, Aug 21 2009, 12:22:21)
[GCC 4.4.1 20090818 (Red Hat 4.4.1-6)] on linux2
Type "help", "copyright", "credits" ...
>>> import layerdb
>>> D = layerdb.ReadData("a.data")
>>> layerdb.PrintLayerData( D )
Layer = 46
 0 0 10 10
 20 10 30 20
 5 5 10 5>>>
```

As the above example shows, SWIG and Python can maintain the object structure of complex C++ classes and structure, and pass them to and from functions.

9.8 Conclusion

In this chapter we have presented application development libraries which aid in the development of remote procedure call enabled applications. The libraries presented included RPC (remote procedure call), XDR (extensible data representation), checksum, SHA1, MD5, OpenSSL. The use of XML files for data representation was presented with examples using C++. Persistent storage of data using BerkeleyDB, and its use in C++ and Python code, was also discussed. Lastly, we presented the Simplified Wrapper Interface Generator (SWIG) tool which can automate the wrapping of C and C++ APIs for use in Python, Scheme, and other interactive languages.

Chapter 10
Hierarchical Data Format 5 : HDF5

Abstract In this chapter we introduce the Hierarchical Data Format 5 (HDF5) specification. We present a number of examples which use the HDF5 API for reading
and writing HDF files. We present C/C++ API examples of the API including examples of writing, reading compound types in HDF5. For large scientific data, HDF5 is
representative of the APIs which are used in CDF and other data formats containing
simulation or experimentally collected data.

Contents

Scientific datasets can be enormously large in size (comprising of many terabytes
of data). It is thus essential that large datasets are transported and accessed in an
efficient and standardized manner (otherwise multiple tools will have to reinvent
the wheel of doing performance optimized IO on large datasets). Since scientific
datasets also have a lot of structure to them, efficient storage policies can be adopted.

HDF5 is a hierarchical data format specification and supporting library implementation. HDF version 5 addresses some of the limitations of the older HDF formats, such as (i) 2 Gb file size restriction, (ii) cap on maximum number of objects
per file, and (iii) library source of the previous HDF was old and outdated. HDF5
also includes the following improvements to the previous version: (i) new file format
(which removes the restrictions on file size), (ii) simpler, more comprehensive data
model (which incidentally has only two basic structures: a multi-dimensional array
of record structure and a grouping structure, and (iii) simpler library with improved
support of parallel I/O and threading.

10.1 HDF5 files

HDF5 are organized in a hierarchical structure around two basic objects: (i) groups
and (ii) datasets. *Datasets* contain the actual multi-dimensional data elements (along-

S. Koranne, *Handbook of Open Source Tools*,
DOI 10.1007/978-1-4419-7719-9_10, © Springer Science+Business Media, LLC 2011

with supporting metadata), while *groups* provide the data with organizational skeleton; a group contains instances of other groups or datasets (alongwith supporting metadata). The *metadata* is actually data about the data. In addition to metadata, groups and datasets may have an associative *attribute* list. Attributes are user-defined HDF5 structures containing extra information about the HDF5 object.

HDF5 groups and group members are deliberately organized in a manner analogous to the UNIX file system hierarchy, with / representing the *root* group, and /foo denoting a group foo which is a member of the root group. An HDF5 group comprises of: (i) group header (consisting of group name and list of attributes), and (ii) group symbol table (which is a list of the HDF5 objects which belong to the group).

An HDF5 dataset comprises of: (i) dataset header, and (ii) data array. The header contains information needed to interpret the array; it contains the name of the object (header name), its dimensionality (header dataspace), number-type (header datatype), and disk storage information (header storage layout). The name of the object is an ASCII string.

The dimensionality of the dataset is stored in the dataspace in the header. The dimensions can be static or *unlimited* (extensible). Properties of the dataspace consist of its *rank* (number of dimensions) of the data array, actual sizes of the dimensions, and the maximum size of the dimensions. Importantly, a dataspace can also describe portions of the dataset, allowing input-output operations on a *hyperslab* of the dataset. Selection of such a region is supported in HDF5 using the dataspace interface (H5S). The region(s) can be (i) contiguous, (ii) non-contiguous hyperslab, (iii) union of hyperslabs, and (iv) list of independent points.

The HDF5 supported data-types are (i) integer, (ii) floating point (IEEE 32-bit and 64-bit) in both endian formats, (iii) references and (iv) strings. Symbolically the data-types can be referenced using the API enumerations, e.g., for int the corresponding HDF5 data-type is H5T_NATIVE_INT, similarly for long double, H5T_NATIVE_LDOUBLE. A C language struct can be represented as an HDF5 *compound* data-type.

The *storage* of the HDF5 data on disk is possible in a number of ways, with contiguous storage being the default. In this default mode, there is a one-to-one correspondence between data items in memory and disk. The other types of storage are: (ii) compact (useful when data is small and can be stored alongwith the header), and (iii) chunked (data is divided into equal sized chunks, which are stored separately). Chunked data increases performance, especially when accessing subsets of datasets. The HDF5 appears to the user as a directed graph whose nodes are HDF5 objects such as: (i) groups, (ii) datasets, (iii) datatypes and (iv) dataspaces. At the lowest level, an HDF5 file is comprised of:

1. A super block,
2. B-tree nodes: containing either symbol, nodes or raw data chunks,
3. Object headers:
4. Local heaps:
5. Free space.

Although the APIs provide opaque access to the underlying data, it is useful to understand how the underlying data is stored on disk file for improving efficiency. As an example, a HDF5 *group* is an object header in the HDF5 file. This object header contains a message that points to a local heap and to a B-tree node pointing to symbol nodes.

10.1.1 HDF5 API Naming Conventions

The naming convention of the HDF5 API are given below:

1. H5F: file-level access, e.g., `H5Fopen`,
2. H5G: group functions, e.g., `H5Gset`,
3. H5T: datatype functions, e.g., `H5Tcopy`,
4. H5S: dataspace functions, e.g., `H5Screate_simple`,
5. H5D: dataset functions, e.g., `H5Dread`,
6. H5P: property list functions, e.g., `H5Pset_chunk`,
7. H5A: attribute access, e.g., `H5Aget_name`,
8. H5Z: compression registration, e.g., `H5Zregister`,
9. H5E: error handling routine, e.g., `H5Eprint`,
10. H5R: reference function, e.g., `H5Rcreate`,
11. H5I: identifier routing, e.g., `H5Iget_type`.

10.2 Example of HDF5 API

We present two simple examples of using the HDF5 API. In Listing 10.1 we have presented a HDF5 dataset writer, which writes a two-dimensional table of 32-bit integers to a disk file.

```
    // \file hdf_example.cpp
    // \author Sandeep Koranne (C) 2010
    // \description Example of using HDF5 for data storage
    #include <cstdio>              // for C FILE
5   #include <cassert>            // assertion checking
    #include <cstdlib>            // exit
    #include <iostream>           // program IO
    #include <hdf5.h>             // HDF5

10  static const char fileName[] = "example_A.h5";
    static const char datasetName[] = "ASTR_10";
    static const unsigned int DIM0 = 3;
    static const unsigned int DIM1 = 3;

15  static void InitializeDataSet( int  D[DIM0][DIM1] ) {
      for( unsigned int i=0; i < DIM0; ++i )
        for( unsigned int j=0; j < DIM1; ++j )
          D[i][j] = i+j;
    }
20
    int main( int argc, char *argv[] ) {
```

```
      hid_t file, space, dataset;            // Opaque HDF5 handles
      herr_t status;
      hsize_t dims[2] = { DIM0, DIM1 };
25    int     write_data[DIM0][DIM1];        // write buffer
      InitializeDataSet( write_data );
      file = H5Fcreate( fileName, H5F_ACC_TRUNC, H5P_DEFAULT, H5P_DEFAULT );
      space= H5Screate_simple( 2, dims, NULL );
      dataset = H5Dcreate( file, datasetName, H5T_STD_I32LE, space,
30            H5P_DEFAULT, H5P_DEFAULT, H5P_DEFAULT );
      status = H5Dwrite( dataset, H5T_NATIVE_INT, H5S_ALL, H5S_ALL,
            H5P_DEFAULT, write_data[0] );
      status = H5Dclose( dataset ); // first close the dataset
      status = H5Sclose( space );   // then close the space
35    status = H5Fclose( file );    // and lastly close the file.
      return (0);
    }
```

Listing 10.1 Example of using HDF5 library API for writing dataset

The corresponding *reader* example is presented in Listing 10.2. The same file is opened using H5F_ACC_RDONLY mode, and the contents are verified and printed to the screen.

```
    // \file hdf_example.cpp
    // \author Sandeep Koranne (C) 2010
    // \description Example of using HDF5 for data storage
    #include <cstdio>              // for C FILE
5   #include <cassert>            // assertion checking
    #include <cstdlib>            // exit
    #include <iostream>           // program IO
    #include <hdf5.h>             // HDF5

10  static const char fileName[] = "example_A.h5";
    static const char datasetName[] = "ASTR_10";
    static const unsigned int DIM0 = 3;
    static const unsigned int DIM1 = 3;

15  static void CheckAndPrintDataSet( int  D[DIM0][DIM1] ) {
      for( unsigned int i=0; i < DIM0; ++i ) {
        std::cout << std::endl;
        for( unsigned int j=0; j < DIM1; ++j ) {
          assert( D[i][j] == (int)(i+j) );
20        std::cout << " " << D[i][j];
        }
      }
    }

25  int main( int argc, char *argv[] ) {
      hid_t file, dataset;                   // Opaque HDF5 handles
      herr_t status;
      int     read_data[DIM0][DIM1];         // read buffer
      file = H5Fopen( fileName, H5F_ACC_RDONLY, H5P_DEFAULT );
30    dataset = H5Dopen( file, datasetName, H5P_DEFAULT );
      status = H5Dread( dataset, H5T_NATIVE_INT, H5S_ALL, H5S_ALL,
            H5P_DEFAULT, read_data[0] );
      // check and print dataset
      CheckAndPrintDataSet( read_data );
35    status = H5Dclose( dataset ); // first close the dataset
      status = H5Fclose( file );    // and lastly close the file.
      std::cout << std::endl;
      return (0);
    }
```

Listing 10.2 Example of using HDF5 library API for reading dataset

The SCons file for compiling these applications is shown for reference:

```
Program('hdf_example', ['hdf_example.cpp'],
        CXXFLAGS="-Wall -O0 -ggdb",
        LIBS=['hdf5','m','pthread'])
Program('hdf_read_example', ['hdf_read_example.cpp'],
        CXXFLAGS="-Wall -O0 -ggdb",
        LIBS=['hdf5','m','pthread'])
```

Compiling and executing the writer/reader programs gives:

```
$ ./hdf_example
$ ls -l example_A.h5
-rw-rw-r--. 1 skoranne example_A.h5
$ ./hdf_read_example

 0  1  2
 1  2  3
 2  3  4
```

In the above Listing 10.1, the use of HDF5 API functions is demonstrated. In HDF5, *datatype* and *dimensionality* are independent objects, which are created independently from any particular dataset they may be attached to. A dataspace can be created once the *rank* and *dimension* of the space are known; in the above example we use the H5Screate_simple function to create the dataspace.

The datatype is created using the H5Tcopy(H5T_NATIVE_INT) function (in the example it is simple integer), which creates a datatype using the native integer type as a template. The endian behavior of the datatype can be set using the H5Tset_order function; we can choose little-endian (H5T_ORDER_LE) or big-endian. Once the dataspace and datatype handles have been created the actual dataset can be created. At this point in time, the dataset is connected to the underlying data file, the datatype and the dataspace. The H5D function for creating a dataset is: H5Dcreate. In addition to the filename, the name of the dataset has to be given. Once the dataset has been created, actual data can be read or written to the disk file using the H5Dread and H5Dwrite functions. These functions take the pointer to the data for reading and writing.

The datatype, dataspace and dataset object handles should be released once they are no longer needed. Each handle must be released separately using: H5Tclose (for datatype), H5Dclose (for dataset), and H5Sclose (for dataspace).

Using the public API functions of the HDF5 library we can write a HDF inspection tool as shown in Listing 10.3.

```
     // \file hdf_info.cpp
     // \author Sandeep Koranne (C) 2010
     // \description HDF5 info program
     #include <cstdio>              // for C FILE
 5   #include <cstring>             // for string function
     #include <cassert>            // assertion checking
     #include <cstdlib>            // exit
     #include <iostream>           // program IO
     #include <hdf5.h>             // HDF5
10
     static const unsigned int MAX_NUMBER = 1024;
```

```cpp
   static char fileName[ MAX_NUMBER ];
   static char datasetName[ MAX_NUMBER ];

15 int main( int argc, char *argv[] ) {
     if( argc != 3 ) {
       std::cerr << "Usage: hdf_info <h5> <dataset>..\n";
       exit(1);
     }
20   strcpy( fileName, argv[1] );
     strcpy( datasetName, argv[2] );
     hid_t file, dataset;                      // Opaque HDF5 handles
     file = H5Fopen( fileName, H5F_ACC_RDONLY, H5P_DEFAULT );
     // analyze the file object
25   hsize_t filesize;
     (void) H5Fget_filesize( file, &filesize );

     hsize_t objc[ 5 ];
     for( int i=0; i < 5; ++i ) objc[i] = 0;
30   objc[0] = H5Fget_obj_count( file, H5F_OBJ_FILE ); // no. file objects
     objc[1] = H5Fget_obj_count( file, H5F_OBJ_DATASET ); // no. datasets
     objc[2] = H5Fget_obj_count( file, H5F_OBJ_GROUP ); // no. group objects
     objc[3] = H5Fget_obj_count( file, H5F_OBJ_DATATYPE ); // no. datatype
     objc[4] = H5Fget_obj_count( file, H5F_OBJ_ATTR ); // no. attributes
35   std::cout << "File " << fileName
               << " File size = " << filesize
               << " free space = " << H5Fget_freespace( file ) << "\n"
               << " has " << objc[0] << " file objects\n"
               << " has " << objc[1] << " dataset objects\n"
40             << " has " << objc[2] << " group objects\n"
               << " has " << objc[3] << " datatype objects\n"
               << " has " << objc[4] << " attribute objects\n";
     // in case we dont know the dataset name we can query it
     size_t max_dataset_number = objc[1];
45   hid_t *dataset_objects = new hid_t[ max_dataset_number ];
     (void) H5Fget_obj_ids( file, H5F_OBJ_DATASET,
                            max_dataset_number, dataset_objects );

     for( size_t i=0; i < max_dataset_number; ++i ) {
50     char temp_dataset_name[ MAX_NUMBER ];
       H5Fget_name( dataset_objects[i], temp_dataset_name, MAX_NUMBER );
       std::cout << "Dataset name = " << temp_dataset_name;
     }
     std::cout << std::endl;
55   dataset = H5Dopen( file, datasetName, H5P_DEFAULT );
     hsize_t storage_size = H5Dget_storage_size( dataset );
     haddr_t offset = H5Dget_offset( dataset );
     std::cout << "\nStorage size : " << storage_size
               << "\nOffset       : " << offset
60             << std::endl;

     hid_t datatype = H5Dget_type( dataset );
     hid_t hdf5class= H5Tget_class( datatype );
     switch( hdf5class ) {
65   case H5T_INTEGER:{std::cout << "Data set class = INTEGER\n"; break;}
     case H5T_FLOAT:{std::cout << "Data set class = FLOAT\n"; break;}
     case H5T_TIME:{std::cout << "Data set class = TIME\n"; break;}
     case H5T_STRING:{std::cout << "Data set class = STRING\n";break;}
     case H5T_BITFIELD:{std::cout << "Data set class = BITFIELD\n";break;}
70   case H5T_OPAQUE:{std::cout << "Data set class = OPAQUE\n";break;}
     case H5T_COMPOUND:{std::cout << "Data set class = COMPOUND\n";break;}
     case H5T_REFERENCE:{std::cout << "Data set class = REFERENCE\n";break;}
     case H5T_ENUM:{std::cout << "Data set class = ENUM\n";break;}
     case H5T_VLEN:{std::cout << "Data set class = VLEN\n";break;}
75   case H5T_ARRAY:{std::cout << "Data set class = ARRAY\n";break;}
     case H5T_NO_CLASS:
     default:
       { std::cout << "Data set class = NO CLASS\n";  break; }
```

```
      }
 80   if( hdf5class == H5T_FLOAT ) {
        // check point normalization
        H5T_norm_t norm = H5Tget_norm( datatype );
        if( norm == H5T_NORM_NONE )
          std::cout << "Floating point is not normalized.\n";
 85     else
          std::cout << "Floating point is normalized.\n";
      } else if( hdf5class == H5T_INTEGER ) {
        // check integer signed type
        H5T_sign_t sign_t = H5Tget_sign( datatype );
 90     if( sign_t == H5T_SGN_2 )
          std::cout << "Integer is 2's complement.\n";
        else if( sign_t == H5T_SGN_NONE )
          std::cout << "Integer is UNSIGED.\n";
      }
 95   hid_t order = H5Tget_order( datatype );
      switch( order ) {
      case H5T_ORDER_LE:{std::cout << "Data order is Little Endian\n"; break;}
      case H5T_ORDER_BE:{std::cout << "Data order is Big Endian\n"; break;}
      case H5T_ORDER_VAX:{std::cout << "Data order is VAX\n"; break;}
100   case H5T_ORDER_NONE:{std::cout << "Data order is NONE\n"; break;}
      }

      hid_t size = H5Tget_size( datatype );
      std::cout << "Data size = " << size << "\n";
105
      hid_t dataspace = H5Dget_space( dataset );
      hid_t dataspace_class = H5Sget_simple_extent_type( dataspace );
      switch( dataspace_class ) {
      case H5S_NO_CLASS:{std::cout<<"Dataspace class = H5S_NO_CLASS\n";break;}
110   case H5S_SCALAR:{std::cout << "Dataspace class = H5S_SCALAR\n"; break;}
      case H5S_SIMPLE: {std::cout << "Dataspace class = H5S_SIMPLE\n";break;}
      case H5S_NULL: {std::cout << "Dataspace class = H5S_NULL\n"; break;}
      }
      hssize_t num_points = H5Sget_simple_extent_npoints( dataspace );
115   hid_t rank = H5Sget_simple_extent_ndims( dataspace );
      hsize_t *dims = new hsize_t[ rank ];
      hsize_t *maxdims = new hsize_t[ rank ];
      for( int i=0; i < rank; ++i ) maxdims[i] = 32;
      (void)H5Sget_simple_extent_dims( dataspace, dims, maxdims );
120   std::cout << "Num points = " << num_points << "\n";
      std::cout << "Is Simple ? = " << H5Sis_simple( dataspace ) << "\n";
      std::cout << "Rank = " << rank << "\n";
      for( int i=0; i < rank; ++i ) {
        std::cout << "dim[" << i << "] = " << dims[i] << "\n";
125   }
      std::cout << std::endl;

      delete[] dims;
      H5Sclose( dataspace );
130   H5Tclose( datatype );
      H5Fclose( file );
      H5Dclose( dataset );
      std::cout << std::endl;
      return (0);
135 }
```

Listing 10.3 HDF5 API example

When using the C++ API the following libraries should be linked with the application program:

```
Program('hdfcxx', ['hdfcxx.cpp'],
        CXXFLAGS="-Wall -O0 -ggdb",
        LIBS=['hdf5','hdf5_hl','hdf5_cpp',
```

```
'hdf5_hl_cpp','pthread'])
```

Compiling and running this 'info' program on an HDF5 file gives:

```
./hdf_info example_A.h5 ASTR_10
File example_A.h5 File size = 2084 free space = 0
 has 1 file objects
 has 0 dataset objects
 has 0 group objects
 has 0 datatype objects
 has 0 attribute objects

Storage size : 36
Offset       : 2048
Data set class = INTEGER
Integer is 2's complement.
Data order is Little Endian
Data size = 4
Dataspace class = H5S_SIMPLE
Num points = 9
Is Simple ? = 1
Rank = 2
dim[0] = 3
dim[1] = 3
```

An example of using C++ API is shown in Listing 10.4.

```
   // \file hdfcxx.cpp
   // \author Sandeep Koranne (C) 2010
   // \description HDF API in C++
   #include <cstdlib>
 5 #include <H5Cpp.h>
   #include <iostream>

   using namespace H5;

10 int main( int argc, char *argv[] ) {
     H5File file( argv[1], H5F_ACC_RDONLY );
     hid_t dataset_id = H5Dopen( file.getId(), argv[2], H5P_DEFAULT );
     if( dataset_id == -1 ) {
       std::cerr << "Dataset : " << argv[2] << " not found in file.\n";
15     exit(1);
     }
     DataSet dataset( dataset_id );
     std::cout << "Storage size = " << dataset.getStorageSize();

20   dataset.close();
     file.close();
     std::cout << std::endl;
     return (0);
   }
```

Listing 10.4 HDF5 C++ API example

The primary use of HDF5 is in maintaining large datasets, for example, those coming from simulations of physical processes, or astronomy. In these cases the size of the dataset exceeds the installed physical memory on the computer system by orders of magnitude. For example, an 8 Gb machine may be expected to analyze a terabyte size file containing temperature and pressure readings from a simulation. HDF5 includes the ability to load into memory *sections* of data from disk.

The sections are arranged into *hyperslabs* which is a dimensional generalization of multi-dimensional arrays. See Figure 10.1 for a depiction of how data organized on disk can be loaded into memory. The data can be loaded from a contiguous slab, or accessed using *strides*. The function H5Sselect_hyperslab can be used to select regions from the dataspace, both in dataset space, as well as the memory space.

Fig. 10.1 Hyperslab data
reading in HDF5

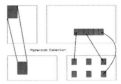

10.2.1 Writing and Reading compound datatype in HDF5

Writing and reading compound datatype (such as a C language structure) can be performed using HDF5 *groups*. Consider the example of temperature and pressure (as floating point values) readings taken at various coordinates (X,Y pairs). This organization can be written in C as:

```
  typedef struct {
    int X,Y;
  } Coord;
  typedef struct {
5   float temperature, pressure;
    Coord location;
    hobj_ref_t group;
  } Reading;
```

The HDF5 API calls needed to implement writing and reading this compound type can be written as:

```
   hid_t file = H5Fcreate( FILE, H5F_ACC_TRUNC, H5P_DEFAULT, H5P_DEFAULT );
   space = H5Screate_simple( 2, dims, NULL );
   dataset = H5Dcreate( file, ``LOC_READING'', H5T_NATIVE_DOUBLE,..);
   group = H5Gcreate(file, ``SensorA'', H5P_DEFAULT, H5P_DEFAULT, H5P_DEFAULT);
5  (void)H5Gclose( group );
   group = H5Gcreate(file, ``SensorB'',H5P_DEFAULT,H5P_DEFAULT,H5P_DEFAULT);
   (void)H5Gclose( group );
   Reading *readings = new Reading[4];
   reading[0]->temperature = 100.0; reading[0]->pressure = 65.1;
10 reading[0]->X = 6, reading[0]->Y = 8;
   reading[1]->temperature = 92.4; reading[1]->pressure = 47.2;
   reading[1]->X = 12, reading[0]->Y = 18;
   status = H5Rcreate( &reading[0].group, file, ``SensorA'', H5R_OBJECT, -1);
```

The *group* handle is stored in the data structure representing the location based readings. A group can be created using the H5Gcreate function. A hierarchical group can be created by providing the absolute name of the group to this function, e.g.:

```
group = H5Gcreate( file, ''/Data/Another_Group'');
```

creates a group "Another_Group" inside the group "Data" which is in the root group. A dataset can be placed in a group by giving the name of the group as the argument to the HD5create function. To open the dataset in a group, two methods can be used: (i) use the absolute name of the group in the HD5open function, or (ii) open the group, and pass the group handle to the HD5open function.

10.2.2 HDF5 Attributes

HDF5 attributes are small named datasets that are attached to primary datasets, or named datatypes. An attribute has two components: (i) name, and (ii) value. The value part contains one or more data entries of the same datatype. As mentioned above, H5A is the API prefix for all attribute related HDF5 functions. All the attributes of an object can be iterated using the *index value* identification of attributes. Since attributes are expected to be small, they are stored in the header of the object they are attached to. Attributes can be created using H5Acreate, an example is shown below:

```
id = H5Screate( H5S_SCALAR );
attr = H5Acreate( dataset, ''Integer'', H5T_NATIVE_INT, id, H5P_DEFAULT);
..
H5Aopen_name( dataset, ''Integer'' ); // reads same attribute back
```

10.2.3 References to objects

In HDF5 object references are based on the relative file address of the object header in the file and is deemed constant for the life of the object.

10.2.4 Conclusion

HDF5 is a versatile and efficient mechanism of data storage and access. Since it is widely supported on a number of high-performance computing platforms, availability of tools to analyze HDF5 files, and the use of multi-threaded I/O for reading and writing large datasets in HDF5 format can be expected. In the above section we have given a brief overview of the capabilities of HDF5.

Chapter 11
Graphics and Image Processing Libraries

Abstract In this chapter we discuss the Cairo graphics library. Cairo graphics library is used for device independent graphics rendering. Thereafter image formats such as PNG and JPEG are interfaced with the help of libraries libPNG and jpegLib, and these libraries are also discussed in this chapter. We conclude with a discussion of the GraphicsMagick and ImageMagick software.

Contents

11.1 Cairo: A Vector Drawing Library

Cairo is a device independent vector drawing library which supports a number of backends including X11 (Xlib), Windows, PDF, PostScript, PNG, and MacOS (Quartz). The same rendering code (using Cairo drawing primitives) can generate the required output using the appropriate backend.

Cairo graphics supports rendering of fonts, arcs, points, polylines, as well as bitmap images. Drawing with Cairo involves the use of the `cairo_surface` and the `cairo_t` state. The surface models the backend, while the `cairo_t` state models the current drawing state machine. Attributes such as backing store of the surface, size, color space depth, number of visuals on the surface can be controlled using Cairo functions. The state machine attributes (including current color, pen type) can be changed on the `cairo_t` data-structure using Cairo functions. A simple example of using the X11 Xlib backend is shown in Listing 11.1.

```
// \file cairo_example.cpp
// \author Sandeep Koranne (C) 2010
// \description Use of Cairo graphics library
#include <iostream>          // for program IO
#include <cstdlib>           // for exit
```

S. Koranne, *Handbook of Open Source Tools*,
DOI 10.1007/978-1-4419-7719-9_11, © Springer Science+Business Media, LLC 2011

```
     #include <cairo/cairo.h>      // Cairo library
     #include <cairo/cairo-xlib.h> // X-lib surface for Cairo

     static const unsigned int SIZE = 400;
10
     static void EraseWindow( cairo_surface_t *surface ) {
       cairo_t *cr = cairo_create( surface );
       cairo_set_source_rgb( cr, 1.0, 1.0, 1.0 );
       cairo_paint( cr );
15     cairo_destroy( cr );
     }

     static void DrawData( cairo_surface_t *surface ) {
       cairo_t *cr = cairo_create( surface );
20     while( true ) {
         EraseWindow( surface );
         cairo_set_source_rgb( cr, 0.5, 0.1, 0.5 );
         cairo_rectangle( cr, rand()%SIZE,rand()%SIZE, 100, 100 );
         cairo_stroke( cr );
25     }
       cairo_destroy( cr );
     }

     static void InitializeWindow( Display *dpy ) {
30     cairo_surface_t *surface ;
       Drawable drawable;
       int screen = DefaultScreen( dpy );
       XSetWindowAttributes xwa;
       drawable = XCreateWindow( dpy, DefaultRootWindow( dpy ),
35             100,100, SIZE, SIZE, 0,
               DefaultDepth( dpy, screen ),
               InputOutput,
               DefaultVisual( dpy, screen ),
               CWOverrideRedirect, &xwa );
40     XMapWindow( dpy, drawable );
       surface = cairo_xlib_surface_create( dpy, drawable,
                   DefaultVisual( dpy, screen ),
                   SIZE, SIZE );
       cairo_xlib_surface_set_size( surface, SIZE, SIZE );
45     DrawData( surface );

       EraseWindow( surface );
       XDestroyWindow( dpy, drawable );
     }
50
     int main( int argc, char *argv[] ) {
       Display *dpy = XOpenDisplay( NULL );
       if( dpy == NULL ) {
         std::cerr << "Unable to open X display...\n";
55       exit(1);
       }
       InitializeWindow( dpy );

       std::cout << std::endl;
60     XCloseDisplay( dpy );
       return (0);
     }
```

Listing 11.1 Example of using Xlib with Cairo

The following functions are available for use when using Xlib surface with Cairo:

```
     cairo_public cairo_surface_t *
     cairo_xlib_surface_create (Display      *dpy,
             Drawable drawable,
             Visual        *visual,
5            int      width,
```

```
                    int    height);

 cairo_public cairo_surface_t *
 cairo_xlib_surface_create_for_bitmap (Display  *dpy,
10                Pixmap  bitmap,
                  Screen  *screen,
                  int width,
                  int height);

15 cairo_public void
 cairo_xlib_surface_set_size (cairo_surface_t *surface,
                int             width,
                int             height);

20 cairo_public void
 cairo_xlib_surface_set_drawable (cairo_surface_t *surface,
           Drawable   drawable,
                int             width,
                int             height);
25
 cairo_public Display *
 cairo_xlib_surface_get_display (cairo_surface_t *surface);

 cairo_public Drawable
30 cairo_xlib_surface_get_drawable (cairo_surface_t *surface);

 cairo_public Screen *
 cairo_xlib_surface_get_screen (cairo_surface_t *surface);

35 cairo_public Visual *
 cairo_xlib_surface_get_visual (cairo_surface_t *surface);

 cairo_public int
 cairo_xlib_surface_get_depth (cairo_surface_t *surface);
40
 cairo_public int
 cairo_xlib_surface_get_width (cairo_surface_t *surface);

 cairo_public int
45 cairo_xlib_surface_get_height (cairo_surface_t *surface);
```

To draw using Cairo surfaces, a Cairo *context* `cairo_t` must be created using the Cairo surface as the target. The function `cairo_create()` is used for this purpose. The various attributes of the Cairo context can be changed using Cairo functions, e.g., the default line width can be changed using `cairo_set_line_width` function, and the *fill rule* can be set to either the *winding number*, or the even-odd overlap count using the `cairo_set_fill_rule` function. Even the *join types* of lines, their *stroke* (whether to use solid lines or dashes) can be set. In this regards, Cairo is similar to OpenGL (see Section 19.2) and GD (see Section 19.10).

Similar to OpenGL, a Cairo context can be given a matrix transform to be applied. The transform may contain rotation, translation, and scaling. The drawing primitives include (i) points, (ii) rectangles, (iii) paths, (iv) curves and arcs.

The advantage of using Cairo is manifest when the same rendering code has to generate multiple output formats such as PDF, PostScript, PNG, and SVG. Cairo provides functions to create *surfaces* for each of these backends. A page compositing program may use multiple backends of Cairo at the same time, Xlib for on-screen editing, PDF and PostScript for paper printing, and PNG and SVG for web

deployment. For example, the SVG (scalable vector graphics) backend functions
are:

```
cairo_public cairo_surface_t *
cairo_svg_surface_create (const char   *filename,
        double  width_in_points,
        double  height_in_points);
```

Fig. 11.1 Cairo graphics with
SVG surface, example

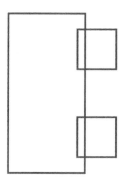

An example of using SVG with Cairo is shown in Listing 11.2.

```
    // \file cairo_svg.cpp
    // \author Sandeep Koranne, (C) 2010
    // \description Example of using SVG with Cairo
    #include <cstdio>            // C FILE*
5   #include <cstdlib>           // exit
    #include <iostream>          // program IO
    #include <cairo/cairo.h>     // Cairo
    #include <cairo/cairo-svg.h> // SVG support
    static unsigned int SIZE = 500;
10  static void DrawExample( const char* filename ) {
      cairo_surface_t *surface =
        cairo_svg_surface_create( filename, SIZE, SIZE );

      cairo_t *cr = cairo_create( surface );
15    cairo_set_source_rgb( cr, 0.5, 0.1, 0.5 );
      #if 0
      cairo_translate( cr, 200, 100 );
      cairo_rotate( cr, 45.0 );
      #endif
20    cairo_rectangle( cr, 10, 10, 100, 200 );
      cairo_rectangle( cr, 100, 30, 50, 50 );
      cairo_rectangle( cr, 100, 140, 50, 50 );

      cairo_stroke( cr );
25    cairo_destroy( cr );
      cairo_surface_destroy( surface );
    }

    int main( int argc, char *argv [] ) {
30    if( argc != 2 ) {
        std::cerr << "Usage: cairo_svg <filename>..\n";
        exit(1);
      }
      DrawExample( argv[1] );
35    std::cout << std::endl;
      return (0);
```

```
}
```

Listing 11.2 Using SVG surface with Cairo

Compiling and running this program as:

Fig. 11.2 Example of using
`cairo_rotate`

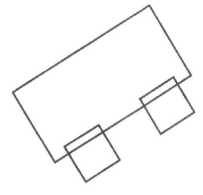

```
$g++ cairo_svg.cpp -lcairo
$./a.out a.svg
```

generates the SVG file as shown in Figure 11.1.

By adding `cairo_translate` and `cairo_rotate` to the Cairo context we can perform translation and rotation of the scene. An example is shown in Figure 11.2.

11.2 Graphics File Formats

Portable Network Graphics (PNG) is the modern loss-less image format which is widely used on the Internet. JPEG is loss-less and even though it is prevalent in Internet and amateur photography, its role in scientific image processing is thus limited only to preview generation. We concentrate on PNG as its design, and the software library are instructive in designing efficient file formats for two-dimensional bitmap graphics. The equivalent for vector images is Scalable Vector Graphics (SVG) (see Section 11.2.2) for more details.

11.2.1 libPNG: library for Portable Network Graphics

Portable Network Graphics (PNG) is a standard for loss-less representation of bit-mapped image data. PNG supports 24-bit and 32-bit RGBA color images (but does not support print color spaces such as CMYK). The PNG format uses the zlib *deflate* compression (see Section 8.1) and achieves good compression. Since two-dimensional image data is often involved in scientific experiments; reading, writing, and otherwise manipulating PNG files is often required. We discuss the use of the `libPNG` library for this purpose. A more general purpose image manipulation library is discussed in the next section, see Section 11.2.3. Tools to programmatically generate PNG files are described in Chapter 19, see Section 19.10 and Section 19.11.

11.2.1.1 Format of a PNG file

PNG files are divided into *chunks*, which help in maintaining backwards compatibility. A PNG reader when confronted with a chunk which it cannot decode can simply skip the chunk (if it is not defined as critical). Each chunk is represented as:

1. Chunk length in bytes: this field is 4 bytes long,
2. Chunk type: also 4 bytes long. The case of the character in these 4 bytes denotes information that a decoder can use. This information includes whether the chunk is *critical*, *public*, *reserved*, and *safe to copy*,
3. Actual chunk data: length in bytes is given in the first field above,
4. CRC checksum: 4 bytes long.

The color depth of the image, compression, and transparency are also stored.

The compression used in PNG is divided into multiple phases. In the first phase a compression (using predictor functions) based on the value of neighboring pixels is performed. Then interlacing and *Adam7* interlacing (to quickly regenerate an approximation of the image during decoding) is done. This 7-pass scheme (Adam7) not only requires computation time, but also reduces compression since not all adjacent bytes of the images are compressed together. The raw byte stream comprising the chunk is then compressed using `deflate`, the standard ZLIB compression (see Section 8.1).

11.2.1.2 JPEG file format

The library for reading and writing JPEG files is `libjpeg` and it provides functions for reading (decompressing) and writing (compressing) JPEG image data from/to files. It provides a JPEG compress structure `jpeg_compress_struct` which can be used to write JPEG files, (part of the structure are shown) as shown in the listing below.

```
    struct jpeg_compress_struct {
      /* partial structure see libjpeg.h */
      struct jpeg_destination_mgr * dest;
      JDIMENSION image_width; /* input image width */
5     JDIMENSION image_height;  /* input image height */
      int input_components;
      J_COLOR_SPACE in_color_space; /* colorspace of input image */
      double input_gamma;     /* image gamma of input image */
      int data_precision;     /* bits of precision in image data */
10    int num_components;     /* # of color components in JPEG image */
      J_COLOR_SPACE jpeg_color_space; /* colorspace of JPEG image */
```

```
     jpeg_component_info * comp_info;
     JHUFF_TBL * dc_huff_tbl_ptrs[NUM_HUFF_TBLS];
     JHUFF_TBL * ac_huff_tbl_ptrs[NUM_HUFF_TBLS];
15   };
```

Correspondingly, `jpeglib` also has a decompression structure for reading JPEG data as shown in the next listing (only some part of the structure are shown):

```
   struct jpeg_decompress_struct {
     /* partial structure see libjpeg.h */
     struct jpeg_source_mgr * src;
     JDIMENSION image_width; /* nominal image width  */
5    JDIMENSION image_height;  /* nominal image height */
     int num_components;    /* # of color components */
     J_COLOR_SPACE jpeg_color_space; /* colorspace of JPEG image */
     J_COLOR_SPACE out_color_space; /* colorspace for output */
     unsigned int scale_num, scale_denom;
10   JDIMENSION output_width;  /* scaled image width */
     JDIMENSION output_height; /* scaled image height */
   };
```

A list of function from the `jpeglib.h` header file is shown:

```
   /* Destruction of JPEG compression objects */
   jpeg_destroy_compress JPP((j_compress_ptr cinfo));
   jpeg_destroy_decompress JPP((j_decompress_ptr cinfo));

5  jpeg_stdio_dest JPP((j_compress_ptr cinfo, FILE * outfile));
   jpeg_stdio_src JPP((j_decompress_ptr cinfo, FILE * infile));

   jpeg_set_defaults JPP((j_compress_ptr cinfo));
   jpeg_set_colorspace JPP((j_compress_ptr cinfo,
10                   J_COLOR_SPACE colorspace));
   jpeg_default_colorspace JPP((j_compress_ptr cinfo));
   jpeg_set_quality JPP((j_compress_ptr cinfo, int quality,
                     boolean force_baseline));
   jpeg_set_linear_quality JPP((j_compress_ptr cinfo,
15                       int scale_factor,
                        boolean force_baseline));
   jpeg_add_quant_table JPP((j_compress_ptr cinfo, int which_tbl,
                      const unsigned int *basic_table,
                      int scale_factor,
20                     boolean force_baseline));
   jpeg_quality_scaling JPP((int quality));
   jpeg_simple_progression JPP((j_compress_ptr cinfo));
   jpeg_suppress_tables JPP((j_compress_ptr cinfo,
                         boolean suppress));
25 jpeg_alloc_quant_table JPP((j_common_ptr cinfo));
   jpeg_alloc_huff_table JPP((j_common_ptr cinfo));

   /* Main entry points for compression */
   jpeg_start_compress JPP((j_compress_ptr cinfo,
30                    boolean write_all_tables));
   jpeg_write_scanlines JPP((j_compress_ptr cinfo,
                         JSAMPARRAY scanlines,
                         JDIMENSION num_lines));
   jpeg_finish_compress JPP((j_compress_ptr cinfo));
35
   jpeg_write_raw_data JPP((j_compress_ptr cinfo,
                        JSAMPIMAGE data,
                        JDIMENSION num_lines));

40 /* Same, but piecemeal. */
   jpeg_write_m_header
   JPP((j_compress_ptr cinfo,
       int marker,
```

```
                unsigned int datalen));
45   jpeg_write_m_byte
     JPP((j_compress_ptr cinfo, int val));

     /* Alternate compression function:
        just write an abbreviated table file */
50   jpeg_write_tables JPP((j_compress_ptr cinfo));

     /* Decompression startup:
        read start of JPEG datastream to see what's there */
     jpeg_read_header JPP((j_decompress_ptr cinfo,
55                          boolean require_image));

     /* Main entry points for decompression */
     jpeg_start_decompress JPP((j_decompress_ptr cinfo));
     jpeg_read_scanlines JPP((j_decompress_ptr cinfo,
60                             JSAMPARRAY scanlines,
                               JDIMENSION max_lines));
     jpeg_finish_decompress JPP((j_decompress_ptr cinfo));

     /* Replaces jpeg_read_scanlines when reading raw downsampled data. */
65   jpeg_read_raw_data JPP((j_decompress_ptr cinfo,
                             JSAMPIMAGE data,
                             JDIMENSION max_lines));

     /* Additional entry points for buffered-image mode. */
70   jpeg_has_multiple_scans JPP((j_decompress_ptr cinfo));
     jpeg_start_output JPP((j_decompress_ptr cinfo,
                            int scan_number));
     jpeg_finish_output JPP((j_decompress_ptr cinfo));
     jpeg_input_complete JPP((j_decompress_ptr cinfo));
75   jpeg_new_colormap JPP((j_decompress_ptr cinfo));
     jpeg_consume_input JPP((j_decompress_ptr cinfo));
```

11.2.2 Scalable Vector Graphics (SVG)

In the above sections we have looked at image representation using PNG and JPEG. These file formats are designed for *bit-mapped* data. Another way to represent geometric information in images is using a *vector approach*, where the image is composed of well defined primitives. The primitives can be combined hierarchically, transformed (scaled and rotated, as well as translation). The advantage of vector graphics is: (i) faithful reproduction independent of scaling, (ii) much better compression for line art type drawings, (iii) programmatic generation, and (iv) elegant mathematical model underlying the drawing principles. Scalable Vector Graphics (SVG) is an international standard for representing vector image data using the XML format.

Since SVG files are XML files, they can be parsed using standard XML processing tools (see Section 9.4).

```
<?xml version="1.0" encoding="UTF-8"?>
<svg xmlns="http://www.w3.org/2000/svg"
        xmlns:xlink="http://www.w3.org/1999/xlink"
        width="500pt" height="500pt"
        viewBox="0 0 500 500" version="1.1">
<g id="surface1">
```

```
<path style="fill:none;stroke-width:2;stroke-linecap:butt;
            stroke-linejoin:miter;stroke:rgb(50%,10%,50%);
            stroke-opacity:1;stroke-miterlimit:10;"
            d="M 10.000546
  " transform="matrix(0.525322,0.850904,
                      -0.850904,0.525322,200,100)"/>
</g>
</svg>
```

The SVG specification defines the semantics of the XML elements; XML is used as the transmission medium and container. To properly interpret an SVG files, the semantics of the SVG primitives such as: (i) path, (ii) fill, (iii) width, (iv) stroke, (v) join-styles, etc. must be understood.

Thus, it is possible for an application to generate SVG directly, but most applications use a higher order abstraction. Generating XML from geometric primitives can be done using Cairo (see Section 11.1), and see Figure 11.2 for an example of using Cairo to generate a shape in SVG, then also perform a rotational transformation on the shape. Inkscape (see Section 19.8) uses SVG as its native file format.

11.2.3 GraphicsMagick and ImageMagick

GraphicsMagick is described as the "swiss army knife" of image processing. It provides tools and libraries which support reading and writing image of many formats including JPEG, PNG, and TIFF. Source code of GraphicsMagick is portable and recently has added OpenMP (see Section 12.2) support for parallel processing. GraphicsMagick has been used with large images (of gigapixel-size). GraphicsMagick is originally derived from ImageMagick but now is developed independently.

ImageMagick is an open-source software suite to create, edit, and compose bitmap images from a multitude of file formats including JPEG, PNG, and TIFF. ImageMagick can perform the following functions on files:

1. Format conversion: convert an image from one format to another,
2. Transform: resize, rotate, crop, and flip,
3. Transparency, Draw, and Decorate,
4. Special effects: blur, sharpen and tint,
5. Image identification,
6. Discrete Fourier Transform,
7. High dynamic range image,
8. Thread safe and implemented with OpenMP.

ImageMagick is mostly used with the command-line tool `convert` which has command-line options to perform many of the above tasks. Images used in this book were often processed with `convert` for resizing and format conversion.

11.3 Conclusion

In this chapter we discussed the Cairo graphics library, and presented example of its usage. Cairo is used for device independent graphics rendering in many open-source applications including Firefox. Image formats such as PNG and JPEG are interfaced with the help of libraries (libPNG and jpegLib). The Scalable Vector Graphics (SVG) file format (in XML) was explained with the help of an example, and we used Cairo to generate an SVG file. We concluded with a discussion of the GraphicsMagick and ImageMagick software for image manipulation on the command-line.

Part III
Parallel and System Programming

Chapter 12
Parallel Programming

Abstract Parallel programming deals with multi-processing and multi-threading. With the advent of multi-core computers, parallel programming has become essential. In this chapter we discuss the POSIX threading library (pthread). User annotated compiler supported parallelism with OpenMP is described with the help of examples in Section 12.2. The new features of OpenMP version 3.0 (task computing) is presented with the help of examples in Section 12.2.0.3. In addition to multi-threading, parallel computing has been successfully deployed on cluster grids (called the Beowulf class). The most common API used in cluster computing is MPI (message passing interface) which is discussed in this chapter. In addition to these well established parallel programming systems, the rapid rise of many-core and other forms of parallelism has also created new systems which have had less exposure. In particular the Intel Thread Building Block library, and GPGPU computing with NVIDIA CUDA and OpenCL are described.

Contents

In this chapter we discuss the various software tools and libraries available for programming parallel and distributed computers. With the ready availability of dual-core, quad-core and many-core machines in the mainstream, utilizing parallel computation paradigms effectively is important. In this Part we will discuss POSIX threading library (pthreads), OpenMP, MPI, Intel TBB, and general purpose computing on graphics card using CUDA and OpenCL.

S. Koranne, *Handbook of Open Source Tools*,
DOI 10.1007/978-1-4419-7719-9_12, © Springer Science+Business Media, LLC 2011

12.1 POSIX Thread Library (pthreads)

Even before we delve deep into the programming model of pthreads, it is useful to see what are the knobs and tuning parameters available to us to manage pthreads. These management methods are independent of the CPU thread scheduling policies or data-cache touching functions.

Fig. 12.1 Example of pthread spawning 2 threads for functions F1 and F2.

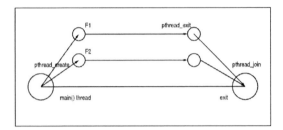

POSIX pthread library allows management of threads by specifying the following:

1. Keys: threads use keys to maintain private copies of shared data items. A single globally defined key points to different memory location for each thread. We can use keys when making function calls and we want thread specific data to be passed along,
2. Attributes: thread attributes allow us to control scheduling behavior, stack size and initial state,
3. pthread_once mechanism: is like the singleton constructor concept, it ensures that a function or action is performed *exactly once* regardless of how many times (or different threads) attempt to call it. Useful for file or network handling, use this mechanism instead of writing static variables for this purpose,
4. Thread cancellation: allows a thread to self-destruct, for example when searching for a data in a binary-tree, once the data is found, the rest of the threads can be canceled,
5. Thread scheduling: specifies to the Operating System which thread you (as a programmer, not as a user, unless you make the thread scheduling visible to the end-user) would like the CPU to be scheduled. CPU has extensive control on the scheduling behavior of threads.

12.1.1 Understanding pthread programming model

Since pthread is an API library, all of its functionality is exposed using standardized function calls and these calls make up the programming model for pthreads.

The pthread_create() function creates a new thread, with attributes specified by attr, within a process. If `attr` is NULL, the default attributes shall be used. If the attributes specified by `attr` are modified later, the thread's attributes shall not be affected. Upon successful completion, pthread_create() store the ID of the created thread in the location referenced by thread.

```
int pthread_create(pthread_t *restrict thread,
      const pthread_attr_t *restrict attr,
      void *(*start_routine)(void*), void *restrict arg);
```

Attributes objects are provided for threads, mutexes, and condition variables as a mechanism to support probable future standardization and customization. Attributes objects provide clean isolation of the configurable aspects of threads. For example, stack size is an important attribute of a thread, but it cannot be expressed portably.

The pthread_attr_init() function shall initialize a thread attributes object attr with the default value for all of the individual attributes used by a given implementation. The pthread_attr_destroy() function shall destroy a thread attributes object.

```
int pthread_attr_destroy(pthread_attr_t *attr);
int pthread_attr_init(pthread_attr_t *attr);
```

```
   #include <stdlib.h>
   #include <stdio.h>
   #include <errno.h>
   #include <pthread.h>
5  int r1 = 0, r2 = 0;
   static void F1(int *pr1) {
     // some compute intensive function
   }
   static void F2(int *pr2) {
10   // some other compute intensive function
   }
   int main(int argc, char **argv) {
     pthread_t t1, t2;
     if(pthread_create(&t1,NULL,(void *) F1,(void *) &r1) != 0)
15     perror("pthread_create"), exit(1);
     if(pthread_create(&t2,NULL,(void *) F2,(void *) &r2) != 0)
       perror("pthread_create"), exit(1);
     if(pthread_join(t1, NULL) != 0)
       perror("pthread_join"),exit(1);
20   if (pthread_join(t2, NULL) != 0)
       perror("pthread_join"),exit(1);
     return (EXIT_SUCCESS);
   }
```

Listing 12.1 Introduction to pthreads

```
int pthread_once(pthread_once_t *once_control,
      void (*init_routine)(void));
pthread_once_t once_control = PTHREAD_ONCE_INIT;
```

Usually the controlling thread in the CPU will perform setup tasks, but this can be sub-optimal, especially when the setup is compute intensive or involves dealing with blocking I/O.

12.1.2 Pthreads Keys: using thread specific data

As threads are created, run, and destroyed, their working space consists of their stack space, global variables, and memory allocated from the heap. During the course of program execution if you want to associate thread specific data to each thread, so that the same variable in different threads points to different memory, you can use *keys* to achieve this. Consider the example of searching for a value in a tree, every thread function will store its own fragment of the binary tree on which its operating. This can be done using keys:

```
static pthread_key_t tree_key;
int search_init() {
  pthread_key_create( &tree_key, (void*)free_key );
}
```

This function needs the variable for the key as well as a destructor function. Since thread cancellations and exits can leak thread specific data pointed to by keys, implementing this destructor correctly helps prevent memory leaks. Now we can use this key as follows:

```
pthread_setspecific( tree_key, (int*)(binary_tree+offset));
```

We set this threads `tree_key` to `binary_tree`offset+, later on we can use this key to retrieve the binary tree offset as follows:

```
int *tree_offset;
pthread_getspecific( tree_key, (void**)&tree_offset);
```

```
void *pthread_getspecific(pthread_key_t key);
int pthread_setspecific(pthread_key_t key, const void *value);
```

12.1.3 Pthreads Summary

Detailed function description of pthread functions can be found using the `man pthread_create` manpage and then following the related sections. In this section we present the most often used pthread functions, and their description for ready reference.

```
    #include <stdlib.h>
    #include <stdio.h>
    #include <errno.h>
    #include <pthread.h>
 5  #define N 100
    typedef int matrix_t[N][N];
    typedef struct {
      int        id,size,Arow,Bcol;
      matrix_t   *MA, *MB,*MC,dummy;
10  } arg_t;
    matrix_t MA,MB,MC;
    void mult(int size,int row,int col,
             matrix_t A, matrix_t B,matrix_t C) {
```

Table 12.1 Pthread Functions

Name	Description
pthread_create	creates thread (pthread_t, attr, function, arg)
pthread_exit	exits thread (status)
pthread_attr_init	initializes attribute (pthread_attr_t*)
pthread_attr_destroy	destroys attribute (pthread_attr_t*)
pthread_join	blocks parent till T terminates (pthread_t T, **void****)
pthread_detach	storage can be claimed when thread terminates (pthread_t),
pthread_self	return the thread id of calling thread (pthread_t)
pthread_equal	compare thread id (pthread_t t1, pthread_t t2)
pthread_control	allows 1 call to function (pthread_once_t *, **void*** ())
pthread_mutex_init	initializes mutex with attributes (pthread_mutex_t*, pthread_mutexattr_t*)
pthread_mutex_destroy	destroys mutex (pthread_mutex_t*)
pthread_mutexattr_init	initializes mutex attribute (pthread_mutexattr_t*)
pthread_mutexattr_destroy	destroys mutex attribute (pthread_mutexattr_t*)
pthread_mutex_lock	acquire a lock on a mutex (pthread_mutex_t*)
pthread_mutex_trylock	attemp to acquire a lock (pthread_mutex_t*)
pthread_mutex_unlock	unlock the mutex (pthread_mutex_t*)
pthread_cond_init	initializes cond with attributes (pthread_cond_t*, pthread_condattr_t*)
pthread_cond_destroy	destroys cond (pthread_cond_t*)
pthread_condattr_init	initializes cond attribute (pthread_condattr_t*)
pthread_condattr_destroy	destroys cond attribute (pthread_condattr_t*)
pthread_cond_wait	blocks calling thread until condition (pthread_cond_t*, pthread_mutex_t*)
pthread_cond_signal	wakeun up thread waiting for cond (pthread_cond_t*)
pthread_cond_broadcast	waken up all threads waiting on cond (pthread_cond_t*)

```
        int pos;
15      C[row][col] = 0;
        for(pos = 0; pos < size; ++pos)
          C[row][col] += A[row][pos] * B[pos][col];
      }
      void *mult_worker(void *arg) {
20      arg_t *p=(arg_t *)arg;
        mult(p->size, p->Arow, p->Bcol, *(p->MA), *(p->MB), *(p->MC));
        free(p);
        return(NULL);
      }
25    int main(int argc, char **argv) {
        int       size, row, col, num_threads, i;
        pthread_t *threads;
        arg_t *p;
        unsigned long thread_stack_size;
30      pthread_attr_t *pthread_attr_p, pthread_custom_attr;
        size = N;
        threads = (pthread_t *)malloc(size*size*sizeof(pthread_t));
        for (row = 0; row < size; ++row)
          for (col = 0; col < size; ++col)
```

```
35       MA[row][col] = 1, MB[row][col] = row + col + 1;
     num_threads = 0;
     for(row = 0; row < size; row++)
       for (col=0;col<size;col++,num_threads++) {
         p = (arg_t *)malloc(sizeof(arg_t));
40       p->id = num_threads;p->size = size;
         p->Arow = row;p->Bcol = col;
         (p->MA) = &MA;(p->MB) = &MB;(p->MC) = &MC;
         pthread_create(&threads[num_threads],&pthread_custom_attr,
                        mult_worker, (void *) p);
45     }
     for (i = 0; i < (size*size); i++)
       pthread_join(threads[i], NULL);
     return (EXIT_SUCCESS);
   }
```

Listing 12.2 Parallel matrix multiple

12.2 OpenMP: Open specification for Multi-processing

OpenMP has become the defacto standard for shared-memory parallel programming. The last few years has seen OpenMP emerge as the preferred standard, and with the advent of multi-core computing on every desktop, parallel programming with OpenMP is the method of choice for application developers. OpenMP is an Application Program Interface (API) defined by a standardization body comprised of hardware and software vendors. It provides a simple, flexible, yet efficient model for developing portable and scalable parallel applications. GNU GCC version 4.5.0 was used as the C/C++/FORTRAN compiler which supports the latest OpenMP standard (OpenMP ver 3.0). OpenMP comprises of:

1. Compiler directives (written as # pragmas),
2. Runtime library routines: provided by the compiler or hardware provider,
3. Environment variables: controlling aspects of OpenMP,
4. Programming Model: is based on shared-memory, thread based explicit parallelism with a *fork- join* model.

The compiler directives for OpenMP are written as # pragmas for the compiler. Consider a small program written in C++ using OpenMP as shown in Listing 12.3.

```
   // \file hw_open_mp.cpp
   // \author Sandeep Koranne, (C) 2010
   // \description Initial program of OpenMP
   #include <iostream>          // for program IO
5  #include <cassert>           // assertions
   #include <cstdlib>           // exit
   #include <omp.h>             // OpenMP

   int main( int argc, char *argv [] ) {
10   int num_threads, id;

   #pragma omp parallel private( num_threads, id )
     {
       id = omp_get_thread_num();
15     std::cout << "Thread id " << id << " says hello!\n";
```

```
      if( id == 0 ) { // master thread
        num_threads = omp_get_num_threads();
        std::cout << "Number of threads = " << num_threads;
20    }
    } // join point of all threads

    return ( 0 );
}
```

Listing 12.3 Hello world style program using OpenMP

We can compile and run this program with GCC 4.5.0 as:

```
$g++ -fopenmp hw_open_mp.cpp -o hw_omp
$export OMP_NUM_THREADS=2
$./hw_omp
Thread id 1 says hello!
Thread id 0 says hello!
Number of threads = 2
```

Using the `ldd` program we can check that this binary links to both `libgomp` (which provides the OpenMP run-time library), and `libpthread` (which provides the underlying threading implementation on GNU/Linux, although this is not required by OpenMP).

12.2.0.1 OpenMP directives

OpenMP directives are case sensitive, follow the C/C++ compiler standards for #pragmas. Directives are specified for one structured block by a single line preceding it (if the line becomes too long a line continuation character has to be inserted). Some common OpenMP directives are listed below in Table 12.3.

12.2.0.2 Parallel region construct

defines a block of code which will be executed by multiple threads. It accepts a conditional test clause (to trigger parallel execution), a list of thread-local private variables, shared variables. The number of threads for the block can also be specified, as well as advanced parameters dealing with initialization of variables going into the block (firstprivate) There is an implied barrier at the end of the block.

```
#pragma omp parallel default(shared) private(..)
```

The number of threads can be specified in the parallel construct.

The *data* clause of the parallel directive can be chosen depending on the use of the variable. In OpenMP 3.0 the following data clauses are available:

1. private: variable is localized to each thread,
2. firstprivate: variable is localized and initialized with the value of the variable going into the parallel section,

Table 12.2 GNU libomp runtime functions

Directive	Description
omp_get_active_level	Number of active parallel regions
omp_get_ancestor_thread_num	Ancestor thread ID
omp_get_dynamic	Dynamic teams setting
omp_get_level	Number of parallel regions
omp_get_max_active_levels	Maximal number of active regions
omp_get_max_threads	Maximal number of threads of parallel region
omp_get_nested	Nested parallel regions
omp_get_num_procs	Number of processors online
omp_get_num_threads	Size of the active team
omp_get_schedule	Obtain the runtime scheduling method
omp_get_team_size	Number of threads in a team
omp_get_thread_limit	Maximal number of threads
omp_get_thread_num	Current thread ID
omp_in_parallel	Whether a parallel region is active
omp_set_dynamic	Enable/disable dynamic teams
omp_set_max_active_levels	Limits the number of active parallel regions
omp_set_nested	Enable/disable nested parallel regions
omp_set_num_threads	Set upper team size limit
omp_set_schedule	Set the runtime scheduling method
omp_init_lock	Initialize simple lock
omp_set_lock	Wait for and set simple lock
omp_test_lock	Test and set simple lock if available
omp_unset_lock	Unset simple lock
omp_destroy_lock	Destroy simple lock
omp_init_nest_lock	Initialize nested lock
omp_set_nest_lock	Wait for and set simple lock
omp_test_nest_lock	Test and set nested lock if available
omp_unset_nest_lock	Unset nested lock
omp_destroy_nest_lock	Destroy nested lock
omp_get_wtick	Get timer precision.
omp_get_wtime	Elapsed wall clock time.

3. lastprivate: variable is localized, the last iteration copies the value of the variable to the corresponding variable outside the parallel block,
4. shared: all threads in the team share the variable (same memory reference),
5. default: changes the default for all variables in the parallel block,
6. reduction: each thread has a local copy which is combined (*reduced*) using the specified binary operator (only simple binary operator such as +, - are supported),
7. copying: the value of the master thread variable is copied to all threads.

Table 12.3 OpenMP directives

Directive	Description
parallel	parallel region construct
for	for-loop with implied barrier at end
sections	code executed in parallel
single	executed by exactly one thread
task	finer schedulable entity (ver 3.0)
master	synchronization directive
critical	code block with critical section
barrier	synchronization primitive
taskwait	synchronization primitive
atomic	memory location updated atomically
flush	specific point at which memory is consistent
ordered	loop iteration should execute in serial order
threadprivate	variables become thread specific

If *nested regions* are enabled (use the library function omp_get_nested() to check this), then parallel sections can be nested. The parallel directive can be combined with the worksharing directives shown in the next section.

Work sharing directives are the work-horses of OpenMP. Using these directives the programmer directs the compiler to create parallel threads. The work sharing construct divides the code block execution between the threads on the team. Work sharing directives can be *iterative* (such as **do**, **while** and **for**), or they can be non-iterative SECTIONS, which are listed in the code block. Each section is executed once by a thread (although the same thread may execute multiple sections).

```
#pragma omp parallel shared(sum,A,count) private(i)
{
   #pragma omp for schedule(static) nowait
   for( i=0; i < count; ++i ) {
5       sum += A[i];
   }
} /* end of parallel section */
```

As an example of SECTIONS:

```
#pragma omp parallel shared(sumA,A,subB,B,count) private(i)
{
   #pragma omp sections nowait
   {
5       #pragma omp section
      for( i=0; i < count; ++i ) sumA += A[i];

      #pragma omp section
      for( i=0; i < count; ++i ) sumB += B[i];
10   } /* end of sections */
} /* end of parallel section */
```

For FORTRAN language there is a special WORKSHARE directive which treats all statements in the enclosed structured code block as independent work units.

There also exists a directive **single** which ensures that the enclosed code is executed by only one thread. The work-sharing directives accept arguments which control the scheduling of the threads and the division of the work between the threads on the team. The schedules are:

1. Static: fixed schedule, depends on the size of the work unit in the iteration, number of threads and chunk size,
2. Dynamic: as threads become free they execute on available data.

A comparison of the scheduling methods available in OpenMP was performed. The workload used was the evaluation of the return-count of the *Collatz* function of a number. The Collatz function is defined as a recurrence:

```
collatz(n)  = 1      : n == 1
            = n/2    : n is even
            = 3n+1   : n is odd
```

The return-count of the Collatz function is the number of times the function would need to be evaluated given a positive integer as input. Although defined as a recurrence, we implement it as a `while` loop as shown in Listing 12.4.

```
    volatile unsigned long CountCollatzReturn( unsigned long N ) {
      static const unsigned int LOOP_COUNT = 1;
      unsigned long count = 0;
      unsigned long storedN = N;
5     for( unsigned long j=0; j < LOOP_COUNT; ++j ) {
        N = storedN;
        count=j;
        while( 1 ) {
          count++;
10        if( N <= 1 ) break;
          if( N % 2 ) N = 3*N+1;
          else N = N/2;
        }
      }
15    return count;
    }
```

Listing 12.4 Return count of the Collatz function

We write several OpenMP functions, each differing in the scheduling method. These functions are listed in Listing 12.5. The performance comparison is shown in Figure 12.2. In Figure 12.2(a), the performance benefit of adding another processor to the thread pool is shown. Since the Collatz function is compute bound, we get linear speedup. As stated above, the scheduling of threads can be left as static (using the static schedule in OpenMP). The Collatz function has the property that the return count of any integer is not completely dependent on the size of the integer (e.g., return count for all even numbers is simply $\lg(n)$), thus the static schedule performs adequately. But if the workload was dependent on the size of the integer, then a static schedule would exhibit the *tail syndrome*, where the initial values would be computed quickly, but as the length of the integers increases, the time would increase (e.g., this manifests itself in number theoretic functions checking whether a given number is prime or not). The static schedule divides the iteration count of the

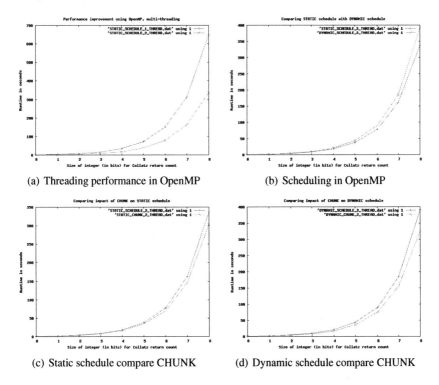

(a) Threading performance in OpenMP (b) Scheduling in OpenMP

(c) Static schedule compare CHUNK (d) Dynamic schedule compare CHUNK

Fig. 12.2 Comparing performance in OpenMP using various scheduling and data chunking strategies.

`for` loop by the number of threads (as specified in the parallel section) or based on the `OMP_NUM_THREADS` environment value.

In *dynamic* scheduling, as and when threads get free they take new work from the pool of not yet finished loop iterations. At the expense of managing this dynamic queue, the tail syndrome can be mitigated. The runtime of the function using static schedule is compared with dynamic schedule as shown in Figure 12.2(b). As we discussed above, in this example the static schedule is reasonably *fair*, thus the dynamic schedule's queue management overhead is visible.

By default OpenMP processes a single loop iteration per-thread. If the workload computation is small (as in this case), it is more efficient to process larger *chunks* per thread. This can be specified using the `chunk` construct in the OpenMP pragma. We compare the addition of chunks to static schedule in Figure 12.2(c), and to dynamic schedules in Figure 12.2(d).

```
   // experiment to try different schedule
   int dynamic_chunk( unsigned long scount ) {
     std::cout << __PRETTY_FUNCTION__ << std::endl;
     unsigned long i;
5    unsigned long ans=0;
     int chunk = 100;
```

```
      #pragma omp parallel private(i) shared( scount,chunk,ans )
      {
        #pragma omp for schedule( dynamic,chunk ) nowait
10      for( i=0; i < scount; ++i ) {
          unsigned long retCount = CountCollatzReturn( i );
          if( retCount > 400 ) ans++;
        } /* end of parallel section */
      } // join point of all threads
15    std::cout << "Ans = " << ans << std::endl;
      return 0;
    }

    int dynamic_schedule( unsigned long scount ) {
20    std::cout << __PRETTY_FUNCTION__ << std::endl;
      unsigned long i;
      unsigned long ans=0;
      #pragma omp parallel private(i) shared( scount,ans )
      {
25      #pragma omp for schedule( dynamic ) nowait
        for( i=0; i < scount; ++i ) {
          unsigned long retCount = CountCollatzReturn( i );
          if( retCount > 400 ) ans++;
        } /* end of parallel section */
30    } // join point of all threads
      std::cout << "Ans = " << ans << std::endl;
      return 0;
    }

35  int static_chunk( unsigned long scount ) {
      std::cout << __PRETTY_FUNCTION__ << std::endl;
      unsigned long i;
      unsigned long ans=0;
      int chunk = 100;
40    #pragma omp parallel private(i) shared( scount,chunk,ans )
      {
        #pragma omp for schedule( static, chunk) nowait
        for( i=0; i < scount; ++i ) {
          unsigned long retCount = CountCollatzReturn( i );
45        if( retCount > 400 ) ans++;
        } /* end of parallel section */
      } // join point of all threads
      std::cout << "Ans = " << ans << std::endl;
      return 0;
50  }


    int static_schedule( unsigned long scount ) {
      std::cout << __PRETTY_FUNCTION__ << std::endl;
55    unsigned long i;
      unsigned long ans=0;
      #pragma omp parallel private(i) shared( scount,ans )
      {
        #pragma omp for schedule( static ) nowait
60      for( i=0; i < scount; ++i ) {
          unsigned long retCount = CountCollatzReturn( i );
          if( retCount > 400 ) ans++;
        } /* end of parallel section */
      } // join point of all threads
65    std::cout << "Ans = " << ans << std::endl;
      return 0;
    }

    int static_wait( unsigned long scount ) {
70    std::cout << __PRETTY_FUNCTION__ << std::endl;
      unsigned long i;
      unsigned long ans=0;
      #pragma omp parallel private(i) shared( scount,ans )
```

```
     {
75     #pragma omp for schedule( static )
       for( i=0; i < scount; ++i ) {
         unsigned long retCount = CountCollatzReturn( i );
         if( retCount > 400 ) ans++;
       } /* end of parallel section */
80   } // join point of all threads
     std::cout << "Ans = " << ans << std::endl;
     return 0;
   }

85 int dynamic_wait( unsigned long scount ) {
     std::cout << __PRETTY_FUNCTION__ << std::endl;
     unsigned long i;
     unsigned long ans=0;
     #pragma omp parallel private(i) shared( scount,ans )
90   {
       #pragma omp for schedule( dynamic )
       for( i=0; i < scount; ++i ) {
         unsigned long retCount = CountCollatzReturn( i );
         if( retCount > 400 ) ans++;
95     } /* end of parallel section */
     } // join point of all threads
     std::cout << "Ans = " << ans << std::endl;
     return 0;
   }
100

   int main( int argc, char * argv[] ) {
     int shift_count = atoi( argv[1] );
     unsigned long scount = 1UL << shift_count;
105  int which_algo = atoi( argv[2] );
     switch( which_algo ) {
     case 0: { return static_schedule( scount ); }
     case 1: { return dynamic_schedule( scount); }
     case 2: { return static_chunk( scount ); }
110  case 3: { return dynamic_chunk( scount); }
     case 4: { return static_wait( scount); }
     case 5: { return dynamic_wait( scount); }
     default:{ return static_schedule( scount ); }
     }
115  return 0;
   }
```

Listing 12.5 Experimenting with OpenMP scheduling

The *shared* and *private* lists of data in the parallel sections control the data visibility of variables to the threads. A variable present in the 'private' list implies that all threads receive a new object of the same type (value of this variable is uninitialized, unless *copying* or *firstprivate* is specified). Shared variables, on the other hand, refer to the same location in memory, and all threads can access the same variable. A variable may also be passed as a *reduction* variable with an associated operator. OpenMP currently limits the operator to be arithmetic operations such as addition, subtraction, and bitwise Boolean. Although, sufficient for basic purposes such as scans and prefix-sums, a generic binary reduction operator as a C++ template type can easily be written. An example is shown in Listing 12.6, which computes the minimum value of a given vector.

```
// \file min_omp.cpp
// \author Sandeep Koranne, (C) 2010
// \description Calculate minimum value of vector
#include <omp.h>
```

```
 5  #include <iostream>
    #include <cstdlib>
    #include <limits>

    template <typename T>
10  T MyOp( T a, T b ) { return std::min( a, b ); }

    // Although OpenMP does not support reduction on
    // arbitrary functions, it is simple to code it.

15  int main( int argc, char *argv [] ) {
      if( argc != 2 ) {
        std::cerr << "Usage: <N>\n";
        exit(1);
      }
20    int i;
      int N = atoi( argv[1] ); // number of elements
      int *data = new int[N];
      for( i=0; i < N; ++i ) data[i] = rand() ^ 177317;
      int chunk = N/10; // chunk size
25    int result = std::numeric_limits<int>::max();

      #pragma omp parallel private(i) shared(N,chunk,result)
      {
        int local_min = std::numeric_limits<int>::max();
30      #pragma omp for schedule(static,chunk) nowait
          for(i=0; i < N; ++i) local_min = MyOp( local_min, data[i] );

        #pragma omp critical
        {
35        result = MyOp( result, local_min );
        }
      }
      std::cout << "Minimum value = " << result << std::endl;
      delete [] data;
40    return (0);
    }
```

Listing 12.6 OpenMP reduction style min operator

The use of a local variable to accumulate partial results per thread which are then *merged* with the final result under a critical section is a powerful idiom. This can reduce locking on the shared variable, and thus improve performance.

12.2.0.3 Task parallelism in OpenMP ver 3.0

New in OpenMP 3.0 is the concept of task parallelism. This was added in response to the increasing demand of application writers who had to deal with irregular parallelism. The *Task* construct defines an explicit task (which is then executed by some thread), which shares data according to the data sharing attributes (as discussed above, see *shared*, and *private*).

The *fork* and *join* concept of OpenMP is exemplified in the design of the TASK construct. See Figure 12.3; an example of using task parallelism in OpenMP is shown in Listing 12.7.

```
// \file task_example.cpp
// \author Sandeep Koranne, (C) 2010
// \description Example of task based parallelism
#include <iostream>          // for program IO
```

Fig. 12.3 Task parallelism in
OpenMP 3.0

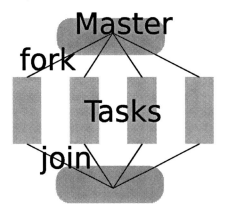

```
 5   #include <vector>              // data storage
     #include <algorithm>           // for copy
     #include <iterator>            // ostream iterator
     #include <omp.h>               // OpenMP

10   static void AnalyzeData( std::vector<int>& D,
                              int index,
                              const unsigned int N,
                              const unsigned int chunk ) {
     // each chunk = [N/chunk]*index, [N/chunk]*(index+1)
15   unsigned int start_index = (N/chunk) * index;
     unsigned int stop_index = start_index + (N/chunk);
     for( unsigned int i=start_index; i < stop_index; ++i )
       D[i] = index;
     }

20   void ExecuteFunction(void) {
       static const unsigned int N = 48;
       std::vector<int> Data( N );
       #pragma omp parallel
25     {
         #pragma omp single nowait
         {
           for( int i=0; i < N; ++i ) Data[i]=i;
         }
30       // create 16 tasks, each responsible for doing part of the work
         for( int i=0; i < 16; ++i ) {
         #pragma omp task if( i > 10 )
           {
             AnalyzeData( Data, i, N, 16 );
35           #ifdef _OPENMP
             std::cout<<"\nComputing with "<<i<<"\t"<<omp_get_thread_num();
             #endif
           }
         }
40       #pragma omp taskwait
       }
       std::copy( Data.begin(), Data.end(),
                  std::ostream_iterator<int>( std::cout, " " ) );
     }
45

     int main( int argc, char *argv [] ) {
       ExecuteFunction();
```

```
     std::cout << std::endl;
50   return (0);
     }
```

Listing 12.7 Example of TASK parallelism in OpenMP

The TASK construct has an optional scalar expression which controls whether the task should be executed on a new thread or executed on the existing thread. On entry upon the parallel sections, already threads have been created, and if the scalar expression returns false, the task code block is executed in-situ.

OpenMP also supports a *Critical* construct which should enclose a block of code which must be executed only by one thread at any given time. Example can be a function to write messages to log file. Task synchronization is achieved using the *Barrier* directive which requires threads to wait at the barrier until all other threads also reach that barrier, at which time all the threads resume execution.

12.3 MPI: Message Passing Interface

Before multicore computing became commonplace in the last 2-3 years, large scale high performance computing projects relied on *cluster computing* or *distributed* computing. A cluster of 100 nodes proved more beneficial (flops per dollar cost) to the end user, as compared to an expensive shared memory multi-CPU machine with 100 CPUs. As the number of vendors deploying clusters increased, a need for standardized APIs for messaging arose. The MPI (message passing interface) API was deigned in response to this need and it comprises of a number of functions as listed below. The functions are divided into categories of:

1. Environment functions: to initialize the MPI execution environment, determine the size of the process group, and compute the rank,
2. C/C++ Library API: includes the point-to-point communication, `MPI_Send` and `MPI_Recv`.
3. Run time library,
4. Configuration file.

12.3.0.4 MPI Environment functions

The environment functions include `MPI_Init()`, `MPI_Comm_size()`, `MPI_Abort()`, and `MPI_Finalize()`.

12.3.0.5 C/C++ Library API

The MPI library contains functions for blocking and non-blocking message passing. As a communication library, MPI ensures (a) ordering of messages and (b) fairness of messages. The MPI functions of `MPI_Send()`, and `MPI_Recv()` can be used. Messages to the same host can be differentiated using *tags* included with the message, as:

```
MPI_Send( &buf, count, datatype, dest, tag, communication);
```

Consider an example:

```
    MPI_Init(&argc,&argv);
    MPI_Comm_size(MPI_COMM_WORLD, &numtasks);
    MPI_Comm_rank(MPI_COMM_WORLD, &rank);

5   if( rank == 0 ) {
      rc = MPI_Send(&outmsg, 1, MPI_CHAR, dest,
                    tag, MPI_COMM_WORLD);
      rc = MPI_Recv(&inmsg, 1, MPI_CHAR, source,
                    tag, MPI_COMM_WORLD, &Stat);
10  }
```

12.3.0.6 Run-time library

The compiler options to compile the C/C++ program and link it with the appropriate MPI library can be done automatically using the `mpicc` command.

12.3.0.7 Configuration file

The `mpd.hosts` file lists the hosts on which MPI programs can run. The port range and password for authentication is listed in `.mpd.conf`. Thereafter, the `mpd` program can be launched, either using `mpdboot` command or direct invocation of the `mpd` program. The servers can be linked together into the same MPI ring. Consider the program as shown in Listing 12.8.

```
    /*
     * \file mpi_hw.c
     * \author Sandeep Koranne (C) 2010
     * \description MPI hello_world program
5    */
    #include <mpi.h>              /* for MPI */
    #include <stdio.h>            /* for program IO */
    #include <stdlib.h>          /* for exit */

10  int main( int argc, char *argv[] ) {
      int number_tasks, rank, rc;
      rc = MPI_Init( &argc, &argv );
      if( rc != MPI_SUCCESS ) {
        perror("MPI init failed");
15      MPI_Abort( MPI_COMM_WORLD, rc );
        exit( 1 );
      }

      MPI_Comm_size( MPI_COMM_WORLD, &number_tasks );
20    MPI_Comm_rank( MPI_COMM_WORLD, &rank );
      printf("\n Num tasks = %d, rank = %d",
       number_tasks, rank );
      MPI_Finalize();
```

```
        return (0);
25   }
```

Listing 12.8 Example program using MPI

Running this program on our MPI cluster prints the number of tasks and the rank of the process. It should be noted that MPI itself is only a specification, and an *implementation* is provided by a specific vendor (hardware or software), or a generic version such as MPICH can be used on the existing cluster. Since MPI was designed for primarily message passing, it has the concept of a *communicator group* which is an object that specifies the collection of MPI processes which may communicate with each other. Moreover, each process has an unique integer identifier denoted *rank* which can be used for communication.

12.3.1 Using Boost.MPI

In addition to using the default MPI API, we can also write MPI programs in C++ using Boost. The Boost.MPI is a library for message passing built on top of MPI. Boost.MPI is a C++ interface which provides an alternative style of programming and development style including (a) support for user-defined data types, (b) C++ STL, (c) arbitrary function objects, and (d) template functions. The functions of MPI.Boost can be classified the same as MPI:

1. Communicators: creation, cloning, and splitting of MPI communicators, manipulation of process groups,
2. Point-to-point communication: communication of primitive and user-defined data types with send and receive operation with blocking and non-blocking interfaces,
3. Collective communications: including *reduce* and *gather*,
4. MPI Datatypes.

Boost.MPI can be directly used through C++ or using Python interface.

12.4 Other libraries and tools

In this section we discuss some of the new parallel programming libraries and software.

12.4.1 Thread Building Blocks

Thread Building Blocks is an C++ library by Intel designed to improve the efficacy of expressing parallelism using C++. Intel TBB is written using modern C++ techniques such as *templates*, and object oriented programming. TBB is written to allow the expression of task based, as well the *plain-old* thread based parallelism. It supports parallel design patterns and promotes scalable data parallel programming. As we discussed above in the case of OpenMP (see Section 12.2), TBB also specifies tasks, as opposed to threads, TBB manages the scheduling, which in the general case results in better performance; however, TBB has excellent support for C++, as shown in Listing 12.9.

```cpp
// \file tbb_example.cpp
// \author Sandeep Koranne, (C) 2010
// \description Example of using Thread Building Block library

#include <iostream>            // for program IO
#include <string>             // for std::string
#include <cassert>            // assertion checking
#include <cstdlib>            // exit
#include <algorithm>          //
#include "tbb/parallel_for.h"
#include "tbb/blocked_range.h"

class NoWorkDone {
public:
  void operator()( const tbb::blocked_range<size_t>& R ) const {
    //std::cout << std::endl << "Range = " << R.begin() << "\t" << R.end();
  }
};


class TakeuchiBenchmark {
  static unsigned int Takeuchi( int x, int y, int z) {
    if( y >= x ) return z;
    // extra looping constructs, to add computation time
    #if 0
    for( unsigned int i=0; i < z; ++i )
      for( unsigned int j=0; j < x; ++j )
        x = ( x & i) | x, y = y & j | y;
    #endif
    return Takeuchi( Takeuchi( x-1, y, z ),
                     Takeuchi( y-1, z, x ),
                     Takeuchi( z-1, x, y ));
  }

public:
  void operator()( const tbb::blocked_range<size_t>& R ) const {
    //std::cout << std::endl << "Range = " << R.begin() << "\t" << R.end();
    unsigned long int sum = 0;
    for( size_t i = R.begin(); i < R.end(); ++i ) {
      sum += Takeuchi(i, i+1, i+2 );
    }
    //std::cout << "\n Sum = " << sum << std::endl;
  }
};


int main(int argc, char *argv[]) {
  unsigned int N = 16;
  if( argc == 2 ) N = atoi( argv[1] );
  std::cout << "Calculate schedule.\n";
  tbb::parallel_for( tbb::blocked_range<size_t>( 0, N ),NoWorkDone() );

  std::cout << "Takeuchi benchmark...\n";
  tbb::parallel_for( tbb::blocked_range<size_t>( 0, N ),
                     TakeuchiBenchmark() );
  return ( EXIT_SUCCESS );
```

```
}
```

Listing 12.9 Example of using Intel TBB

As shown in Listing 12.9, the `operator()` function is defined for the class; the TBB library splits the workload using `tbb::blocked_range<size>`. The partitioning scheme can be changed by passing a custom partitioner to the `parallel_for` function. The `parallel_for` function introduces parallelism in the code. TBB also supports *nested parallelism*. A nice use of C++ templates is the *lambda* expression to encapsulate a code block without creating a dummy class with the `operator()`. This features uses recent features of the C++ 00x standard, but is supported in GCC 4.5.0. An example is shown below in Listing 12.10.

```
   int main(int argc, char *argv[]) {
     unsigned int N = 16;
     if( argc == 2 ) N = atoi( argv[1] );

5    parallel_for( blocked_range<size_t>( 0, N ),
       [=]( const blocked_range<size_t>& R ) {
         unsigned long int sum = 0;
         for( size_t i = R.begin(); i < R.end(); ++i )
           sum += TakeuchiBenchmark::Takeuchi(i, i+1, i+2 );
10     }
     );
     return ( EXIT_SUCCESS );
   }
```

Listing 12.10 Lambda expressions in TBB

As with OpenMP, TBB also defines *reductions* to be performed in parallel. Complete C++ support for binary operator is available, and using this, arbitrary complex binary reductions can be performed in parallel. Another key facility provided by TBB is a memory allocator optimized for multi-threaded allocations.

12.4.2 CUDA : C Unified Device Architecture

A recent entrant to the parallel computing stage is the GPGPU (general purpose computing on graphics processing unit) concept. Graphics chips (such as those by NVIDIA and AMD) have enormous compute power in the form of streaming processors designed for high-performance graphics. As compared to the CPU architecture (shown in Figure 12.4(a)) the GPUs have most of the silicon area dedicated to ALUs (as shown in Figure 12.4(b)). The branch control logic and memory latency hiding caches are smaller. This presents both an opportunity and a challenge, as traditional C/C++ programs are written for execution on a CPU and thus cannot directly take advantage of the massive data-parallelism which is present in the GPU.

Thus, programming GPUs is accomplished by special compilers and tools, although recent progress in auto-parallelizing (at least FORTRAN code) has been made. One such programming system for GPUs is CUDA by NVIDIA. CUDA is an acronym for C Unified Device Architecture. A more standardized (able to execute

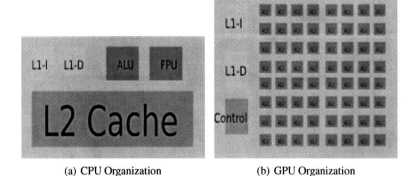

      (a) CPU Organization   (b) GPU Organization

Fig. 12.4 Difference between CPU and GPU organization

on NVIDIA as well as AMD chips) method is to use OpenCL. In this section we describe the use of CUDA and OpenCL to program GPUs.

Our test hardware for the CUDA setup is a GeForce GT 240 graphics card plugged into a PCI-Express slot on an Opteron x86-64 host running GNU/Linux (Fedora Core 12). NVIDIA CUDA ships with a `deviceQuery` utility which can report the various properties of the underlying hardware, this is required when writing programs which are expected to run on generic CUDA hardware. Running this program on our CUDA setup we get:

```
./bin/linux/release/deviceQuery Starting...

CUDA Device Query (Runtime API) version (CUDART static linking)

There is 1 device supporting CUDA

Device 0: "GeForce GT 240"
  CUDA Driver Version:                      3.10
  CUDA Runtime Version:                     3.10
  CUDA Capability Major revision number:    1
  CUDA Capability Minor revision number:    2
  Total amount of global memory:            1073020928 bytes
  Number of multiprocessors:                12
  Number of cores:                          96
  Total amount of constant memory:          65536 bytes
  Total amount of shared memory per block:  16384 bytes
  Total number of registers available per block: 16384
  Warp size:                                32
  Maximum number of threads per block:      512
  Maximum sizes of each dimension of a block: 512 x 512 x 64
  Maximum sizes of each dimension of a grid: 65535 x 65535 x 1
  Maximum memory pitch:                     2147483647 bytes
  Texture alignment:                        256 bytes
  Clock rate:                               1.34 GHz
  Concurrent copy and execution:            Yes
```

```
Run time limit on kernels:                     Yes
Integrated:                                    No
Support host page-locked memory mapping:       Yes
Compute mode:                                  Default
Concurrent kernel execution:                   No
Device has ECC support enabled:                No
```

deviceQuery, CUDA Driver = CUDART, CUDA Driver Version = 3.10,
CUDA Runtime Version = 3.10, NumDevs = 1, Device = GeForce GT 240

PASSED

12.4.3 SIMT in CUDA

The CUDA stream multiprocessors create, manage, schedule and execute threads in groups called *warps*. Threads belonging to a warp have the same instruction stream (program code) and start off from the same location. Each thread has its own program counter (PC), register state and thus can branch independently of each other; however, as long the threads execute the same instruction path full efficiency is realized, but as soon as a thread diverges, the warp serializes the execution of thread until the threads synchronize to the same point. This behavior, where a group of threads execute the same instruction, but may diverge out of sequence is termed SIMT (single instruction multiple thread). As long as the programmer can optimize that code path divergence is minimized, efficiency can be optimized, but unlike SIMD, branches can be taken by the threaded code. The ability of the GPU architecture to launch threads and schedule them in hardware makes the runtime penalty of this thread management small compared to execution time of the data-parallel portions of the kernel.

12.4.4 Compute Kernels in CUDA

The key concept in CUDA is the idea of *compute kernels*. Compute kernels are C functions which are executed in parallel by N different CUDA threads. Consider a compute kernel as shown below:

```
   __global__ void AddVector( float* A, float* B, float* SUM ) {
     int i = threadIdx.x;
     SUM[i] = A[i] + B[i];
   }

5
   int main() {
     AddVector<<<1,N>>>( A, B, SUM );
     // Parallel invocation of N threads
     return (0);
10 }
```

In the above listing *N* threads execute the `AddVector` compute kernel in parallel. The `__global__` declaration specifier denotes the function as a CUDA kernel function. Inside the kernel the variable `threadIdx` and `blockIdx` are defined by the CUDA system. These are actually three dimensional vector. Calling the kernel with different blocks is done as:

```
{
    AddVector<<< numBlocks, numThreadsPerBlock >>>(A,B,SUM);
}
```

There is a system imposed limit on the number of threads per block, due in part to the CUDA constraints on limited memory resources on the core. Moreover, thread blocks are required to execute independently, as well as in any order. Threads within a block can cooperate by sharing data through *shared memory* and using synchronization functions.

12.4.5 Compiling CUDA code with NVCC

NVCC is the CUDA compiler driver which compiles C code to a form which can be executed on the GPU. The compiler outputs PTX (assembly) and/or binary instructions from C code containing a mixture of host code and GPU code (kernels).

CUDA functions to allocate memory on the host and device allow the programmer to transfer data for computation on the GPU. Texture memory on the GPU can also be used as a lookup memory for constant data. Consider the following example:

```
   // \file cuda_memory.cu
   // \author Sandeep Koranne, (C) 2010
   // \description CUDA example file

 5 #include <cuda.h>
   #include <cuda_runtime_api.h>
   #include <iostream>

   __global__ void DepictBlockDim( float* SUM, int N ) {
10   int i = blockDim.x * blockIdx.x + threadIdx.x;
     if( i < N ) SUM[i] = blockDim.x;
   }

   __global__ void DepictBlockId( float* SUM, int N ) {
15   int i = blockDim.x * blockIdx.x + threadIdx.x;
     if( i < N ) SUM[i] = blockIdx.x;
   }

   __global__ void DepictIteration( float* SUM, int N ) {
20   int i = blockDim.x * blockIdx.x + threadIdx.x;
     if( i < N ) SUM[i] = i;
   }

   __global__ void AddVector( float* A, float* B, float* SUM, int N ) {
25   int i = threadIdx.x;
     SUM[i] = A[i] + B[i];
   }

   int main() {
30   cuInit(0);
```

```
     int deviceCount = 0;
     cudaGetDeviceCount( &deviceCount );
     if( deviceCount == 0 ) {
       std::cerr << "Unable to find any CUDA devices..\n";
35     return (1);
     }
     const unsigned int N = 16;
     const size_t size = N*sizeof(float);
     std::cout << deviceCount << " CUDA device(s) found.\n";
40   float* A = (float*) malloc( size );
     float* B = (float*) malloc( size );
     float* SUM = (float*) malloc( size );
     for( unsigned int i=0; i < N; ++i ) {
       A[i] = B[i] = 1.0 * i;
45     SUM[i] = 0.0;
     }
     float *d_A = NULL, *d_B = NULL, *d_SUM = NULL;
     cudaMalloc( (void**) &d_A, size ); // device allocate
     cudaMalloc( (void**) &d_B, size ); // device allocate
50   cudaMalloc( (void**) &d_SUM, size ); // device allocate

     cudaMemcpy( d_A, A, size, cudaMemcpyHostToDevice );
     cudaMemcpy( d_B, B, size, cudaMemcpyHostToDevice );

55   //AddVector<<<1,N>>>( d_A, d_B, d_SUM ); // SUM = A + B
     int numThreadsPerBlock = 256;
     int blocksPerGrid = ( N + numThreadsPerBlock -1 )/numThreadsPerBlock;
     std::cout << "Blocks per grid = " << blocksPerGrid << "\n";
     //DepictBlockDim<<<8,8>>> (d_SUM,N);
60   //DepictBlockId<<<8,8>>> (d_SUM,N);
     //DepictIteration<<<8,8>>>( d_SUM, N );
     AddVector<<<blocksPerGrid,numThreadsPerBlock>>>( d_A, d_B, d_SUM, N );
     cudaMemcpy( SUM, d_SUM, size, cudaMemcpyDeviceToHost );
     cudaFree( d_A );
65   cudaFree( d_B );
     cudaFree( d_SUM );
     for( unsigned int i=0; i < N; ++i ) {
       std::cout << SUM[i] << " ";
     }
70   std::cout << std::endl;
     return (0);
   }
```

Listing 12.11 Kernel dimensions in CUDA

We compile this CUDA file as:

```
nvcc -I /usr/local/cuda/include/ cuda_memory.cu  -lcuda
```

Listing 12.11 has many different compute kernels. Except for the `AddVector` kernel, the purpose is to demonstrate the effect of kernel *dimensions*, parameters to the CUDA runtime listed within the <<<B,N>>> angled brackets. The first argument is the `blocksPerGrid`, and the second argument is `threadsPerBlock`. The kernel implicit variable `blockDim` is defined as the *dimension* of the block, while individual values of the block id are available in the implicit variable `blockIdx`. We selectively run each of these compute kernels, the results are shown below:

Kernel is `DepictBlockDim`

```
$ ./a.out
1 CUDA device(s) found.
Blocks per grid = 2
```

```
8 8 8 8 8 8 8 8 8 8 8 8 8 8 8 8
```

The block dimension is indeed 8, as that was passed as the argument to the CUDA runtime.

Kernel is `DepictBlockId`

```
$ ./a.out
1 CUDA device(s) found.
Blocks per grid = 2
0 0 0 0 0 0 0 0 1 1 1 1 1 1 1 1
```

Since there are a total of 16 values, and 8 blocks per grid, there will be a total of 2 blocks, of id 0 and id 1. This is what we see from the above output.

Kernel is `DepictIteration`

```
$ ./a.out
1 CUDA device(s) found.
Blocks per grid = 2
0 1 2 3 4 5 6 7 8 9 10 11 12 13 14 15
```

In this kernel we compute the iteration index for each member of the array. Typically, this formulation is used in most kernels, the id of the variable is computed as a function of blockDim, blockId, and threadId. If a multi-dimensioned thread block is used, conventional row-column matrix index conversion formulas can be used to compute the index.

Kernel is `AddVector`

```
$ ./a.out
1 CUDA device(s) found.
Blocks per grid = 2
0 2 4 6 8 10 12 14 16 18 20 22 24 26 28 30
```

This kernel performs the actual addition. Since both vectors *A* and *B* are sequentially filled, we expect the *SUM* array to be comprising of sequential even numbers. For high-efficiency, number of threads per block is set to 256.

12.4.5.1 PyCUDA

While CUDA API and the `nvcc` compiler provide complete access to the GPU hardware, programming in CUDA is low-level and time consuming. For numerical applications which involve operations on array, an alternative is to use a Python software library called PyCUDA which is a higher level alternative.

Consider the following example in PyCUDA as shown in Listing 12.12.

```
  # \file pycuda_props.py
  # \author Sandeep Koranne (C) 2010
  # \description Example file using PyCUDA for GP GPU Computing
  import pycuda.driver as drv
5 import pycuda.autoinit
  import numpy
```

```
     import sys

     from pycuda.compiler import SourceModule
10
     mod = SourceModule("""
       __global__ void VectorAdd( float* A, float* B, float* SUM ) {
         const int i = threadIdx.x;
         SUM[i] = A[i] + B[i];
15     }

       __global__ void DepictBlockId( float *SUM ) {
          const int i = blockDim.x * blockIdx.x + threadIdx.x;
          SUM[i] = i;
20     }
     """)

     VectorAdd = mod.get_function("VectorAdd")

25   numCudaDevices = drv.Device.count()

     if numCudaDevices == 0:
         print "No CUDA devices detected..\n"
         sys.exit(1)
30   else:
         print "%d CUDA devices detected." % numCudaDevices
         dev = drv.Device(0)
         print "Compute Capability %d.%d" %dev.compute_capability()
         print "Device has %s MB memory" % (dev.total_memory()//(1024*1024))
35
     # we have the CUDA device
     N = 16
     a = numpy.random.randn(N).astype(numpy.float32)
     b = numpy.random.randn(N).astype(numpy.float32)
40   sum = numpy.zeros_like( a )
     #VectorAdd( drv.In( a ), drv.In( b ), drv.Out( sum ),block=(N,2,1))
     DepictBlockId = mod.get_function("DepictBlockId")
     DepictBlockId( drv.Out( sum ), block=(N,2,1) )

45   print sum

     sys.exit(0)
```

Listing 12.12 Example of PyCUDA

Running the code in Listing 12.12 we get:

```
$ python pycuda_props.py
1 CUDA devices detected.
Compute Capability 1.2
Device has 1023 MB memory
[   0.   1.   2.   3.   4.   5.   6.   7.   8
...
]
```

Using GPUarrays, the task of running simple compute kernels can be simplified
even further as shown in Listing 12.13.

```
     # \file gpuarray.py
     # \author Sandeep Koranne (C) 2010
     # \description Example of using GPUarray in PyCUDA with CUDA
     import pycuda.gpuarray as gpuarray
5    import pycuda.driver as drv
     import pycuda.autoinit
```

```
     import numpy
     import sys

10   numCudaDevices = drv.Device.count()

     if numCudaDevices == 0:
         print "No CUDA devices detected..\n"
         sys.exit(1)
15
     N=4
     h_a = numpy.arange(N).astype(numpy.float32)
     d_a = gpuarray.to_gpu( h_a )
     a_2x = (2*d_a).get()
20   print "Host Array = "
     print h_a
     print "After computing"
     print a_2x

25
     sys.exit(0)
```

Listing 12.13 Example of GPUarrays in PyCUDA

Running the code in Listing 12.13, we get:

```
$python gpuarray.py
Host Array =
[ 0.   1.   2.   3.]
After computing
[ 0.   2.   4.   6.]
```

Comparing the code in Listing 12.13 to the C code compiled with nvcc, one can immediately see the reduction in code that can be done using PyCUDA. The use of NumPy and general Python language can be used to write the non compute-intensive part of the program, while CUDA based kernels can perform the compute intensive code blocks. The low level CUDA APIs are exposed using the Driver and Device class in PyCUDA.

12.4.6 OpenCL (Open Compute Language)

OpenCL (Open Compute Language) is an open standard for parallel programming of heterogeneous systems. OpenCL standard is managed by the Khronos Group (same group who manages OpenGL standard). OpenCL is designed to run embedded systems, personal computers, and high-performance computing systems. OpenCL runs on CUDA for NVIDIA, and implementations of OpenCL for other GPUs are also available.

It is instructive to compare the compute kernel for vector addition (see Listing 12.4.4) with the OpenCL kernel shown below:

```
__kernel void AddVector( __global const float *A,
                         __global const float *B,
                         __global     float *SUM ) {
    int elementIndex = get_global_id(0);
```

```
5   SUM[elementIndex] = A[elementIndex] + B[elementIndex];
}
```

The key difference in the OpenCL kernel is the computation of thread index equivalent, elementIndex is done using function calls as opposed to implicit variables (as done in CUDA). OpenCL has functions to return the (i) work group index, (ii) work group size, (iii) thread index. The function used above `get_global_id` is used to get the id globally, while `get_local_id` is used to get the id within the work group. Starting with version 3.1 of the CUDA software, OpenCL is supported.

12.5 Conclusion

Parallel, especially multicore computing has become the mainstay of high-performance computing. More and more desktop computers also come equipped with multicore CPUs. In this chapter we have briefly introduced open-source tools and techniques to write efficient and scalable parallel programs. In this chapter we have discussed POSIX Threading library, OpenMP, MPI, Intel Thread Building Blocks, and GPU computing with CUDA and OpenCL. We expect parallel and multi-core computing to increase its impact on scientific as well as general computing in the years ahead.

Chapter 13
Compiler Construction

Abstract A compiler refers to a software tool or system which performs automatic conversion from one computer language to another. Along the way, the compiler tries to optimize the program while maintaining the semantics of the computation. This chapter discusses the compiler construction tools `flex`, `bison`, and LLVM. Related text processing tools such as `m4`, `gperf` and `readline` are also presented. We also discuss the various GNU binutils tools, including `ar`, `nm` and `ld`. Examples using the various compiler construction tools are presented.

Contents

13.1 Introduction

In this chapter we discuss language processors and compiler construction tools. In Section 13.2 we introduce the compiler construction concept. The flowchart for language processing and translation is presented, and appropriate software tools for each part of the flow are introduced. In Section 13.3 we discuss lexical analysis. This problem has elegant mathematical roots and solutions from finite automata are discussed. Utilities such as `m4`, `readline`, `getopt` are also introduced. In Section 13.4 venerable parsing tool YACC and the ANTLR compiler construction toolkit are described. The Boost SPIRIT framework is also described in 13.4.1.

In Section 13.5 we describe the code generation, and instruction representation libraries available through GNU binutils. Other system tools such as linkers, archive managers, ELF inspection are also discussed.

S. Koranne, *Handbook of Open Source Tools*,
DOI 10.1007/978-1-4419-7719-9_13, © Springer Science+Business Media, LLC 2011

The Low Level Virtual Machine (LLVM) intermediate representation, and compiler optimization system is discussed in Section 13.6. GCC plugin `dragonegg`, the LLVM bit-code analyzer, and compiler optimization passes are discussed.

13.2 Anatomy of a Compiler

A compiler refers to a software tool or system which performs automatic conversion from one computer language to another. Along the way, the compiler tries to optimize the program while maintaining the semantics of the computation. The history of compiler construction tools is rich and varied. Classical texts such as Aho, Ullman and Sethi's Dragon Book have defined the field. As shown in Figure 13.1 the process of compiling a source text can be broken into (i) lexical analysis, (ii) parsing, (iii) intermediate code representation, (iv) optimization (which includes register allocation, scheduling and rewriting), and (v) finally, instruction generation.

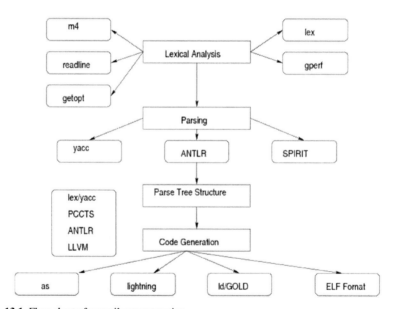

Fig. 13.1 Flow chart of compiler construction

Alongwith the compiler proper, the act of compiling also includes ancillary activities such as (i) assembler, (ii) loaders and linkers, (iii) binary object file manipulation, and (iv) library archive maintenance. Towards this end, in this part we have also discussed the important and ubiquitous GNU `bin-utils`. More details are given below in Section 13.5.1 Intermediate representation (IR) and compiler optimizations are discussed later in Section 13.6.

13.3 Lexical Analysis

Lexical analysis is defined as the task of *tokenizing* a given input stream according to pre-defined rules. Examples of lexical analysis abound in computer literature, from command line parsing, macro substitution, UNIX `grep`, to actual programming language compilers. The problem can be modeled as a halting problem on finite automata. The rules for lexical analysis are input by the user, and most commonly a software tool is used to construct a lexical scanner. The software usually also optimizes the scanner so that it reduces memory usage during scanning, and optimizes the performance. The common lexical analysis generator is `lex`, and its GNU implementation is called `flex`. This is discussed below in Subsection 13.3.1.

As we mentioned above, lexical analysis is also used in `grep` (see Chapter 1.3), macro processor (`m4`), GNU `readline`, and the GNU perfect hash generator (`gperf`). These are discussed below.

13.3.1 GNU flex

GNU `flex` automatically generates programs whose control flow is directed by instances of regular expressions in the input stream. It is well suited for generating lexical analyzers and token processors, and for segmenting input in preparation for a parsing routine.

GNU `flex` source is a table of regular expressions (called *rules*) and corresponding program fragments, called *actions*. The table is translated to a program by `flex` which reads an input stream, copying it to an output stream and partitioning the input into strings which match the given expressions. Each such string is recognized using deterministic finite automata (DFA); when the string is matched the corresponding program fragment is executed. GNU `flex` generates as output a C source file named 'lex.yy.c' (by default), which defines a routine `yylex()`. This file is compiled and linked with the '-lfl' library to produce an executable. The following flex input specifies a scanner which whenever it encounters the string "username" will replace it with the user's login name:

```
%%
username     printf( "%s", getlogin() );
```

By default, any text not matched by a flex scanner is copied to the output, so the net effect of this scanner is to copy its input file to its output with each occurrence of "username" expanded.

13.3.1.1 Format of the input file

The GNU `flex` input file consists of three subsections, separated by a line with just %% in it:

```
definitions
%%
rules
%%
user code
```

The definitions subsection contains declarations of simple name definitions to simplify the scanner specification, and declarations of start conditions. The command-line usage of GNU `flex` is:

```
Usage: flex [OPTIONS] [FILE]...
Generates programs that perform pattern-matching on text.

Debugging:
  -p, --perf-report write performance report
  -v, --verbose     write summary of scanner statistics

Files:
  -o, --outfile=FILE      specify output filename
  -S, --skel=FILE         specify skeleton file
  -t, --stdout            write scanner on stdout
      --yyclass=NAME      name of C++ class
      --header-file=FILE  create a C header file
      --tables-file[=FILE] write tables to FILE
  -i, --case-insensitive  ignore case in patterns
      --yylineno          track line count in yylineno

  -+, --c++               generate C++ scanner class
  -Dmacro[=defn]
        #define macro defn  (default defn is '1')
```

A more complete list of command line options can be found by running `man flex`. Now we present some example of using GNU `flex` to generate a scanner.

```
   /* A simple lexical scanner
      (C) Sandeep Koranne, 2010.
   */
   %{
5  #include <stdio.h>
   #include <unistd.h>
     int num_lines = 0;
     char login_name[1024];
   %}
10
   %%
   \n    { ++num_lines; printf("\n"); }
   username { cuserid( login_name ); printf( "%s", login_name ); }
   %%
15 int main()
   {
     yylex();
     printf( "# of lines = %d\n", num_lines );
   }
```

Listing 13.1 Lex exampe file for username analysis

We generate a scanner using GNU `flex` as:

```
$flex -o username.c username.l
$gcc -o substitute_username username.c -lfl
$ cat scanner_example.txt
This is a simple report generator.
username is writing this.
$ cat scanner_example.txt | ./substitute_username
This is a simple report generator.
skoranne is writing this.
# of lines = 2
```

A slightly more involved example (coming from a Pascal compiler) is shown below:

```
";" { return SEMI; }
":" { return COLON;}
"," { return COMA; }
"("      { return LP; }
")"      { return RP ; }
"[" { return LB ; }
"]" { return RB ; }
":="     { return ASSGN ; }
and { return AND_TOK ; }
array { return ARRAY;   }
[bB][eE][gG][iI][nN]{ return BEGIN_TOK; }
case { return CASE; }
div { return DIV_TOK; }
do { return DO; }
downto { return DOWNTO; }
else { return ELSE; }
end { return END; }
for { return FOR; }
function{ return FUNCTION ; }
if { return IF ; }
```

This conversion of fixed size and constant strings (mostly coming from the reserved words and keywords of programming languages) into tokens is so common that a special tool GNU gperf has been developed which performs *perfect hashing*. This makes the lexical scanner perform much faster at this task of identifying keywords.

13.3.1.2 GNU gperf: the perfect hash generator

GNU gperf reads in a set of keywords from a user-provided file (which typically has a '.gperf' extension, although this is not mandatory)for example, commandoptions.gperfand generates C/C++ sources for the hash table, hashing, and lookup methods. All code is written to standard output (by default).

The input file for `gperf` is similar in format to GNU `flex`; it contains

```
%{
/* C code */
%}
declarations
%%
keywords
%%
functions
```

The declarations subsection is optional and can be omitted as long as GNU `gperf`
is not invoked with the -t option. In case the -t option is used, the last component in
the declaration subsection must be a structure whose first field must be a char* or
const char* identifier called name.

The next construct in the file is the *keywords* subsection. Each line in this subsection that begins with the number sign (#) in the first column is a comment. The
keywords are the first field of each non-comment line in the keywords subsection.
Using the Pascal example from above:

```
    /*
     GNU gperf input file for the Pascal language
     (C) Sandeep Koranne, 2010
    */
 5  %{
    #include <stdio.h>
    #include <string.h>
    %}
    %%
10  #Pascal keywords
    and, AND_TOK
    array, ARRAY
    begin, BEGIN_TOK
    case, CASE
15  div, DIV_TOK
    do, DO
    downto, DOWNTO
    else, ELSE
    end, END
20  for, FOR
    function, FUNCTION
    if, IF
    of, OF
    or, OR_TOK
25  procedure, PROCEDURE
    program, PROGRAM
    then, THEN
    to, TO
    var, VAR_TOK
30  while, WHILE
    write, WRITE_TOK
    integer, INT_TYPE
    real, REAL_TYPE
    read, READ_TOK
35  repeat, REPEAT
    until, UNTIL
    goto, GOTO
    %%
```

Listing 13.2 Example `gperf` input file for Pascal keywords

A simple test harness to exercise the GNU `gperf` generated keyword checker compared to a string comparison function was also written.

```
/*
 Test harness for GNU gperf, performance characterization also
 (C) Sandeep Koranne, 2010
 */

#include <stdio.h>
#include <stdlib.h>
#include <string.h>

extern const char *in_word_set (const char *str, unsigned int len );

static const char* PASCAL_KW[] = { "and", "array", "begin",
            "case", "div", "do", "downto",
            "else", "end", "for",
            "function", "if", "of", "or",
            "procedure", "program","then",
            "to", "var", "while", "write",
            "integer", "real", "read",
            "repeat", "until", "goto"
};

int main( int argc, char *argv[] )
{
  FILE *fp = NULL;
  int kw_found = 0;
  if( argc != 2 ) {
    fprintf( stderr, "usage: pascal_kw <filename>\n" );
    exit(1);
  }
  fp = fopen( argv[1], "rt" );
  if( !fp ) {
    fprintf( stderr, "Unable to open Pascal program file : %s", argv[1] );
    exit(1);
  }

  while( 1 ) {
    int i;
    char kw[1024];
    int  num_read = fscanf( fp, "%s", kw );
    if( num_read == 0 ) break;
    if( strcmp( kw, "end." ) == 0 ) { kw_found++; break; }
    #ifndef USE_GPERF
    for( i=0; i < 27;++i ) {
      if( strcmp( kw, PASCAL_KW[i] ) == 0 ) { kw_found++; break; }
    }
    #else
    if( in_word_set( kw, strlen( kw ) ) ) kw_found++;
    #endif
  }
  fclose( fp );
  printf("%d Pascal keywords found.\n", kw_found );
  return 0;
}
```

Listing 13.3 Test harness for `gperf` for Pascal keywords

On a collection of Pascal programs totalling 4.9 million lines of code containing 5.7 million Pascal keywords, the `gperf` based functions are almost 2 times faster than the string comparison function.

13.3.2 GNU m4

GNU m4 is a macro processor, meaning that it copies its input to the output, expanding macros along the way. In this regard it's similar to the C pre-processor 'cpp'. Like cpp, m4 originally was written as the pre-processor for a programming language (Rational FORTRAN); however, m4 is much more powerful and feature-rich than 'cpp', which makes it much more useful than just defining constants in programs. Consider the simple example:

```
$cat expand_hello.m4
define('hello', 'Hello World from here')
```

Note the particular use of quotation mark around the text to be macro processed. The format is intuitive, wherever the pattern exists, m4 will replace that with the defined macro. We run m4 as:

```
$cat m4_example.txt
Are you saying hello
$m4 expand_hello.m4 m4_example.txt
Are you saying Hello World from here
```

GNU m4 writes the processed output on standard output. Some of the commonly used command line options for m4 are:

```
'--help'
'--version'
'-E'
'--fatal-warnings'
    Controls the effect of warnings.
'-i'
'--interactive'
'-e'
    Makes this invocation of 'm4' interactive.

'-p'
'--prefix-builtins'

'-Q'
'--quiet'
```

GNU m4 also allows for defining pre-processor macros on the command line itself.

```
'-D NAME[=VALUE]'
'--define=NAME[=VALUE]'

'-I DIRECTORY'
'--include=DIRECTORY'
```

```
'-U NAME'
'--undefine=NAME'
```

13.3.3 GNU readline

GNU `readline` reads a line from the terminal and returns it as a prompt. If prompt is NULL or the empty string, no prompt is issued. The line returned is allocated with `malloc`; the caller must free it when finished. The line returned has the final newline removed, so only the text of the line remains. GNU `readline` offers editing capabilities while the user is entering the line. By default, the line editing commands are similar to those of GNU `emacs`. Using GNU `readline` is simple as shown in the following API example:

```
#include <stdio.h>
#include <readline/readline.h>
#include <readline/history.h>
char * readline (const char *prompt);
```

13.3.4 getopt

The `getopt()` function parses the command-line arguments. Its arguments `argc` and `argv` are the argument count and array as passed to the `main()` function on program invocation. An element of `argv` that starts with '-' is an option element. The characters of this element (aside from the initial '-') are the option characters. If `getopt()` is called repeatedly, it returns successively each of the option characters from each of the option elements. The variable optind is the index of the next element to be processed in argv. If there are no more option characters, `getopt()` returns -1. Then optind is the index in argv of the first argv-element that is not an option. Consider the following API description:

```
    #include <unistd.h>
    int getopt(int argc, char * const argv[],
            const char *optstring);

5   extern char *optarg;
    extern int optind, opterr, optopt;
    #include <getopt.h>
```

The `getopt\_long()` function works like getopt() except that it also accepts long options, started with two dashes. A more detailed example from a real example is shown below:

```
int main( int argc, char* argv [] ) {
    int polymake_mode=0, graph_extraction_mode=0,latex_mode=0,reduced_mode=1;
    char *cano_file_name = NULL;
    int c;
5   dimension = 3; // by default
```

```
      while (( c = getopt( argc, argv, "hprgGlf:d:m:") ) != -1 ) {
        switch( c ) {
        case 'f':
          cano_file_name = new char[ strlen(optarg) + 1];
10        strcpy( cano_file_name, optarg );
          break;
        case 'd':
          dimension = atoi( optarg );
          break;
15      case 'g':
          graph_extraction_mode = 1;
          break;
        case 'G':
          graph_extraction_mode = 2;
20        break;
        case 'r':
          reduced_mode = 0;
          break;
        case 'l':
25        latex_mode = 1;
          break;
        case 'm':
          SINGLE_SEPARATION_MODE = atoi( optarg );
          break;
30      case 'p':
          polymake_mode = 1;
          break;
        case 'h':
        case '?':
35        std::cerr << "pm_cd -f <cano> -d <dim> [ -p | -g ]\n";
          return 1;
        }
      }
```

Listing 13.4 Example of using getopt

13.4 YACC: Yet Another Compiler Compiler

Yacc provides a general tool for describing the input to a computer program. The
user specifies the structures of his input, together with code to be invoked as each
such structure is recognized. Yacc turns such a specification into a subroutine that
handles the input process. The GNU implementation of yacc is called bison. Bison is a general-purpose parser generator that converts a grammar description for
an LALR(1) context-free grammar into a C (as well as C++) program to parse
that grammar. Bison is upward compatible with Yacc. Input files should follow the
yacc convention of ending in '.y'. Unlike yacc, the generated files do not have fixed
names, but instead use the prefix of the input file. Usage:

```
Usage: bison [OPTION]... FILE
Generate LALR(1) and GLR parsers.
Operation modes:
  -h, --help                display this help and exit
  -V, --version             output version information
  -y, --yacc                emulate POSIX Yacc
```

```
Output:
  --defines[=FILE]              also produce a header file
  -d                            likewise
  -v, --verbose                 same as '--report=state'
  -b, --file-prefix=PREFIX      specify a PREFIX for output files
  -o, --output=FILE             leave output to FILE
  -g, --graph[=FILE]            also output a graph of the automaton
```

A more comprehensive listing of command line options can be found by running man bison. An example of a simple library checkin, and checkout system can be constructed using GNU flex and GNU bison. The grammar of the library database control language is very simple; consider an example transcript shown below:

```
checkout book 98
checkout book 200
checkin book 76
checkin book 98
checkout book 23
checkout book 465
```

The grammar of the library database control language in BNF notation is the following:

```
commands: /* empty */
        | commands command
        ;
command:
        book_check_in | book_check_out
        ;
book_check_in: CHECKIN_TOKEN BOOK_TOKEN NUMBER_TOKEN
        { gTotalBooksCheckedOut--; }            ;
book_check_out: CHECKOUT_TOKEN BOOK_TOKEN NUMBER_TOKEN
        { gTotalBooksCheckedOut++; }            ;
```

The language consists of production rules which refer to tokens and other rules. To implement the lexical scanner which will translate the keywords checkin, checkout to a valid representation we use GNU flex with the following input file.

```
/*
 Simple library checkin, checkout controller.
 (C) Sandeep Koranne, 2010
*/
%{
#include <stdio.h>
#include "y.tab.h"
%}

%%
[0-9]+  return NUMBER_TOKEN;
book    return BOOK_TOKEN;
checkin return CHECKIN_TOKEN;
```

```
checkout return CHECKOUT_TOKEN;
\n       ;
[ \t]+ ;
%%
```

The GNU `bison` input file is shown in Listing 13.5.

```
 1   /* \file lib.y
        \author Sandeep Koranne, (C) 2010
        \description Simple library checkin, checkout controller.

 5    Library Example:
      simple example is
      checkin book 76
      checkout book 80
      */
10   %{
     #include <stdio.h>
     #include <string.h>
       int gTotalBooksCheckedOut = 0;
       void yyerror( const char *str ) {
15       fprintf( stderr, "syntax error: %s\n", str );
       }
       int yywrap() { return 1; }
       int main() {
         yyparse();
20       printf("\n %d books still checked out.\n", gTotalBooksCheckedOut );
         return 0;
       }
     %}

25   %token NUMBER_TOKEN BOOK_TOKEN CHECKIN_TOKEN CHECKOUT_TOKEN

     %%
     commands: /* empty */
             | commands command
30           ;
     command:
             book_check_in | book_check_out
             ;
     book_check_in: CHECKIN_TOKEN BOOK_TOKEN NUMBER_TOKEN
35           { gTotalBooksCheckedOut--; }        ;
     book_check_out: CHECKOUT_TOKEN BOOK_TOKEN NUMBER_TOKEN
             { gTotalBooksCheckedOut++; }        ;
```

Listing 13.5 Example of GNU `bison` Yacc file

We generate the lexical scanner and parser using the above tools as:

```
$flex lib.l
$yacc -d lib.y
$gcc lex.yy.c y.tab.c -o library_system -lfl
$cat lib.txt | ./library_system
2 books still checked out.
```

At this time it is instructive to compare the EBNF grammar with the Yacc description we input to GNU `bison` (as shown in Listing 13.5). Mid-rule actions can also be specified in the rules sections; these can be used to call user defined function when part of a rule has been successfully matched.

13.4.1 Boost SPIRIT Framework

The Spirit Parser Framework is an object oriented recursive descent parser generator framework implemented using template meta-programming techniques. Expression templates allow users to approximate the syntax of Extended Backus Naur Form (EBNF) completely in C++. Parser objects are composed through operator overloading and the result is a backtracking LL(∞) parser that is capable of parsing rather ambiguous grammars. Spirit can be used for both lexing and parsing, together or separately. Consider the following example of parsing a file in the '.eqn' format:

```
a=b#c;
o=(c&d)$(e#!f);
```

The above '.eqn' file describes two equations, with output a as the boolean OR of b and c, and the output o as the Boolean XOR of two sub-expressions; c AND d and e OR NOT f. Such a format is common in VLSI CAD tools. The SPIRIT implementation of this parser is shown below:

```
struct EqnParser : public grammar<EqnParser> {
  template <typename ScannerT>
  struct definition  {
    definition(EqnParser const& /*self*/)  {
      expression
        = str_p("INPUT()") [&do_input]
        | str_p("OUTPUT()")[&do_output]
        | term
        >> *(   ('#' >> term)    [&do_or]
             |  ('$' >> term)[&do_xor]
             |  ('-' >> term)[&do_subt]
             )
        ;

      term
        =   factor
        >> *(   ('&' >> factor)[&do_and]
             |   ('/' >> factor)[&do_div]
             )
        ;

      factor
        =   lexeme_d[(+digit_p)[&do_int]]
        |   lexeme_d[(+(alnum_p|'_'|['|']'))[&do_literal]]
        |   '(' >> expression[&do_expr] >> ')'
        |   ('!' >> factor)[&do_neg]
        |   ('#' >> factor)
        ;

      assign
        =   lexeme_d[(+(alnum_p|'_'|['|']'))[&do_lhs]]
        >> '=' >> expression[&do_final] >> ';'
        ;
  }
```

```
    rule<ScannerT> expression, term, factor, assign;
    rule<ScannerT> const&
    start() const { return assign; }
  };
};
```

The parser is modeled after a classic recursive descent parser for parsing mathematical expressions of the form:

```
start <- assignment;
assignment <- lhs '=' rhs;
lhs <- literal;
rhs <- expression;
expression <- '(' expression ')' | term op expression;
op <- XOR | NEG;
term <- term AND factor;
factor <- literal | literal OR factor;
```

Using Boost SPIRIT framework, the formal grammar is translated to C++ code using the following conventions: (i) production choices are delineated using the |
operator, (ii) Kleene star closures are defined using the '*' operator, (iii) the stream redirection operator `>>` is overloaded to define production rules, and (iv) the array index operator '[]' is overloaded to define the function to be called when the production is matched. In the above example, we see that when the operator '#' is matched for example, the do_or function is called. The function to be called is defined to accept the *left* and *right* operand, i.e.:

```
  void do_xor(char const* l, char const* r) {
    // take the 2 top elements from the stack
    // and make their tree and add to sub tree
    Tree* pT = new Tree( XOR, GetName() );
5   pT->left = treeStack.top();   treeStack.pop();
    pT->right = treeStack.top();  treeStack.pop();
    treeStack.push(pT);
  }
```

Consider a '.eqn' file comprising of a single equation:

```
o1 = !a # !b;
```

When we run our equation parser program with this input we get:

```
$./eqn < b.eqn
LHS = (o1)
PUSH LIT(a)
NEG
 PUSH LIT(b)
NEG
 ADD
Parsing succeeded

InSet  has 2 inputs
```

```
OutSet has 1 outputs
INORDER = a b  ;
OUTORDER = o1  ;

o1 = T2;
T2 = T1 + T0;
T1 = !b;
T0 = !a;
```

The parser automatically generates the NEG function calls to compute the complement of a to get a', similarly for b. These are then ORed together to produce $T2$ which is assigned to the output $o1$. The complete equation parser program includes logic to figure out the input set (any literal which is not assigned), and the output set (any literal which is assigned, but is itself not an operand in any expression). The final set of outputs is written out to standard out. A more complex example is shown below:

```
a=INPUT();
b=INPUT();
c=INPUT();
d=INPUT();
o1=OUTPUT();
o2=OUTPUT();
o1 = (!a # !b) & (!c $ (a&d) );
o2 = (!e $ f) $ ( a # !b);
o3 = (f $ e) & (a # b);
```

The generated equations are:

```
INORDER = a b c d e f  ;
OUTORDER = o1 o2 o3  ;

o1 = T6;
T6 = T5 * T2;
T5 = T4 * !T3 + ! T4 * T3 ;
T4 = d * a;
T3 = !c;
T2 = T1 + T0;
T1 = !b;
T0 = !a;

o2 = T11;
T11 = T10 * !T8 + ! T10 * T8 ;
T10 = T9 + a;
T9 = !b;
T8 = f * !T7 + ! f * T7 ;
T7 = !e;

o3 = T14;
T14 = T13 * T12;
T13 = b + a;
```

```
T12 = e * !f + ! e * f ;
```

13.5 Code Generation

In this chapter we discuss the code representation used in the back-end of the compiler. System utilities related to compilers, such as assemblers, linkers, archive managers, ELF inspection utilities are also discussed.

13.5.1 GNU Binutils

The GNU Binutils are a collection of binary tools. The main ones are:

1. `ld` - the GNU linker.
2. `as` - the GNU assembler.

All the tools within the GNU binutils toolset are listed below:

1. `addr2line`: converts addresses into filenames and line numbers,
2. `ar`: a utility for creating, modifying and extracting from archives,
3. `c++filt`: filter to demangle encoded C++ symbols,
4. `gold`: a new, faster, ELF only linker,
5. `gprof`: displays profiling information,
6. `nm`: lists symbols from object files,
7. `objcopy`: copys and translates object files,
8. `objdump`: displays information from object files,
9. `ranlib`: generates an index to the contents of an archive,
10. `readelf`: displays information from any ELF format object file,
11. `size`: lists the subsection sizes of an object or archive file,
12. `strings`: lists printable strings from files,
13. `strip`: discards symbols,
14. `BFD`: binary file descriptor,
15. `opcodes`: opcode generalization and indirection library.

We discuss each of them in the following sections.

13.5.1.1 GNU Binutils : addr2line and libunwind

We first show the usage of `addr2line`.

```
Usage: addr2line [option(s)] [addr(s)]
 Convert addresses into line number/file name pairs.
  -b --target=<bfdname>  Set the binary file format
  -e --exe=<executable>  Set the input file name
```

```
-i --inlines              Unwind inlined functions
-f --functions            Show function names
-C --demangle[=style]     Demangle function names
-h --help                 Display this information
-v --version              Display the program's version
```

Consider the C language program:

```c
#include <stdio.h>
#define UNW_LOCAL_ONLY
#include <libunwind.h>

void show_backtrace (void) {
  unw_cursor_t cursor; unw_context_t uc;
  unw_word_t ip, sp;

  unw_getcontext(&uc);
  unw_init_local(&cursor, &uc);
  while (unw_step(&cursor) > 0) {
    unw_get_reg(&cursor, UNW_REG_IP, &ip);
    unw_get_reg(&cursor, UNW_REG_SP, &sp);
    printf ("ip = %lx, sp = %lx\n", (long) ip, (long) sp);
  }
}

int function_one( int x, int y ) {
  if ( y == 0 ) { show_backtrace(); return 0; }
  return (x+2)/y;
}

int function_two( int y ) {
  return function_one( y,0 ) + 2;
}

int main( ) {
  function_two( 12 );
  return 0;
}
```

Listing 13.6 Example of using `addr2line`

We compile this program with debug as follows:

```
$gcc -ggdb -c adr_example.c -o adr_example
```

The advantage of using `addr2line` is present when debugging a crash where the stacktrace contains only function addresses. Using this program these function addresses can be translated into file name and location. By using the `libunwind` library we can debug the divide-by-zero problem. Running the executable gives us the back trace:

```
ip = 80485df, sp = bfe22490
ip = 804860f, sp = bfe224a0
ip = 8048629, sp = bfe224c0
ip = 126bb6, sp = bfe224e0
ip = 8048491, sp = bfe22560
```

On a simple executable compiled with debug symbols, running `addr2line` returns:

```
$addr2line -e adr_example -f 80485df 804860f 8048629
function_one
/home/skoranne/MYBOOK/adr_example.c:19
function_two
/home/skoranne/MYBOOK/adr_example.c:24
main
/home/skoranne/MYBOOK/adr_example.c:29
```

Another excellent usage of `addr2line` is in using the new *advice* functionality in GCC 4.5.0. See Section 3.1.4 for more details and an example use.

13.5.1.2 GNU Binutils ar : archive manager

The GNU archive manager `ar` program creates, modifies, and extracts from archives. An archive is a single file holding a collection of other files in a structure that makes it possible to retrieve the original individual files (called members of the archive). The original files contents, mode (permissions), timestamp, owner, and group are preserved in the archive, and can be restored on extraction.

```
commands:
  d           delete file(s) from the archive
  m[ab]       move file(s) in the archive
  p           print file(s) found in the archive
  q[f]        quick append file(s) to the archive
  r[ab][f][u] replace or insert new file(s) into the archive
  t           display contents of archive
  x[o]        extract file(s) from the archive
  c           do not warn if new archive was created
  s           perform indexing
  v           be verbose
  V           display version information and exit.
```

13.5.1.3 GNU Binutils c++filt : Name de-mangler for C++

Anyone who has every programmed in C++ and used STL knows the reams of messages and long complex mangled names generated by the compiler to support everything from operator overloading to lookup resolution. Fortunately, the system also comes with a name demangler to perform the reverse translation using exactly the same rules (ABI permitting). Consider the following code fragment:

```
   // example code for C++ de-mangler
   #include <vector>
   #include <set>
   typedef std::vector< std::set<int> > CPLX;
5  int function(void) {
     CPLX A;
     CPLX::const_iterator it = A.begin();
     return ( it != A.end() );
   }
```

Listing 13.7 Example of using `c++filt` for demangling C++ names

Now consider a break-point associated with the symbol:

```
_ZNSaISt3setIiSt4lessIiESaIiEEEC2Ev
```

Using the `c++filt` command we can easily translate this to:

```
$c++filt _ZNSaISt3setIiSt4lessIiESaIiEEEC2Ev
std::allocator<std::set<int, std::less<int>,
   std::allocator<int> > >::allocator()
```

This informs us that the symbol is actually the memory allocator for the STL set.

13.5.1.4 GNU Binutils gprof

This utility is covered in Section 3.12.1.

13.5.1.5 GNU Binutils nm

The `nm` utility prints symbols present in object and executable files.

```
Usage: nm [option(s)] [file(s)]
 List symbols in [file(s)] (a.out by default).
 The options are:
  -a, --debug-syms
  -A, --print-file-name
  -C, --demangle[=STYLE]
      --no-demangle
  -g, --extern-only
  -l, --line-numbers

  -n, --numeric-sort
  -S, --print-size
  -V, --version
```

The utility prints the function names of referenced functions as well. Consider the `nm` utility run on an executable:

```
$nm adr_example
...
08048500 t frame_dummy
080485ce T function_one
080485f6 T function_two
         U getcontext@@GLIBC_2.1
08048614 T main
         U printf@@GLIBC_2.0
```

The nm command can be used to create a map file (a map file lists the address of functions in the binary and the corresponding function name). A map file can be passed as an input to the linker program (see Section 13.5.3). which will respect the addressing information. It can also be reordered, based on profile data (see Section 3.12.1). To create a map file using nm:

```
nm --extern-only --defined-only -v --print-file-name <object-file>
```

The produced map file has the following syntax:

```
cano_proc:08049690 T _init
cano_proc:08049cc0 T _start
cano_proc:08049cf0 T __gmon_start__
cano_proc:0804a1cb T _Z1sRSoRK22PolyRep
cano_proc:0804a97c T _ZN22PolyRep13RKSt6vectorIiSaIiEERSs
cano_proc:0804b0e4 T _ZN22PolyRep11ComputeRankEv
cano_proc:0804b12a T _ZN22PolyRep19CalculatePropertiesEv
cano_proc:0804dccc T _ZN13DP14parse_verticesEj
cano_proc:0804eb1c T _ZN13DP22parse_vertices_reducedEj
cano_proc:0804f406 T _ZN13DP12new_polytopeEv
cano_proc:0804f424 T _ZN13DP3RunEv
cano_proc:0804fe64 T main
```

13.5.1.6 GNU Binutils objcopy

The GNU objcopy utility copies the contents of an object file to another. GNU objcopy uses the GNU BFD Library to read and write the object files. It can write the destination object file in a format different from that of the source object file. The exact behavior of GNU objcopy is controlled by command-line options. Note that GNU objcopy should be able to copy a fully linked file between any two formats. However, copying a relocatable object file between any two formats may not work as expected.

```
Copies a binary file, possibly transforming it in the process
The options are:
 -I --input-target <bfdname>
 -O --output-target <bfdname>
 -p --preserve-dates
 -j --only-subsection <name>
    --add-gnu-debuglink=<file>
 -R --remove-subsection <name>
 -S --strip-all
```

A more complete set of command line options can be found by running man objcopy.

13.5.1.7 GNU Binutils objdump

GNU objdump displays information about one or more object files. The options control what particular information to display. This information is mostly useful to system programmers who are working on the compilation tools, as opposed to programmers who just want their program to compile and work.

```
Usage: objdump <option(s)> <file(s)>
 Display information from object <file(s)>.
 At least one of the following switches must be given:
  -a, --archive-headers
  -f, --file-headers
  -p, --private-headers
  -h, --[subsection-]headers
  -x, --all-headers
  -d, --disassemble
  -D, --disassemble-all
  -S, --source
  -s, --full-contents
  -g, --debugging
  -e, --debugging-tags
  -G, --stabs
  -W[lLiaprmfFsoR] or
  -t, --syms
  -T, --dynamic-syms
  -r, --reloc
  -R, --dynamic-reloc
  @<file>
  -v, --version
  -i, --info
  -H, --help
```

Running GNU objdump on an existing archive gives the (i) permission, (ii) file size, (iii) file modification time stamp, and (iv) name of object file for all objects in the archive. Example:

```
In archive /home/skoranne/OSS/lib/libelf.a:

begin.o:     file format elf32-i386
rw-rw-r-- 500/500   18308 Jun 12 11:15 2010 begin.o

cntl.o:      file format elf32-i386
rw-rw-r-- 500/500   10984 Jun 12 11:15 2010 cntl.o
```

13.5.1.8 GNU Binutils ranlib

GNU ranlib generates an index to the contents of an archive and stores it in the archive. The index lists each symbol defined by a member of an archive that is a relocatable object file. You may use nm -s or nm --print-armap to list this index. An archive with such an index speeds up linking to the library and allows routines in the library to call each other without regard to their placement in the archive. Usage:

```
-a,    archive header information
-f,    display file header
-p,    display format specific object header
-h,    display subsection headers
-x,    display all headers
-d,    show assembler section
-S,    show source code as well
-s,    show full contents of subsections.
```

A more complete list of command line options can be found by running man
ranlib.

13.5.1.9 GNU Binutils readelf

The readelf command displays information about one or more ELF format object
files. The options control what particular information to display. For example, if we
run readelf adr_example we get:

```
$readelf -h adr_example
ELF Header:
  Magic:    7f 45 4c 46 01 01 01 03 00 00 00 00 00 00 00 00
  Class:                             ELF32
  Data:                              2's complement, little endian
  Version:                           1 (current)
  OS/ABI:                            UNIX - Linux
  ABI Version:                       0
  Type:                              EXEC (Executable file)
  Machine:                           Intel 80386
  Version:                           0x1
  Entry point address:               0x8048470
  Start of program headers:          52 (bytes into file)
  Start of subsection headers:          5640 (bytes into file)
  Flags:                             0x0
  Size of this header:               52 (bytes)
  Size of program headers:           32 (bytes)
  Number of program headers:         8
  Size of subsection headers:           40 (bytes)
  Number of subsection headers:         38
  Subsection header string table index: 35
```

The command line options for readelf are:

```
Usage: readelf <option(s)> elf-file(s)
 Display information about the contents of ELF format files
 Options are:
  -a --all               Equivalent to: -h -l -S -s -r -d -V -A -I
  -h --file-header       Display the ELF file header
  -l --program-headers   Display the program headers
     --segments          An alias for --program-headers
  -S --subsection-headers   Display the subsections' header
     --subsections          An alias for --subsection-headers
  -e --headers           Equivalent to: -h -l -S
```

A more complete list of command line options can be found by running `man readelf`.

13.5.1.10 GNU Binutils size

The GNU `size` utility lists the subsection sizes—and the total size—for each of the object or archive files objfile in its argument list. By default, one line of output is generated for each object file or each module in an archive.

```
Usage: size [option(s)] [file(s)]
Displays sizes of subsections
  -t --totals            Display the total sizes
     --common            Display total size
     --target=<bfdname> Set binary file format
```

For example if we run `size adr_example` we get:

```
$size -d adr_example
   text      data      bss      dec      hex filename
   1616       276        8     1900      76c adr_example
```

13.5.1.11 GNU Binutils strings

For each file given, GNU strings prints the printable character sequences that are at least 4 characters long (or the number given with the options below) and are followed by an unprintable character. By default, it only prints the strings from the initialized and loaded subsections of object files; for other types of files, it prints the strings from the whole file.

```
The options are:
  -a - --all    Scan the entire file
  -f            Print name of file as well
  -n --bytes=[n] change minimum len to 4
  -<number> least [number] characters (default 4).
```

A more complete list of all command line options can be generated by running `man strings`. Consider the following program fragment:

```
const char RCS_ID[] = "IDversion1.234";
int function( int x ) { return x+1; }
```

Listing 13.8 Example of using the `strings` program

This program when compiled into object file still contains the string containing the version string. Using the `strings` command we can find the version string embedded inside it.

```
$strings rcs.o
$ID version 1.234$
```

13.5.1.12 GNU Binutils strip

GNU `strip` discards all symbols from object files objfile. The list of object files may include archives. At least one object file must be given. GNU `strip` modifies the files named in its argument, rather than writing modified copies under different names. It is useful in reducing program size before production, which makes the program load faster. Usage:

```
Usage: strip <option(s)> in-file(s)
 Removes symbols and subsections from files
 The options are:
  -I --input-target=<bfdname>
  -O --output-target=<bfdname>
  -p --preserve-dates
  -R --remove-subsection=<name>
  -s --strip-all
  -K --keep-symbol=<name>
  -o <file>
```

13.5.2 GNU Binutils libelf and elfutils

GNU libelf lets you read, modify or create ELF files in an architecture-independent way. The library takes care of size and endian issues, e.g. you can process a file for SPARC processors on an Intel-based system.

```
    /*
     * Example of libelf
     * (C) Sandeep Koranne, 2010
     */
5   #include <fcntl.h>
    #include <stdio.h>
    #include <stdlib.h>
    #include "libelf.h"

10  int main( int argc, char* argv[] ) {
      Elf *elf_ptr;
      Elf_Kind elf_k;
      int fd;
      char *c;
15    if( argc != 2 ) {
        fprintf( stderr, "usage: poke_elf <filename>" );
        exit(1);
      }
      if( elf_version( EV_CURRENT ) == EV_NONE ) {
20      fprintf( stderr, "ELF library init failed");
        exit(1);
      }

      fd = open( argv[1], O_RDONLY, 0 );
25    if( fd < 0 ) {
        fprintf( stderr, "Unable to open file %s", argv[1] );
        exit(1);
      }

30    elf_ptr = elf_begin( fd, ELF_C_READ, NULL );
      if( elf_ptr == NULL ) {
```

```
            fprintf( stderr, "elf_begin() failed : %s", elf_errmsg(-1) );
            exit(1);
        }
35
        elf_k = elf_kind( elf_ptr );

        switch( elf_k ) {
        case ELF_K_ELF: fprintf( stdout, "ELF object" ); break;
40      case ELF_K_AR : fprintf( stdout, "ar(1) archive" ); break;
        case ELF_K_NONE:fprintf( stdout, "data" ); break;
        default: fprintf( stdout, "_NONE_" ); break;
        }
        elf_end( elf_ptr );
45      close( fd );
        exit( 0 );
    }
```

Listing 13.9 Example of using `libelf` library

We can compile this file as:

```
$gcc elf_poke.c -o elf_poke -lelf
```

And now we have an ELF detection tool which can distinguish between ELF objects and 'ar' archives. We run the executable on itself:

```
$./elf_poke elf_poke
ELF object[
$./elf_poke libelf.a
ar(1) archive
```

As shown in Figure 13.2 ever ELF has the following data:

1. ELF executable header: describes the *class* (whether the file is 32-bit or 64-bit), *type* (whether the file is relocatable, shared or executable), and the *endianess* of the file,
2. Optional ELF program header table (PHDR)
3. Subsection Data and the subsection header table are used for relocatable data.

ELF Header   PHDR   Section data   ELF Section Header Table

Fig. 13.2 Layout of an ELF file

The `libelf` library contains many examples which show how to analyze the subsections and headers present inside the ELF file. Using this library interesting and useful programs can be built for program maintenance and analysis.

13.5.3 GNU Binutils ld

A compiler translates program code (given as human readable text) into into a machine specific but still readable text called assembler code. Assembly code is a readable form of the machine code (CPU instructions) which the computer can execute directly. A linker converts object files into executables and shared libraries. GNU linker (or GNU ld) is the GNU Project's implementation of the Unix command ld. GNU `ld` runs the linker, which creates an executable file (or a library) from object files. A linker script may be passed to GNU `ld` to exercise greater control over the linking process. GNU `ld` combines a number of object and archive files, relocates their data and ties up symbol references. Usually the last step in compiling a program is to run `ld`. If the linker is being invoked indirectly, via a driver such as `gcc` then all the linker command line options should be prefixed by -Wl, as shown below:

```
$gcc -Wl,--start-group foo.o bar.o -Wl,--end-group
```

GNU `ld` command supports a number of options which control its behavior. The common options are listed below:

```
'-E'
'--export-dynamic'
'--no-export-dynamic'

'-L SEARCHDIR'
'--library-path=SEARCHDIR'
    Add SEARCHDIR to list of paths that 'ld' will search for

'-M'
'--print-map'

'-o OUTPUT'
'--output=OUTPUT'

'-r'
'--relocatable'
    Generate relocatable output

'-Bdynamic'
'-dy'
'-call_shared'
    Link against dynamic libraries.

'-Bstatic'
'-dn'
'-non_shared'
'-static'
    Do not link against shared libraries.

'--cref'
    Output a cross reference table.

'--demangle[=STYLE]'
```

'--no-demangle'

For a complete list of command line options for ld, run man ld, or info ld.

13.5.3.1 Linker script and map files

When passing arguments to the linker, the GNU gcc commandline -Wl as follows:

```
$gcc file.o -o file -Wl, -M
```

will generate the linker map for the file. The format of the linker map is discussed below, an example is:

```
.group          0x0000000000000000        0x8 cano_proc.o
_ZN9__gnu_cxx13new_allocatorISt6vectorIiSaIiEEEC5ERKS4_
                0x0000000000000000        0x8 cano_proc.o
.group          0x0000000000000000        0x8 cano_proc.o
.group          0x0000000000000000        0x8 cano_proc.o
.group          0x0000000000000000        0x8 cano_proc.o
.text._ZNSt12__niter_baseIPSt6vectorI7CBTupleSaIS1_EELb0EE3__bES4_
                0x0000000000000000        0xd main.o
.text._ZNSt3setIiSt4lessIiESaIiEE6insertERKi
                0x0000000008050bb4        0x5f cano_proc.o
                0x0000000008050bb4        std::set<int, std::less<int>
 *fill*         0x0000000008050c13        0x1 90909090
...
/DISCARD/
 *(.note.GNU-stack)
 *(.gnu_debuglink)
OUTPUT(cano_proc elf32-i386)
```

The .group and .text sections contain the instruction to the loader to place code and data segments into appropriate sections of the memory. This arrangement of code blocks from disk to memory can be managed and optimized (for cache line performance, see Section 3.12.1.1 for more details), using a *linker script*. The linker will ensure that each output subsection has the required alignment, by increasing the location counter if necessary. In this example, the specified addresses for the '.text' and '.data' subsections will probably satisfy any alignment constraints, but the linker may have to create a small gap between the '.data' and '.bss' subsections. As with all optimizations, linker scripts should be used when the correctness of the program has been verified, and especially when using dynamic loaded shared libraries, linker map scripts should be carefully checked.

13.5.3.2 Linker map cross reference table

The cross reference table is generated using the --cref commandline switch to the linker. It has the form of 'function', 'object file', e.g.:

```
BFS::Run()                      poly_utils.o
BFS::end_visit(unsigned int)    poly_utils.o
BFS::expand(unsigned int)       poly_utils.o
BFS::visit(unsigned int)        poly_utils.o
```

When passing arguments to the linker, the GNU gcc commandline -Wl as follows:

```
$gcc file.o -o file -Wl, --cref
```

will generate the cross reference table for the file.

13.5.4 BFD: Binary File Descriptor Library

The BFD, or Binary File Descriptor library, is the GNU Project's main mechanism for the portable manipulation of object files in a variety of formats. BFD works by presenting a common abstract view of object files. An object file has a "header" with descriptive info; a variable number of "subsections" that each have a name, some attributes, and a block of data; a symbol table; relocation entries; and so forth. To use the library, include 'bfd.h' and link with 'libbfd.a'. BFD provides a common interface to the parts of an object file for a calling application. When an application successfully opens a target file (object or archive), a pointer to an internal structure is returned. This pointer points to a structure called 'bfd', described in 'bfd.h'. Our convention is to call this pointer a BFD, and instances of it within code abfd. All operations on the target object file are applied as methods to the BFD.

```
#include "bfd.h"
unsigned int number_of_subsections (abfd)
bfd *abfd;
{
5   return bfd_count_subsections (abfd);
}
```

The abstraction used within BFD is that an object file has:

1. a header,
2. a number of subsections containing raw data,
3. a set of relocations,
4. some symbol information.

Also, BFDs opened for archives have the additional attribute of an index and contain subordinate BFDs. The design and use of BFD can be organized as follows:

- BFD Front end :

  1. Initialization,
  2. Subsections,
  3. Symbols,
  4. Archives,
  5. File formats,

6. Relocations,
7. Core files,
8. Targets,
9. Architectures,
10. Opening and closing BFDs,
11. File caching,
12. Linker functions,
13. Hash tables.

- BFD Back end:

  1. a.out backends,
  2. coff backends,
  3. elf backends.

13.5.5 GNU lightning

Dynamic code generation is the generation of machine code at runtime. It is typically used to strip a layer of interpretation by allowing compilation to occur at runtime. One of the most well-known applications of dynamic code generation is perhaps that of interpreters that compile source code to an intermediate bytecode form, which is then recompiled to machine code at run-time. For performance, gnu lightning emits machine code without first creating intermediate data structures such as RTL representations traditionally used by optimizing compilers.

13.5.5.1 GNU lightning instruction set

GNU lightning's instruction set was designed by deriving instructions that closely match those of most existing RISC architectures. The library supports a full range of integer types: operands can be 1, 2 or 4 bytes long (64-bit architectures might support 8 bytes long operands), either signed or unsigned. The types are listed in the following table together with the C types they represent: There are at least seven integer registers, of which six are general-purpose, while the last is used to contain the frame pointer (FP). The frame pointer can be used to allocate and access local variables on the stack, using the allocai instruction. A partial list of instruction is given below in Table 13.2.

13.5.5.2 Instructions in GNU lightning

The instruction set of GNU lightning is modeled after a generic RISC machine. The list of instructions categorized by their function is shown in Table 13.2.
  The following example from GNU lightning is shown:

Table 13.1 Data types supported by GNU lightning

Symbol	C Data Type
c	signed char
uc	unsigned char
s	short
us	unsigned short
i	int
ui	unsigned int
l	long
ul	unsigned long
f	float
d	double
p	void *

```
#include <stdio.h>
#include "lightning.h"
static jit_insn codeBuffer[1024];
typedef int (*pifi)(int);
int main() {
    pifi  incr = (pifi) (jit_set_ip(codeBuffer).iptr);
    int   in;
    jit_leaf(1);                      /*      leaf  1            */
    in = jit_arg_i();                 /* in = arg_i              */
    jit_getarg_i(JIT_R0, in);         /*      getarg_i R0        */
    jit_addi_i(JIT_RET, JIT_R0, 1);   /*      addi_i  RET, R0, 1 */
    jit_ret();                        /*      ret                */

    jit_flush_code(codeBuffer, jit_get_ip().ptr);

    /* call the generated code, passing 5 as an argument */
    printf("%d + 1 = %d\n", 5, incr(5));
    return 0;
}
```

Listing 13.10 Example of using GNU lightning

GNU lightning generates the following code on SPARC:

```
save %sp, -96, %sp
mov  %i0, %l0      retl
add  %l0, 1,  %i0  add %o0, 1, %o0
ret
restore
```

Table 13.2 Instruction in GNU lightning

ALU Instructions	
`add[r,i]`	`O1 = O2 + O3`
`sub[r,i]`	`O1 = O2 - O3`
`mul[r,i]`	`O1 = O2 * O3`
`div[r,i]`	`O1 = O2 / O3`
`mod[r,i]`	`O1 = O2 % O3`
and`[r,i]`	`O1 = O2 & O3`
or`[r,i]`	`O1 = O2	O3`
xor`[r,i]`	`O1 = O2 ^ O3`
Compare instructions	
`lt[r,i]`	`O2 < O3`
`le[r,i]`	`O2 <= O3`
`gt[r,i]`	`O2 > O3`
`ge[r,i]`	`O2 >= O3`
`eq[r,i]`	`O2 == O3`
`ne[r,i]`	`O2 != O3`
Load and Store Instructions	
`ld[r,i]`	`O1 = *O2`
`ldx[r,i]`	`O1 = *(O2+O3)`
`st[r,i]`	`*O1 = O2`
`stx[r,i]`	`*(O1+O2) = O3`
Branch Instructions	
`blt[r,i]`	if $(O2 < O3)$ goto O1
`ble[r,i]`	if $(O2 \leq O3)$ goto O1
`bgt[r,i]`	if $(O2 > O3)$ goto O1
Jump and Return Instructions	
`call[r,i]`	function call to O1
`finish`	function call to O1
`finishr`	function call to a register
`jmp[r,i]`	unconditional jump to O1
`ret`	return from subroutine
`retval`	move return value

13.5.6 ANTLR

Before we move on to a compiler optimization and representation framework (LLVM) it is only apt to mention ANTLR, which is a complete suite of compiler construction tools. An example of the ANTRL screen and DFA definition is shown in Figure 13.3.

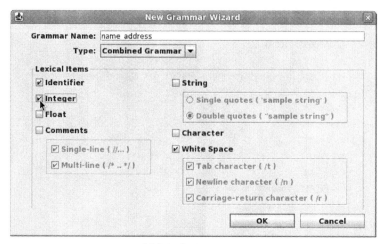

(a) Introductory screen

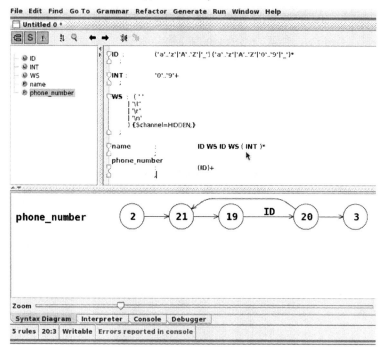

(b) DFA of phone number grammar in ANTLR

Fig. 13.3 ANTLRWorks introductory screen

13.6 LLVM: Low Level Virtual Machine

LLVM (Low Level Virtual Machine), is a compiler framework designed to support transparent, lifelong program analysis and transformation for arbitrary programs, by providing high-level information to compiler transformations at compile-time, link-time, run-time, and in idle time between runs. LLVM defines a common, low-level code representation in Static Single Assignment (SSA) form, with several novel features: a simple, language-independent type-system that exposes the primitives commonly used to implement high-level language features; an instruction for typed address arithmetic; and a simple mechanism that can be used to implement the exception handling features of high-level languages (and setjmp/longjmp in C) uniformly and efficiently. The main projects under the LLVM are (from http://www.llvm.org):

- LLVM Core: the LLVM core provides source and target independent optimization based on SSA, alongwith code generation for modern architectures. The LLVM intermediate representation is well specified and documented,
- Clang: is an LLVM native, C/C++/Objective-C compiler,
- llvm-gcc and dragonegg: integration of LLVM with GCC frontend,
- LLDB: using LLVM and Clang to produce native debugger.

LLVM core and LLVM IR (intermediate representation) are useful in compiler construction, as they provide a generic optimization and instruction scheduling, generation system. Using the front-end tools (Flex and Bison) alongwith LLVM a prototypical compiler can be constructed rapidly, and with ease.

13.6.1 LLVM Core and LLVM IR

Above we saw the GNU lightning (Subsection 13.5.5) and its associated RISC instruction model. In this section we present the LLVM instructions and their intermediate representations.

13.6.1.1 LLVM Intermediate Representation

The LLVM IR can be thought of as the assembly language, or mnemonic instruction set of the low-level virtual machine (LLVM). LLVM is a Single Static Assignment (SSA) representation that provides type safety, flexibility and sufficient breadth to implement all the high level language constructs. LLVM code representation was designed to be used in the following scenarios:

1. in-memory compiler IR,
2. on-disk instruction bit-code, and
3. human readable assembly language.

The use of LLVM instructions are shown below:

```
%result = mul i32 %X, 8
```

Without even going into specifics, it is clear that the above multiplies the content of variable %X with 8, and places the result in %result. LLVM's optimization can do *strength reduction* to convert the above multiply by 8 to a shift left by 3 operation to get:

```
%result = shl i32 %X, i8 3
```

Before we present more examples of LLVM textual IR, and the instructions, we should note that:

- LLVM comments: are delimited by ';' character
- unnamed temporaries: are created when lval is not present, and
- unnamed temporaries: are numbered sequentially.

13.6.1.2 LLVM Program High Level Structure

LLVM programs are composed of *Modules*, each of which corresponds to a translation unit (mostly a single file) of the input program. Each module consists of:

1. functions,
2. global variables,
3. symbol table entries.

Alongwith the module composition, the Linkage types, and calling conventions for the function calls are important considerations when designing an LLVM based IR. LLVM also allows for type aliasing, for example, we can define:

```
%point_type = type { %point_type*, i32 }
```

Functions defined using LLVM have (i) linkage types, (ii) visibility types, (iii) calling convention, (iv) return type, (v) parameter attribute for return type, (vi) function name, (vii) argument list (possibly empty), (viii) optional subsection, (ix) optional alignment, (x) optional garbage collector name, (xi) opening curly brace, (xii) list of basic blocks, and (xiii) closing curly braces. For example:

```
extern int fibo(int n);
=>
define i32 @fibo(i32 %n) nounwind readnone {

static int fibo(int n);
=>
define internal fastcc i32
          @fibo(i32 %n) nounwind readnone {
```

The function attributes and linkage vary depending upon the data present in the translation unit. The actual code of the function:

```
static int fibo( int n ) {
  int i,f0=0,f1=1,f;
  for(i=0; i < n; ++i) {
    f = f0 + f1;
    f0 = f1;
    f1 = f;
  }
  return f;
}
```

is assembled by LLVM into:

```
define internal fastcc i32 @fibo(i32 %n)
   nounwind readnone {
entry:
  %0 = icmp sgt i32 %n, 0                  ; <i1> [#uses=1]
  br i1 %0, label %"4", label %"5"

"4":                                       ; preds = %"4", %entry
  %1 = phi i32 [ %4, %"4" ], [ 1, %entry   ; <i32> [#uses=2]
  %2 = phi i32 [ %1, %"4" ], [ 0, %entry   ; <i32> [#uses=1]
  %3 = phi i32 [ %5, %"4" ], [ 0, %entry   ; <i32> [#uses=1]
  %4 = add nsw i32 %1, %2                   ; <i32> [#uses=2]
  %5 = add nsw i32 %3, 1                    ; <i32> [#uses=2]
  %exitcond = icmp eq i32 %5, %n            ; <i1> [#uses=1]
  br i1 %exitcond, label %"5", label %"4"

"5":                                       ; preds = %"4", %entry
  %.lcssa = phi i32 [ undef, %entry ],
   [ %4, %"4" ] ; <i32> [#uses=1]
  ret i32 %.lcssa
}
```

Function attributes are an important part of the optimization performed by LLVM. The following are some of the function attributes that can be specified per function (their use is self explanatory):

- alignstack
- alwaysinline
- noinline
- optsize
- noreturn
- nounwind
- readnone
- readonly

13.6.1.3 LLVM Instruction Summary: Terminators

The following table presents the terminator instructions in LLVM. Every basic block must end with one of these instruction: The computation instructions in LLVM are standard RISC style instructions; a short summary is presented in the following table.

Table 13.3 LLVM Terminator instructions

ret	ret <void> or ret <type> <value>
br	br <label> or br i1 <cond>, label <iftrue>, label <iffalse>
switch	switch <intty> <value>, label <defaultdest> [..]
indirectbr	indirect branch
invoke	transfer control to specified function
unwind	unwind the stack, transfer control to first callee,
unreachable	optimizer hint

13.6.1.4 LLVM Instruction Summary: Computation

Table 13.4 LLVM Instructions for computation

```
add  = add <ty> <op1> <op2>
sub  = sub <ty> <op1> <op2>
mul  = mul <ty> <op1> <op2>
div  = div <ty> <op1> <op2>
rem  = rem <ty> <op1> <op2>
shl  = shl <ty> <op1> <op2>
lshr = lshr <ty> <op1> <op2>
ashr = ashr <ty> <op1> <op2>
and  = and <ty> <op1> <op2>
or   = or <ty> <op1> <op2>
xor  = xor <ty> <op1> <op2>
```

There are some other special instructions:

- Vector *extractelement,*
- Vector *insertelement,*
- Vector *shufflevector,*
- Memory *alloca,*
- *load,*
- *store,*
- *ptrtoint,*
- *inttoptr,*
- *icmp,*
- *fcmp,*
- *phi*: implement the ϕ node in SSA graph,
- *call*: function call instruction.

LLVM also supports a number of intrinsic functions, a representative set is shown below:

- `llvm.va_start`
- `llvm.gcroot`
- `llvm.returnaddress`
- `llvm.stacksave`
- `llvm.prefetch`
- `llvm.readycyclecounter`
- `llvm.memcpy`
- `llvm.sqrt.*`
- `llvm.bswap.*`
- `llvm.ctpop.* :`
- `llvm.cttz.* :`
- `llvm.memory.barrier llvm.atomic.load.add.* and sub`

13.6.2 LLVM dragonegg

LLVM `dragonegg` is a plugin to GCC, thus it requires a fairly recent GCC version. GCC 4.5.0 works with dragonegg 2.7, and this version is used in the following examples. Consider the following C language program:

```
int function_one( int x ) {
    int i;
    int sum;
    for( i = 0; i < x; ++i ) sum += i;
5   return sum;
}
```

Compiling this program with GCC using the -S option produces the assembly code (this assembly code is for x86 architecture):

```
        .file   "sum.c"
        .text
.globl function_one
        .type   function_one, @function
function_one:
        pushl   %ebp
        movl    %esp, %ebp
        subl    $16, %esp
        movl    $0, -8(%ebp)
        jmp     .L2
.L3:
  movl    -8(%ebp), %eax
  addl    %eax, -4(%ebp)
  addl    $1, -8(%ebp)
.L2:
  movl    -8(%ebp), %eax
  cmpl    8(%ebp), %eax
  jl      .L3
  movl    -4(%ebp), %eax
  leave
  ret
```

```
        .size    function_one, .-function_one
        .ident   "GCC: (GNU) 4.4.2 20091027 (Red Hat 4.4.2-7)"
        .subsection          .note.GNU-stack,"",@progbits
```

We can use the `dragonegg` plugin to use the LLVM intermediate representation, optimization and scheduler. Running:

```
$gcc -fplugin=dragonegg.so -O1 -S sum.c
```

produces:

```
        .file    "sum.c"
# Start of file scope inline assembly
        .ident   "GCC: (GNU) 4.5.0 LLVM: exported"
# End of file scope inline assembly
        .text
        .globl   function_one
        .align   16, 0x90
        .type    function_one,@function
function_one:
        pushl    %ebp
        movl     %esp, %ebp
        pushl    %edi
        pushl    %esi
        subl     $28, %esp
        movl     8(%ebp), %eax
        movl     %eax, -12(%ebp)
        movl     -12(%ebp), %eax
        movl     %eax, -20(%ebp)
        movl     $0, %eax
        movl     %eax, -24(%ebp)
        jmp      .LBB1_3
.LBB1_2:
        movl     -28(%ebp), %eax
        movl     %eax, %ecx
        movl     -32(%ebp), %edx
        addl     %edx, %ecx
        movl     %edx, %esi
        addl     $1, %esi
        movl     %ecx, -36(%ebp)
        movl     %esi, -24(%ebp)
.LBB1_3:
        movl     -24(%ebp), %eax
        movl     %eax, %ecx
        movl     -36(%ebp), %edx
        movl     %edx, %esi
        movl     -20(%ebp), %edi
        cmpl     %edi, %ecx
        movl     %ecx, -32(%ebp)
        movl     %esi, -28(%ebp)
        jl       .LBB1_2
        movl     -28(%ebp), %eax
        movl     %eax, -16(%ebp)
        movl     -16(%ebp), %eax
        addl     $28, %esp
```

```
        popl    %esi
        popl    %edi
        popl    %ebp
        ret
        .size   function_one, .-function_one

        .subsection .note.GNU-stack,"",@progbits
```

To produce the LLVM IR we can add

`-fplugin-arg-dragonegg-emit-ir`

to produce:

```
; ModuleID = 'sum.c'
target triple = "i386-pc-linux-gnu"
module asm "\09.ident\09\22GCC: (GNU) 4.5.0 LLVM: exported\22"
%int = type i32

define i32 @function_one(i32 %x) nounwind {
entry:
  %x_addr = alloca i32                        ]
  %memtmp = alloca i32                        ]
  %"alloca point" = bitcast i32 0 to i32
  store i32 %x, i32* %x_addr
  %0 = load i32* %x_addr, align 32
  %"ssa point" = bitcast i32 0 to i32
  br label %"2"

"2":
  br label %"4"

"3":
  %1 = add nsw i32 %4, %3
  %2 = add nsw i32 %3, 1
  br label %"4"

"4":
  %3 = phi i32 [ %2, %"3" ], [ 0, %"2" ]
  %4 = phi i32 [ %1, %"3" ], [ undef, %"2" ]
  %5 = icmp slt i32 %3, %0
  br i1 %5, label %"3", label %"5"

"5":
  store i32 %4, i32* %memtmp, align 1
  br label %return

return:
  %retval = load i32* %memtmp
  ret i32 %retval
}
```

Thus, using the plugin we can quickly experiment with LLVM, while maintaining the front-end capabilities of GCC. It is indeed possible to compile some parts of an application with LLVM based tools, while compiling the remaining with GCC. In the next section we describe the LLVM compiler and optimization system.

13.6.3 LLVM System

The LLVM system for the end-user (not the developer) comprises of a number of programs and utilities which manipulate, optimize and analyze LLVM '.ll' and '.bc' files. The utilities are:

1. lli: LLVM interpreter and dynamic compiler,
2. llc: LLVM system compiler and optimizer,
3. llvm-ar: LLVM archiver, archives several bit-code files into a single archive. It can optionally compress the members to save space, and it generates a symbol table for efficient lookup during linking,
4. llvm-as: LLVM assembler (.ll to .bc),
5. llvm-bcanalyzer: LLVM bit-code analyzer,
6. llvmc: LLVM compiler driver,
7. llvm-dis: LLVM disassembler. It takes the LLVM bit-code file and converts it to LLVM assembly,
8. llvm-extract: this command takes the name of a function and extracts that function's code from the bit-code file,
9. llvm-ld: LLVM linker, takes a number of LLVM bit-code files and links them together into a single LLVM object. It can also produce native code executables,
10. llvm-nm: GNU nm equivalent for LLVM, reports names of symbols from the LLVM bit-code file,
11. llvm-prof: LLVM profiler, reads in a llvmprof.out file, a bit-code program file, and produces reports, which can be used to deduce program's hotspots,
12. llvm-ranlib: adds or updates the symbol table in an LLVM archive file,
13. llvm-stub:

We discuss some of the important utilities below.

13.6.3.1 LLVM interpreter and dynamic compiler: lli

LLVM lli can directly execute programs in LLVM bitcode ('.bc') format. It uses JIT (just in time) compilation on the bit-code for JIT supported architectures, else it uses an interpreter. LLVM lli has a number of command-line options which control its behavior, including floating-point operation, code model (to choose from (i) small, (ii) kernel, (iii) medium, and (iv) large). It can also be used as a JIT compiler, and for this it accepts the -march and -mcpu command-line options.

13.6.3.2 LLVM System Compiler

Once we have generated the bit-code from the LLVM assembler we can experiment with optimization and instruction scheduling, and other advanced compiler optimizations. One of the design goals of LLVM is that such optimization will happen concurrently with the life of the executable. In the context of mobile applications, this is not as far-fetched as originally thought.

We use llc to recompile the bit-code of our Fibonacci example:

```
$llc -O3  fibo.bc -o fibo.opt.bc -stats \
                  -tailcallopt  -time-passes
```

We generate statistics about the optimization passes as well as the number of machine instructions generated, instructions scheduled.

13.6.3.3 Statistics for fibo.bc

```
 23 asm-printer      - No. machine instrs printed
  1 branchfolding    - No. branches optimized
  1 branchfolding    - No. dead blocks removed
  1 code-placement   - No. intra loop branches moved
  1 code-placement   - No. loop header aligned
  7 dagcombine       - No. dag nodes combined
102 liveintervals    - No. original intervals
  1 loop-reduce      - No. IV uses strength reduced
  1 loop-reduce      - No. PHIs inserted
  1 loop-reduce      - No. loop terminating conds optimized
  2 loopsimplify     - No. pre-header or exit blocks inserted
  1 machine-sink     - No. machine instructions sunk
  4 phielim          - No. atomic phis lowered
  2 regalloc         - No. iterations performed
 22 regcoalescing    - No. identity moves eliminated
  1 regcoalescing    - No. instructions re-materialized
 20 regcoalescing    - No. interval joins performed
  2 regcoalescing    - No. valno def marked dead
  1 scalar-evolution - No. loops with predictable loop counts
  1 twoaddrinstr     - No. instructions commuted to coalesce
  2 twoaddrinstr     - No. two-address instructions
  1 x86-codegen      - No. floating point instructions
```

13.6.3.4 Instruction scheduling

```
  --- Name ---
  DAG Legalization
  Type Legalization
  Instruction Scheduling
  Instruction Creation
  Vector Legalization
  Instruction Selection
```

```
DAG Combining 1
DAG Combining after legalize types
DAG Combining 2
Instruction Scheduling Cleanup
TOTAL
```

13.6.3.5 Compiler optimization passes

1. X86 DAG → DAG Instruction Selection
2. Live Variable Analysis
3. Simple Register Coalescing
4. Live Interval Analysis
5. Linear Scan Register Allocator
6. Loop Strength Reduction
7. X86 AT&T-Style Assembly Printer
8. Induction Variable Users
9. Optimize for code generation
10. Machine Function Analysis
11. Module Verifier
12. Canonicalize natural loops
13. Two-Address instruction pass
14. Virtual Register Map
15. Control Flow Optimizer
16. Eliminate PHI nodes for register allocation
17. Dominator Tree Construction
18. Prolog/Epilog Insertion & Frame Finalization
19. MachineDominator Tree Construction
20. Dominance Frontier Construction
21. Natural Loop Information
22. MachineDominator Tree Construction
23. Machine Natural Loop Construction
24. Remove unreachable machine basic blocks
25. Remove unreachable blocks from the CFG
26. Machine code sinking
27. Machine Natural Loop Construction
28. Exception handling preparation
29. Machine Natural Loop Construction
30. Scalar Evolution Analysis
31. X86 FP Stackifier
32. Code Placement Optimizater
33. Label Folder
34. Machine Instruction LICM
35. Stack Slot Coloring
36. Subregister lowering instruction pass
37. X86 FP_REG_KILL inserter

38. Live Stack Slot Analysis
39. X86 Maximal Stack Alignment Calculator
40. Analyze Machine Code For Garbage Collection
41. Preliminary module verification
42. Insert stack protectors
43. Lower Garbage Collection Instructions
44. Delete Garbage Collector Information

The list of compiler optimization can be better understood by referring to an advanced compiler optimization book such as Muchnik or Kennedy.

13.6.4 Using Clang

Using LLVM Clang, we can dispense with the plugin approach for compiling programs. Again consider the Fibonacci C program, but this time we compile it with clang:

```
$clang -O1 -S -c -emit-llvm -o fibo.ll fibo.c
```

The produced LLVM IR in textual form is similar to the one produced by dragonegg plugin, and it can be assembled using the LLVM assembler, llvm-as,

```
$llvm-as fibo.ll
```

This assembles the textual representation into the bit-code we alluded to earlier in this subsection. This bit-code can be inspected with the help of llvm-bcanalyzer, we present a portion of the 'fibo.bc' below:

```
Summary of fibo.bc:
         Total size: 6432b/804.00B/201W
        Stream type: LLVM IR
  # Toplevel Blocks: 1

Per-block Summary:
  Block ID #0 (BLOCKINFO_BLOCK):
      Num Instances: 1
         Total Size: 637b/79.62B/19W
          % of file: 9.903607e+00
      Num SubBlocks: 0
        Num Abbrevs: 0
        Num Records: 0

  Block ID #8 (MODULE_BLOCK):
      Num Instances: 1
         Total Size: 2270b/283.75B/70W
          % of file: 3.529229e+01
      Num SubBlocks: 7
        Num Abbrevs: 1
        Num Records: 6
        % Abbrev Recs: 1.666667e+01
```

```
        Record Histogram:
                Count    # Bits    % Abv   Record Kind
                    3       207             FUNCTION
                    1        17   100.00    GLOBALVAR
                    1      1557             DATALAYOUT
                    1       303             TRIPLE

    Block ID #9 (PARAMATTR_BLOCK):
        Num Instances: 1
```

Once the bit-code has been generated, it can be analyzed and also optimized. For this we use the LLVM System Compiler which implements the compiler optimizations (see 13.6.3.5). The LLVM system compiler is run using `llc` command, and it has a number of useful command line options:

```
-O=<char>
-asm-verbose
-load=<pluginfilename>
-march -mcpu=<cpu-name>
-o=<filename>
-realign-stack
-stack-protector-buffer-size=<uint>
-stats
-tailcallopt
-time-passes
-unwind-tables
-verify-dom-info
-version
-x86-asm-syntax
  =att
  =intel
```

13.7 Conclusion

Domain specific languages have re-invigorated the need for writing lexical analyzers, parsers, optimizers and instruction analysis engines. From reading configuration files, to writing domanin specific optimized languages for applications, compiler writing remains an important and integral part of scientific computing and engineering. In this chapter we have described several open-source tools for compiler writers. These include lexical analysis generators, macro processing, perfect hash generators, parser generator, and several compiler frameworks. The most important compiler framework discussed is LLVM (low level virtual machine) which provides a complete infrastructure for not only domain specific language development, but also optimization algorithms for existing languages such as C, C++ and Java.

Part IV
Engineering and Mathematical Software

Chapter 14
Scientific Software

Abstract In this chapter we present engineering libraries such as Computer Vision, CImg and FWTools. Geospatial data abstractions are becoming very important with the rise of location aware computing, and several open-source tools such as GDAL and PROJ4 are described in this chapter. Image processing, audio processing, and computational fluid dynamics (CFD) have been part of many engineering applications. More recently, molecular dynamics and simulation programs have also become heavy contenders for the compute time on grids. Molecular dynamics programs (GROMACS, NAMD) as well molecule viewers (JMol) are described in this chapter. Geographical Information Systems (GIS) including GRASS and QGIS are described in Section 14.9.1. Mechanical engineering, as well as use of mechanical CAD software in other disciplines can be accomplished using open-source tools such as QCAD. Solid modeling tools BRL-CAD are described in Section 14.11 as well as Blender 14.12.

Contents

In previous chapters we have discussed the underlying operating system, user shell interaction, various application libraries, compilers and more. In this chapter we focus on integrated open-source software solutions for engineering problems. The domains of engineering include (i) image processing, (ii) audio processing,

(iii) finite-element analysis, (iv) computational biology, (v) geographical information systems, and (vi) VLSI and electronics. This list is not exhaustive, but is representative of the many domains in which open-source software has been used with great positive effect. We first discuss engineering libraries which are used in many open-source software for engineering.

14.1 Computer Vision with OpenCV

OpenCV (Open Source Computer Vision) is a library for real time computer vision. OpenCV has the following modules in its source-code:

1. cxcore: Core functionality including basic structures, array operations, drawing functions, XML, clustering and utility system function. An example of using OpenCV core library is shown in Listing 14.1.

```
// \file cv_example.cpp
// \author Sandeep Koranne, (C) 2010
// \description Example of using cxCore
#include <iostream>          // Program IO
5  #include <cassert>           // assertions
#include <cv.h>              // OpenCV
#include <highgui.h>         // OpenCV GUI

static CvMemStorage *storage;
10 static CvRect gRectA;
static void *data = NULL;
static void InitializeData( ) {
  gRectA.x = 0, gRectA.y = 0;
  gRectA.width = 100, gRectA.height = 100;
15   storage = cvCreateMemStorage( 0 );
  data = cvMemStorageAlloc( storage, 1024 );
}

static void PrintImageInformation(const IplImage* image ) {
20   std::cout << "Image size   = " << image->nSize
    << "Num channels = " << image->nChannels
    << "Depth        = " << image->depth
    << std::endl;
}
25

int main( int argc, char* argv[] ) {
  if( argc != 3 ) {
    std::cerr << "Usage: ./ipl_example <file> <file>...\n";
30     exit(1);
  }
  IplImage *image = cvLoadImage( argv[1] );
  assert( image && "Unable to load image ");
  PrintImageInformation( image );
35   InitializeData();
  CvSize dsize; dsize.width = dsize.height = 100;
  //IplImage *dest = cvCreateImageHeader( dsize, 8, 3 );
  IplImage *dest = cvCreateImage( dsize, 8, 3 );
  CvSeq *contour = NULL;
40   cvThreshold( image, image, 1, 255, CV_THRESH_BINARY );
  cvFindContours( image, storage, &contour,
      sizeof( CvContour ), CV_RETR_CCOMP,
      CV_CHAIN_APPROX_SIMPLE );
  CvScalar color = CV_RGB( 143, 100, 200 );
```

```
45 |    for(; contour; contour = contour->h_next ) {
   |      cvDrawContours( dest, contour, color, color,
   |          -1, CV_FILLED, 8 );
   |    }
   |    cvSaveImage( argv[2], dest );
50 |    cvReleaseImage( &image );
   |    cvReleaseImage( &dest );
   |    cvReleaseMemStorage( &storage );
   |    std::cout << std::endl;
   |    return (0);
55 | }
```

Listing 14.1 OpenCV core functionality

2. cv : Image processing and computer vision: image filtering, geometric image transforms, histograms, feature detection, motion analysis and object tracking, planar subdivisions, object detection and camera calibration,
3. ml : Machine Learning: statistical models, Bayesian classifier, k-nearest neighbors, support vector machines (SVM), decision trees, boosting, random trees and neural networks.

An example of using OpenCV for image processing and display is shown in Listing 14.2.

```
   | // \file ipl_example.cpp
   | // \author Sandeep Koranne, (C) 2010
   | // \description Example of using OpenCV
   | #include <iostream>          // Program IO
5  | #include <cstdlib>           // exit
   | #include <highgui.h>         // OpenCV GUI
   |
   | // Simple program to display an image
   | int main( int argc, char* argv[] ) {
10 |   if( argc != 2 ) {
   |     std::cerr << "Usage: ./ipl_example <file>...\n";
   |     exit(1);
   |   }
   |   IplImage *image = cvLoadImage( argv[1] );
15 |   assert( image && "Unable to load image ");
   |   cvNamedWindow( "IPL_EXAMPLE", CV_WINDOW_AUTOSIZE );
   |   cvShowImage( "IMAGE", image );
   |   cvWaitKey( 0 ); // get user input
   |   cvReleaseImage( &image );
20 |   cvDestroyWindow( "IPL_EXAMPLE" );
   |   std::cout << std::endl;
   |   return(0);
   | }
```

Listing 14.2 Displaying images using OpenCV

Compiling and running this program produces the image as shown in Figure 14.1.

14.2 CImg: C Image Processing Toolkit

The CIMG (C++ Template Image Processing Toolkit) defines classes and method to process image in C++ programs. It defines a single Image class which can represent multi-dimensional datasets upto 4-dimensions with templatized pixel types. It can

Fig. 14.1 OpenCV 'highgui'
image display example

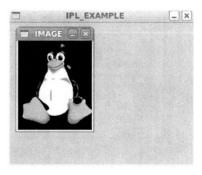

also handle image collections and *sequences*. CIMG is also very efficient and simple
to use since only the single header file cimg.h needs to be included, and all functions
and classes or CIMG are encapsulated in its own namespace. It is self-contained,
however, it can be used alongwith other open-source packages such as libpng (see
Section 11.2.1), libjpeg (see Section 11.2.1.2), FFTW3 (see Section 16.8) and LA-
PACK (see Section 16.3).

A small example to demonstrate CIMG Toolkit is shown in Listing 14.3.

```cpp
   // \file cimg_color.cpp
   // \author Sandeep Koranne (C) 2010
   // \description Example of using CImg toolkit
   #include <cstdlib>            // for exit
5  #include <iostream>           // for program IO
   #include <fstream>            // std::ofstream
   #include <CImg.h>             // CIMG Toolkit
   #include <vector>             // std::vector
   #include <algorithm>          // std::copy
10 #include <iterator>           // ostream iterator
   #include <cassert>            // assertion checking

   using namespace cimg_library;

15 int main( int argc, char *argv [] ) {
     if( argc != 2 ) {
       std::cerr << "Usage: ./cimg_color <file>...\n";
       exit(1);
     }
20   const CImg<unsigned char> image = CImg<>( argv[1] );
     const unsigned int W = image.width();
     const unsigned int H = image.height();
     const unsigned int D = std::min( W, H );
     std::assert( D > 0 );
25   std::cout << "Image H = " << H  << "\t"
         << "Image W = " << W << std::endl;
     // calculate the color profile on line going from (0,0) -> (W,H)
     std::vector<unsigned int> RED( D );
     std::vector<unsigned int> GREEN( D );
30   std::vector<unsigned int> BLUE( D );
     for( unsigned int i=0; i < std::min( W, H ); ++i ) {
       RED[i]   = image( i, i, 0 );
       GREEN[i] = image( i, i, 1 ); // RGB = (0,1,2)
       BLUE[i]  = image( i, i, 2 );
35   }
     std::ofstream r_dat("red.dat"), g_dat("green.dat"), b_dat("blue.dat");
```

```
     std::copy( RED.begin(), RED.end(),
         std::ostream_iterator<unsigned int>( r_dat, "\n" ) );
     std::copy( GREEN.begin(), GREEN.end(),
40       std::ostream_iterator<unsigned int>( g_dat, "\n" ) );
     std::copy( BLUE.begin(), BLUE.end(),
         std::ostream_iterator<unsigned int>( b_dat, "\n" ) );
     std::cout << std::endl;
     return (0);
45   }
```

Listing 14.3 Example of using CImg Toolkit

The program shown in Listing 14.3 calculates the color profile (RGB value) on the diagonal line joining (0,0) to min(W,H) of the given image. The use of C++ templates in CImg is shown as well.

14.3 Binary Decision Diagram (bdd): CUDD Library

Binary Decision Diagrams, are a very useful tool for modeling binary variables and systems. They are primarily used in VLSI CAD for logic synthesis and optimization. They can also be used to perform optimization and Boolean reductions in domains where the problem can be expressed as a Boolean decision problem.

CUDD is an acronym for the Colorado University Decision Diagram Package. It is an open-source C/C++ library for creating and managing BDDs as well as ZBDDs (zero-suppressed BDDs). CUDD implements an internal garbage-collector (see Section 7.3 for a discussion of garbage-collectors), thus the programmer must *reference* and *dereference* BDD nodes during the program computation. By design, CUDD uses an *unique table* to record BDDs; this ensures that identical functions map to the same BDD node and is used for canonical checking on functions for equivalence.

Binary Decision Variables are implemented as a binary tree, where each node has a *then* child, and an *else* child (which are traversed when we assign the value of this BDD node to true, or false, respectively). For each variable listed in the function we traverse the path, and the value of the function is the value of the leaf node we arrive at (true or false). However, for efficiency, the children nodes of a node can be complemented, which complements the value of our assignment. The CUDD library implements Reduced Ordered BDDs, and in this section we present examples of using CUDD to solve problems with BDDs. A short example is shown in Listing 14.4.

```
    // \file bdd_example.cpp
    // \author Sandeep Koranne, (C) 2010
    // \description Example of using BDD with CUDD library
    #include <iostream>          // program IO
5   #include <cassert>           // assertion checking
    #include <cstdio>            // C stdio
    #include <cudd.h>            // CUDD BDD library


10  static DdManager* gManager;   // global DD Manager
    static DdNode**   gNodeArray; // pointer to nodes
```

```
     static unsigned int numVars;  // number of variables

     static void PrintStats(void) {
15     std::cout
         << "CUDD Statistics"
         << "\nNum. vars: " << Cudd_ReadSize( gManager )
         << "\nNum. count:" << Cudd_ReadNodeCount( gManager )
         << "\nNum. order:" << Cudd_ReadReorderings( gManager )
20       << "\nMemory     :" << Cudd_ReadMemoryInUse( gManager )
         << std::endl;
     }

     int main( int argc, char* argv[] ) {
25     // print CUDD version information
       std::cout << "CUDD Version = ";
       Cudd_PrintVersion( stdout );
       numVars = 4; // simple example
       gManager = Cudd_Init( numVars, 0, CUDD_UNIQUE_SLOTS,
30         CUDD_CACHE_SLOTS, 1024*1024 );
       assert( gManager && "Unable to create CUDD Manager" );

       DdNode *x0 = Cudd_bddIthVar( gManager, 0 ); // x0
       DdNode *x1 = Cudd_bddIthVar( gManager, 1 ); // x1
35     DdNode *x0_and_x1 = Cudd_bddAnd( gManager, x0, x1 );
       Cudd_Ref( x0_and_x1 );
       DdNode *x0_and_bar_x1 =
         Cudd_bddAnd( gManager, x0, Cudd_Not( x1 ) );
       Cudd_Ref( x0_and_bar_x1 );
40     DdNode *functionA = Cudd_bddAnd( gManager, x0_and_bar_x1,
             x0_and_x1 );
       DdNode *functionB = Cudd_bddOr( gManager, x0_and_bar_x1,
             x0_and_x1 );

45     Cudd_Ref( functionA );
       Cudd_Ref( functionB );
       Cudd_Ref( x0_and_x1 );
       FILE* f = fopen("bdd.dot","w");
       char **inputNames = new char*[2];
50     inputNames[0] = new char[3]; inputNames[0] = "x0";
       inputNames[1] = new char[3]; inputNames[0] = "x1";
       char **outputNames = new char*[3];
       outputNames[0] = new char[10]; outputNames[0] = "functionA";
       outputNames[1] = new char[10]; outputNames[1] = "functionB";
55     outputNames[2] = new char[10]; outputNames[2] = "x0_and_x1";

       DdNode **outputs = new DdNode*[3];
       outputs[0] = functionA;
       outputs[1] = functionB;
60     outputs[2] = x0_and_x1;
       Cudd_DumpDot( gManager, 3, outputs,
         inputNames, outputNames, f );
       PrintStats();
       std::cout << std::endl;
65     return(0);
     }
```

Listing 14.4 Using CUDD BDD Library

As shown in Listing 14.4 operations on BDDs are managed by the DdManager opaque object. To start the CUDD system we call the Cudd_Init function which sets up the memory space for the required number of variables.

Each DdNode represents a variable in the system. The data-structure for DdNode contains the variable index, reference count, a next pointer for the unique table, and a union type containing either the constant nodes or the BDD children.

The following functions operate on the BDD Node structure `DdNode`:

Table 14.1 CUDD DdNode Functions

Name	Description and Return value
`Cudd_IsConstant`	1 if node is leaf
`Cudd_T`	a pointer to "then" child
`Cudd_E`	a pointer to "else" child
`Cudd_IsComplement`	1 if node is complement
`Cudd_Regular`	1 if node is regular
`Cudd_V`	value of constant node

To build the example shown in Listing 14.4 we use the following SConstruct file:

```
CUDD="/home/skoranne/cudd-2.4.1/"
Program('bdd_example',['bdd_example.cpp'],
        CPPPATH=CUDD+"include",
        LIBPATH=[CUDD+"cudd",
                 CUDD+"st",
                 CUDD+"mtr",
                 CUDD+"epd",
                 CUDD+"util"],
        LIBS=['cudd','util','st','mtr','epd'])
```

Running `scons` gives:

```
$ scons
scons: Reading SConscript files ...
scons: done reading SConscript files.
scons: Building targets ...
g++ -o bdd_example.o -c
    -I/home/skoranne/cudd-2.4.1/include bdd_example.cpp
g++ -o bdd_example bdd_example.o
        -L/home/skoranne/cudd-2.4.1/cudd
        -L/home/skoranne/cudd-2.4.1/st
        -L/home/skoranne/cudd-2.4.1/mtr
        -L/home/skoranne/cudd-2.4.1/epd
        -L/home/skoranne/cudd-2.4.1/util
        -lcudd -lutil -lst -lmtr -lepd
scons: done building targets.

# Running the program
$ ./bdd_example
CUDD Version = 2.4.1
CUDD Statistics
Num. vars: 4
Num. count:5
Num. order:0
Memory    :4225732
```

Using the BDD function we can model the Boolean functions using combinations of AND and OR functions. We can also print the BDD collection to a DOT (see Section 19.4) file. The produced BDD collection is shown in Figure 14.2. It is instructive to see the relationship between the paths, complemented paths from variables in Figure 14.2, to the BDD construction code shown in Listing 14.4.

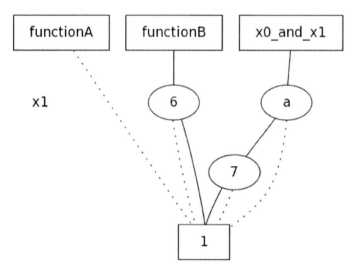

Fig. 14.2 BDD produced using CUDD library

The produced DOT file is shown for reference (and also as an example to writing DOT files, as explained in Section 19.4).

```
digraph "DD" {
size = "7.5,10"
center = true;
edge [dir = none];
{ node [shape = plaintext];
  edge [style = invis];
  "CONST NODES" [style = invis];
" x1 " -> "  " -> "CONST NODES"; }
{ rank = same; node [shape = box]; edge [style = invis];
"  functionA  " -> " functionB  " -> " x0_and_x1  "; }
{ rank = same; " x1 ";
"a";"6";}
{ rank = same; "  ";"7";}
{ rank = same; "CONST NODES";
{ node [shape = box]; "2";}}
"  functionA  " -> "2" [style = dotted];
"  functionB  " -> "6" [style = solid];
"  x0_and_x1  " -> "a" [style = solid];
"a" -> "7";
"a" -> "2" [style = dotted];
```

```
"6" -> "2";
"6" -> "2" [style = dotted];
"7" -> "2";
"7" -> "2" [style = dotted];
"2" [label = "1"];}
```

14.4 FWTools: Open Source GIS

FWTools include OpenEV, GDAL, MapServer, PROJ.4 and OGDI. These tools are described below. More details on Geographical Information Systems (GIS) is given in Section 14.9. OpenEV is a high-performance raster/vector desktop data viewer and analysis tool; an example is shown in Figure 14.3.

Fig. 14.3 OpenEV : Raster/Vector viewer

14.4.1 PROJ4

PROJ.4 is a cartographic projection library. PROJ4 is used by GIS projects (including GRASS, Section 14.9.1) to convert between coordinate systems. Listing 14.5 shows an example of coordinate conversion using PROJ4. It is also possible to perform conversion on the command line using the tools `proj` and `invproj`. Their usage is described below.

```cpp
// \file proj4_example.cpp
// \author Sandeep Koranne, (C) 2010
// \description Example of using PROJ4
// for Cartographic Projection
#include <iostream>        // for program IO
#include <cassert>         // assertion checking
#include <cstdlib>         // exit
#include <proj_api.h>      // PROJ4 library

int main( int argc, char* argv[] ) {
  projPJ transformer, latitude_longitude;
  double gx, gy;

  transformer = pj_init_plus("+proj=utm +lon_0=112w +ellps=clrk66");
  if( !transformer ) {
    std::cerr << "Unable to construct PROJ..\n";
    exit(1);
  }
  latitude_longitude = pj_init_plus("+proj=latlong +ellps=clrk66");
  std::cout << "Enter value:";
  std::cin >> gx >> gy;
  pj_transform( latitude_longitude, transformer, 1, 1, &gx, &gy, NULL );
  std::cout << "Output : " << gx << "\t" << gy;
  std::cout << std::endl;
  return(0);
}
```

Listing 14.5 Example of cartographic projection using PROJ4

PROJ4 includes many projection systems including cylindrical, transverse Mercator, universal transverse Mercator (UTM), central cylinder projection, Airy projection and Miller projection.

The API of PROJ4 includes the `pj_transform` function:

```cpp
int pj_transform( projPJ srcdefn,  // source projection format
                  projPJ dstdefn,  // destination projection format
                  long point_count, // number of points,
                  int point_offset, // which point in the data set,
                  double *x, double *y, // data input-output
                  double *z ); // for 3d transforms.
```

The API is initialized using a string which is the same as the `proj` filter as described below.

14.4.1.1 `proj`: forward cartographic projection filter

The command line tool `proj` implements the PROJ4 library as a command line program and can be used to perform coordinate conversions (e.g., UTM projections).

Consider the following example which calculates Boston's (approximately) UTM coordinates using a standard UTM central meridian of 112 degrees west. The command line option '-r' is used to *reverse* the default ordering of latitude-longitude coordinates.

```
$ proj -E +proj=utm +lon_0=112w +ellps=clrk66 -r << EOF
> 45 111.5W
> 75N -111
> EOF
45 111.5W 460591.19 4982854.80
75N -111 500000.00 8323452.59
```

In the above transcript we used the -E option which prints the input coordinates on the output as well. The related tool `geod` performs direct geodesic conversions. Geodesic (Great Circle) computations for determining latitude, longitude and back azimuth of a point given initial point latitude, longitude and azimuth and distance is supported. For example, to compute the azimuths and distance from Boston to Portland we can use:

```
$geod +ellps=clrk66 <<EOF -I +units=us-mi
            42d15N 71d07W 45d31N 123d41W
            EOF
-180d 0d 2907.509N 71d07W 45d31N 123d41W
```

To calculate Portland's location from Boston use:

```
$geod +ellps=clrk66 <<EOF +units=us-mi
            42d15N 71d07W -66d3150.141" 2587.504
            EOF
45d310.003"N              123d4059.985"W
75d3913.094"
```

The inverse projection tools are also available as `invproj` and `invgeod`.

14.4.2 GDAL : Geospatial Data Abstraction Library and OGR

GDAL is an open-source translator library and command-line utility tools for reading and writing a wide variety of geospatial raster (GDAL) and vector (OGR) formats. GDAL data model prescribes a GDAL Data Set as having a list of raster bands all corresponding to the same physical area, and having the same resolution. GDAL Data also has meta-data, a coordinate system and a geo-referencing transform. The size of the raster is included in the data itself. Using GDAL library we can write a C++ function to print the size of the dataset as:

```cpp
#include <gdal_priv.h>
int WriteDataSetSize( const char* fileName ) {
  GDALDataset *pData;
  GDALAllRegister();
```

```
 5   pData = (GDALDataset*) GDALOpen( fileName, GA_ReadOnly );
     if( pData == NULL ) {
       std::cerr << ''Unable to open file : '' << fileName ;
       exit(1);
     }
10   std::cout << ''Raster X Size = '' << pData->GetRasterXSize()
               << ''\n Raster Y Size = '' << pData->GetRasterYSize()
               << ''\n Raster Count = '' << pData->GetRasterCount()
               << std::endl;
 }
```

Listing 14.6 Example of using GDAL library

The GDAL library provides access to the raster band data, one band at a time using the GDALRasterBand object. Related tools include OGDI, which is a multi-format raster and vector reading technology, and MapServer which is a map server using Web CGI and GeoCode is a geocoding library.

14.5 GNU Image Manipulation Program

GNU Image Manipulation Program (GIMP) is an image processing and paint program which is able to process a number of graphic file formats. GIMP is used to edit and manipulate bitmap images and can be used to convert between formats. It is also possible to use GIMP as a paint program as it features drawing tools such as brush, pencil and clone. An example of an image opened in GIMP for processing is shown in Figure 14.4.

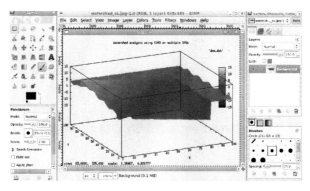

Fig. 14.4 GNU Image Manipulation Program (GIMP)

GIMP has a number of tools built into it, as shown in Figure 14.5.

GIMP also supports *plugins* which provide additional functionality. A plugin can access the bitmap contents of the image currently loaded into the GIMP program and perform analysis, and processing on it. If the image contents are modified, GIMP can then write the resulting output to disk in any supported file format.

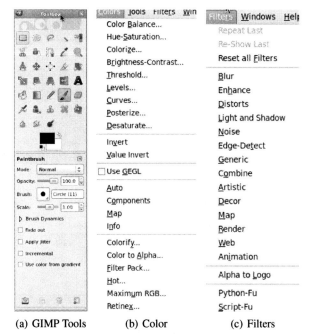

(a) GIMP Tools   (b) Color   (c) Filters

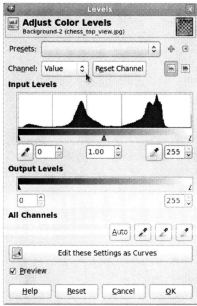

(d) Color levels

Fig. 14.5 GIMP built in tools and filters

14.6 Computational Fluid Dynamics using OpenFOAM

OpenFOAM is an acronym for Open Field Operation and Manipulation. Open-
FOAM is an open-source computational fluid dynamics package and it can support
CFD applications on complex flows involving chemical reactions turbulence and
heat transfers. It has a plugin to ParaView (see Section 19.17), which is used to
view the CFD geometry and perform analysis using the GUI.

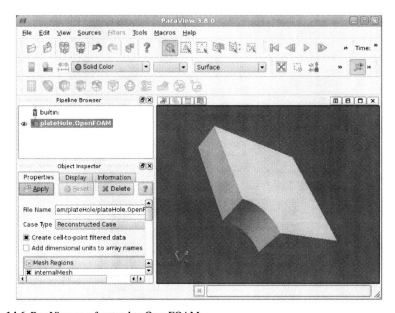

Fig. 14.6 ParaView as a frontend to OpenFOAM

The geometry for analysis can be defined as an ASCII file:

```
FoamFile
{
    version    2.0;
    format     ascii;
    class      dictionary;
    object     blockMeshDict;
}

convertToMeters 1;

vertices
(
    (0.5 0 0)
    (1 0 0)
    (2 0 0)
    (2 0.707107 0)
```

```
    (0.707107 0.707107 0)
...
    (0 2 0.5)
    (0 1 0.5)
    (0 0.5 0.5)
);

blocks
(
    hex (5 4 9 10 16 15 20 21) (10 10 1)
        simpleGrading (1 1 1)
    hex (0 1 4 5 11 12 15 16) (10 10 1)
        simpleGrading (1 1 1)
    hex (1 2 3 4 12 13 14 15) (20 10 1)
        simpleGrading (1 1 1)
    hex (4 3 6 7 15 14 17 18) (20 20 1)
        simpleGrading (1 1 1)
    hex (9 4 7 8 20 15 18 19) (10 20 1)
        simpleGrading (1 1 1)
);

edges
(
    arc 0 5 (0.469846 0.17101 0)
    arc 5 10 (0.17101 0.469846 0)
    arc 1 4 (0.939693 0.34202 0)
    arc 4 9 (0.34202 0.939693 0)
    arc 11 16 (0.469846 0.17101 0.5)
    arc 16 21 (0.17101 0.469846 0.5)
    arc 12 15 (0.939693 0.34202 0.5)
    arc 15 20 (0.34202 0.939693 0.5)
);
```

Figure 14.6 shows the geometry of the plate with a corner cut off. OpenFOAM has a mesh generator which can be invoked with the blockMesh command; this command converts the geometry into *mesh* form. The size of the mesh can be specified.

Once the design has been meshed we can perform analysis using a number of tools provided with OpenFOAM. OpenFOAM by itself is a collection of C++ libraries, but several applications come pre-built with the OpenFOAM release. The applications can be divided into two categories:

1. solvers: each application is designed to solve a very specific problem in CFD,
2. utilities: pre-processors, format converters, data manipulation and post-processing tools.

OpenFOAM allows (and indeed is designed) for runtime linking of user-defined code to provide either new solvers, or new data representations. The shared libraries (which have to be compiled using the OpenFOAM header files) can be placed in the 'controlDict' file for a case using the 'libs' keyword. Then at runtime OpenFOAM will automatically load the shared libraries into the analysis tool for that case.

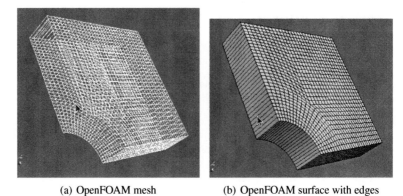

(a) OpenFOAM mesh   (b) OpenFOAM surface with edges

Fig. 14.7 OpenFOAM block mesh

14.6.0.1 Case and File Structure in OpenFOAM

The basic directory structure for an example is shown below:

```
$ ls
0   0.1   0.2   0.3   0.4   0.5   cavity.OpenFOAM
constant   system
$ ls -l system/
total 12
  controlDict
  fvSchemes
  fvSolution

$ ls -l constant/
total 8
 polyMesh
 transportProperties
[skoranne@celex cavity]$ ls -l constant/polyMesh/
total 68
 blockMeshDict
 boundary
 faces
 neighbour
 owner
 points
```

The 'system' directory is used for setting parameters associated with the solution procedure. The 'constant' directory contains a description of the case mesh in the 'polyMesh' sub-directory (as shown above). The physical properties for the application are located in files such as 'transportProperties'. Data in OpenFOAM files is stored as Key-Value dictionaries.

14.6.0.2 CFD Solvers in OpenFOAM

OpenFOAM contains solvers for incompressible flow, solid body stress analysis and combustion analysis. In addition the conventional CFD solvers such as Laplacian solver, scalar transport and laminar flow are also included. In-fact, OpenFOAM also includes a solver for Black-Scholes option pricing (the `financialFoam` solver).

14.6.0.3 Utilities in OpenFOAM

The utilities in OpenFOAM can be classified into the following categories:

1. Pre-processing: box-drawing, initialization of data into OpenFOAM case directories, volume field mapping, and molecular dynamics,
2. Mesh generation and conversion: in the example above we saw the use of `blockMesh` for mesh generation. OpenFOAM also has utilities for converting formats from other CFD packages to the OpenFOAM format,
3. post-processing: these utilities include graphical display and rendering as well as data export from OpenFOAM format to other CFD packages,
4. Sampling Utilities: OpenFOAM `sample` command can be used to probe locations and cells in the case data for a design in OpenFOAM. It can be used to sample field data and generate interpolation data points which are amenable for plotting and external analysis.

14.7 Molecular Dynamics

Computational Molecular Dynamics has become a major driver of high-performance computing. Computer simulations of molecules are used in solving problems in physical chemistry, structure determination in crystallography, and in experimental drug discovery. The latter has significant potential to improve the current treatment options for many of the common ailments, and thus is of considerable social and financial value. In most of the above application domains, a collection of physical structures (atoms, molecules, proteins, etc) are suspended in a fluid of known properties. Starting with a known state (or preconditioned state), the goal of the computer simulation is to calculate the trajectory (physical motion) of the structure's components over a *time step*. Integrating the motion of individual components over a long enough time period can be used to gather information about the physical properties of the compound under study. In this section we describe some of the common molecular dynamics codes and applications.

14.7.1 NAMD

NAMD was designed to support parallel execution on a networked cluster and thus is able to harness the power of a Beowulf class PC-clusters. NAMD supports the following analysis:

1. PME (Particle Mesh Ewald): NAMD implements the PME algorithm for full electrostatic interaction computation. This algorithm reduces the time complexity of the computation to $O(n \log n)$ from $O(n^2)$, which provides significant speedup,
2. Force field compatibility: the force field used in NAMD is the same as those used by other molecular dynamics programs which can help in correlating results, as as well migration of simulations from one software to another,
3. Multiple time stepping: NAMD used Verlet integration method to advance the time step,
4. Interactive simulations
5. Load Balancing.

14.7.2 GROMACS

GROMACS is an open-source package to perform molecular dynamics. Its main features are:

1. Efficiency: GROMACS code base has been systematically optimized for high-performance using latest compilers, and CPU instructions, including 64-bit registers and SSE2,
2. User friendly: GROMACS topology files are plain ASCII text files which are simple to generate and edit,
3. Writing trajectory data using lossy compression :accuracy of compression is user selectable, but provides significant compaction,
4. MPI based parallel programming

14.7.3 Molecular Visualization

Molecules can be viewed using JMol or RasMol, both open-source packages for molecular visualization.

14.7.3.1 JMol

JMol is an open-source molecule viewer which comprises of:

1. JMolApplet: web browser applet which can be integrated into web pages,

2. JMol Application: is a standalone Java application,
3. JMolViewer: is a development tool that can be integrated into other Java programs.

An example of JMol application viewing a PDB file is shown in Figure 14.8.

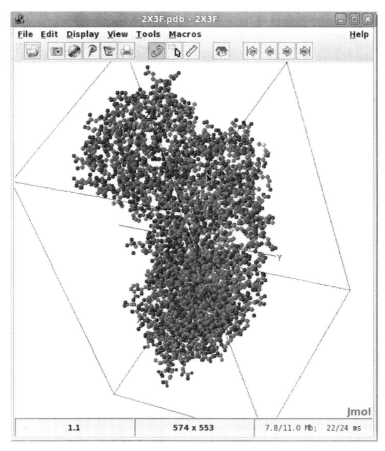

Fig. 14.8 JMol : Molecule Viewer application

14.7.4 Foldng@Home

No section on open-source molecular dynamics can be complete without a reference to Folding@Home. Folding@Home is a distributed computing project, which harnesses the power of idle desktop personal computers from all over the world ti run software to perform protein folding. Work units are distributed to each of the computers, each unit represents some folding computation problem which needs to be solved. Collectively, Folding@Home, represents one of the largest supercomputers implemented using distributed computing. Folding@Home clients are available for a number of operating systems, computers and even supercomputers.

14.8 Audacity

Audacity is a graphical audio editor. It uses the `libsndfile` library to read/write many uncompressed file formats. An example of a WAV file being edited is shown in Figure 14.9. It includes support for filters such as noise removal and stereo editing. Batch processing applications use `sox` which is described in Section 14.8.1.

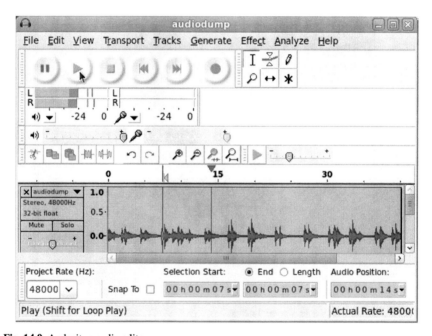

Fig. 14.9 Audacity : audio editor

14.8.1 Sound Exchange : sox

Sound Exchange (sox) is a computer program that can read/write audio files and perform audio processing on them. It can combine various input file, synthesize sounds, and even act as a simple audio player. The `sox` processing chain can be described as:

```
Input -> Combiner -> Effects -> Output
```

The input and output phases can perform conversion to/from a wide variety of file formats. The combiner and effects can include (i) sampling rate change, (ii) sampling size change, (iii) mono down mix, (iv) volume and dither control, and (v) speed control (to speedup or slow down the audio). `sox` can also act as a sound recorder with auto detection of sound.

14.9 Geographical Information Systems

With the advent of localization based services and the widespread use of information technology in urban planning, the role of geographical information systems (GIS) software has burgeoned. In the open-source community the GRASS GIS system, and Quantum GIS (qgis) are popular, and we describe these below.

14.9.1 GRASS GIS

GRASS is an open-source geographical information system (GIS) which can handle raster, topological vector, image processing, and graphic data. An example of using GRASS GIS is shown in Figure 14.10 which shows county data in the USA.

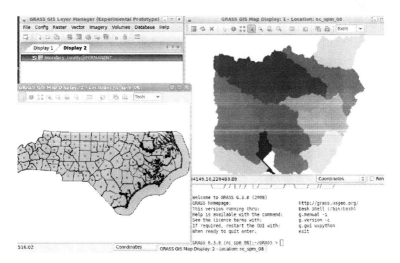

Fig. 14.10 GRASS GIS (geographical information system)

Recently a new topological 2D/3D vector engine has been added to GRASS (version 6.0) which adds support for vector network analysis. Moreover, GRASS can store attributes in SQL-based databases such as MySQL (see Section 20.6), PostgreSQL/PostGIS (see Section 20.5) and SQLite (see Section 20.7).

14.9.2 Quantum GIS

Quantum GIS (QGIS) is another open-source GIS software. QGIS is written in C++ and uses the Qt libraries (see Section 19.1.3), and as such is very portable and runs on GNU/Linux as well as Mac OS X, and Microsoft Windows. An example of QGIS is shown in Figure 14.11. Map data from USA atlas is available at `http://nationalatlas.gov` and we have used the county map, location of dams, and road network in the examples below. The county area of Portland, Oregon is shown in Figure 14.11.

Fig. 14.11 Quantum GIS (geographical information system)

We add another vector shape file, the location of dams, and urban areas. We see that there are a number of dams located near the urban areas as shown in Figure 14.12.

A common use of GIS is in road network planning, and transportation planning. The road network near Portland is loaded from the data file and can be overlayed on the existing maps, as shown in Figure 14.13.

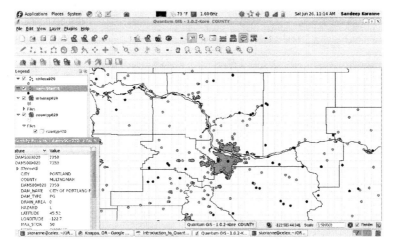

Fig. 14.12 Portland and surrounding area, urban area, county and dams.

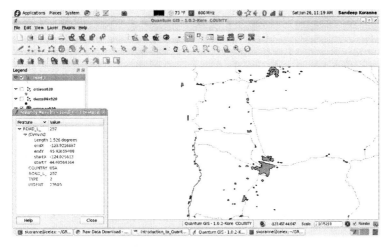

Fig. 14.13 Road network, vector shape file

14.10 QCAD : 2d CAD Tools

Computer Aided Design is the discipline of using computer and information processing to aid in mechanical design. There is a fundamental difference between a CAD program and a general drawing program as shown in Section 19.7, although it is possible to use a general drawing program for initial schematics, the CAD program provides additional features such as ISO compliant dimensioning, exact measurement, wide variety of grid snapping and more. In this section we describe

QCad an open-source CAD program for two-dimensional drafting. An example of
QCad drawing is shown in Figure 14.14.

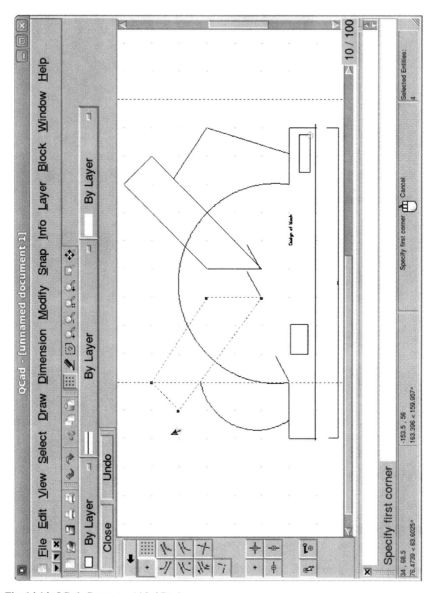

Fig. 14.14 QCad: Computer Aided Design

14.11 BRL-CAD

BRL-CAD is a constructive solid geometry package designed for a wide variety of military and industrial purposes. The software is open-source and as of version 7.16B contains a number of tools including (a) interactive geometry editor, (b) ray-tracing, and (c) network distributed image processing.

The interactive geometry editor is termed the Multi-Device Geometry Editor (mged).

14.11.0.1 Multi-Device Geometry Editor

The editor can be launched using the mged command line tool. An existing geometry database can be specified on the command-line or opened using the GUI. The GUI has an interactive shell window which accepts text commands, while the GUI menus provide the same functionality graphically.

Constructive Solid Geometry (CSG), as the name implies, is the process of model construction through the use and combination of primitive objects such as (a) ellipsoid (generalized sphere), (b) right parallelepiped, (c) toric shapes, and more. These primitives can be placed in the model at appropriate coordinates, scaled and transformed using the editor to form the desired model. The *center-of-view*, scaling and zooming, azimuth view, of the current display can be controlled by the user. Consider an example shown in Figure 14.15.

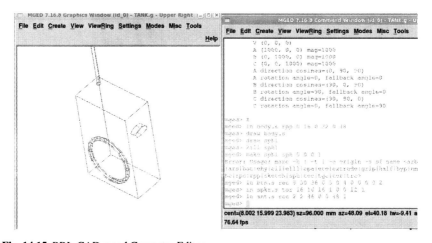

Fig. 14.15 BRL-CAD mged Geometry Editor

The construction of objects can be carried out by entering commands in the console:

```
in rcc3s rcc 3 2 4 0.2 0.2 0.3 0.4
```

```
in rcc4s rcc -3 2 4 0.2 0.2 5.0 0.4
mged> in rcc5s
Enter solid type: rcc
Enter X, Y, Z of vertex: 3 3 4
Enter X, Y, Z of height (H) vector: 0 0 1
Enter radius: 0.4
```

This command in, inserts a right circular cylinder at the given (x,y,z) coordinates as the RCC's center. The height of the RCC is entered next in the current unit, followed by the radius of the cylinder.

Since BRL-CAD is used to order parts for machining directly from the model, the engineering drawing views of the model also need to be generated as shown in Figure 14.16.

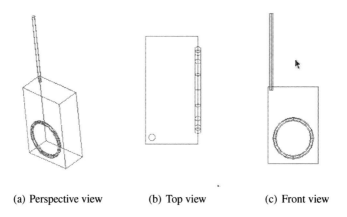

(a) Perspective view   (b) Top view   (c) Front view

Fig. 14.16 Perspective view, top and front view

A realistic image of the constructed model can be generated using the *raytracer* which is included with the software. An example of the raytracer in action and the produced final image is shown in Figure 14.17(a) and (b).

Moreover, BRL-CAD also supports performing Boolean operation on the solid volumes. The Boolean can be *union*, *subtraction* and *intersection*. Consider a machine part comprised of a cylinder in which cutouts for rivet-heads have to be drilled as shown in Figure 14.18(a). The part itself comprises of a RCC, on top of which we add 4 spheres and a small RCC. The content of the current file can be listed using the ls command; while the hierarchical constituents of a model can be analyzed using the tree command. The Boolean operation can be performed as:

```
r mp1.part u rcc1 - sph1 - sph2 - sph3 - sph4 - rcc2
```

The final rendering of the machine part is shown in Figure 14.18(b). The rendering can be influenced by choosing a *material* to apply for the object. The transparency and reflectance of the surface can be changed using this property. Figures

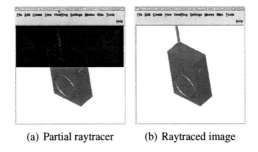

(a) Partial raytracer   (b) Raytraced image

Fig. 14.17 Raytracer in BRL-CAD

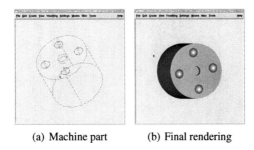

(a) Machine part   (b) Final rendering

Fig. 14.18 Example machine part in BRL-CAD

can be rotated in the editor using the 'Z' command, and using the arrow keys various azimuth and elevation angle views can be generated. Multiple parts of the drawing can be constructed separately and assembled later using the comb function and ray-traced together.

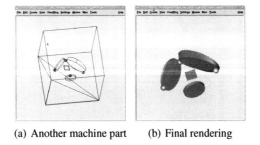

(a) Another machine part   (b) Final rendering

Fig. 14.19 Example machine part in BRL-CAD

An object can be copied to another using the cp command in the console. At-tributes of objects such as scale, translation and individual object property such as *radius*, *height* can also be changed. Another 3d geometry construction software named *Blender* is discussed in Section 14.12. A dedicated raytracing program (POVray) is discussed in Section 19.9.

14.12 Blender

Blender is an open-source software to create 2d and 3d content aimed at media professionals and artists. An example of Blender is shown in Figure 14.20 and Figure 14.21.

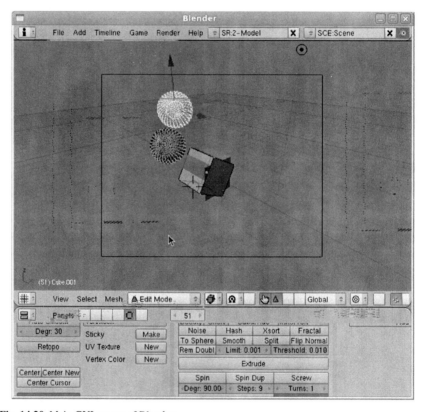

Fig. 14.20 Main GUI screen of Blender

The key features of Blender include:

1. Tools for content creation: including modeling, UV-mapping, texturing, rigging, animation, rendering, compositing, post-production and game creation,
2. Cross platform availability,

3. Small executable size.

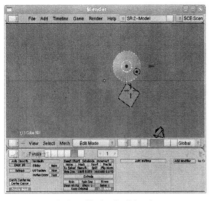

(a) A cylinder in Blender   (b) Raytraced rendering

Fig. 14.21 Rendering in Blender

Blender has its unique GUI with many actions tied to single key shortcuts which increases the productivity of the designer, but has a learning curve.

14.13 Conclusion

In this chapter we presented engineering libraries such as Computer Vision, CImg and FWTools. Geospatial data abstractions are becoming very important with the rise of location aware computing, and several open-source tools such as GDAL and PROJ4 were described in this chapter. Image processing, audio processing, and computational fluid dynamics (CFD) that are part of many engineering were presented. Molecular dynamics simulators and viewers were also discussed. Information Systems (GIS) including GRASS and QGIS are described in Section 14.9.1. Mechanical engineering, as well as use of mechanical CAD software in other disciplines can be accomplished using open-source tools such as QCAD. Solid modeling tools BRL-CAD and Blender were described in this chapter.

Chapter 15
VLSI CAD Tools

In this chapter we describe the various steps in the design flow of an integrated circuit, starting from the algorithmic description of the functionality, to its representation and model in a High Level *hardware description language*(HDL). For each step we present open-source VLSI CAD tools. We limit our attention to standard cell based design methodology, but we should point out that in addition to this style there are these other techniques which are also used:

1. Full-custom: the complete chip is directly drawn as a layout. This is useful for analog and radio-frequency (RF) block design. It is also used in custom bus design where the performance of the block is critical to the chip,
2. Data-path design: used in arithmetic blocks where similar operations happen on a wide-bus.
3. Memory array: memory arrays are often designed as a single *bit*-cell, which is then regularly placed by automated tools,
4. PLA design: control logic can be represented as *sum-of-products*(SOP), and PLA (programmable logic arrays) can be automatically drawn from SOP functions. The array is not run-time programmable, only that the choice of which products to sum is made by switches which are programmed when the PLA is drawn.
5. FPGA design:field-programmable gate arrays use arrays of logic elements (LEs) connected by a fabric of programmable routing. Some of the problems, such as floorplanning, and placement can be put in FPGA context as well.

VLSI design has been the study of automation and computer science, and many tools have been written to solve one or more problems associated with the VLSI design flow. Thus a number of standard (and not so standard) *file-formats* have emerged to communicate data from one step of the flow to another.

We first describe the syntax of the Berkeley Logic Interchange Format, and present several tools to analyze BLIF data, as BLIF is a standardized logic interchange format for synthesis tools. Post synthesis the logic design has been converted into a *gate-level* netlist we discuss the problem of timing estimation, buffer-insertion and logic checking. This part of the flow is traditionally known as *front-end* of the VLSI flow, and the timing correct gate-level netlist was the accepted *handoff* to

S. Koranne, *Handbook of Open Source Tools*,
DOI 10.1007/978-1-4419-7719-9_15, © Springer Science+Business Media, LLC 2011

the physical design flow to follow. The major design flow stages are (i) front-end design and HDL capture, (ii) synthesis to gate level netlist, (iii) floorplanning, (iv) placement, (v) global routing, (vi) detailed routing, and (vii) mask level processing.

15.1 Algorithmic Design and HDL Capture

We state the title of this section as "algorithmic design", which refers to the task of solving a given problem by using an algorithm. Consider the problem of designing an elevator control module which services four elevators in a building with ten floors. The requirement analysis of this problem in detail may produces a list as follows:

1. Interact with sensors, lighting and control switches inside the elevator, in the floor and the main control room of the building,
2. Minimize the number of physical connections to the unit,
3. Minimize wait times,
4. Conserve power by maximizing idle item,

A system architect will perform a Pareto analysis of various solutions to this problem. He may consider using an off-the-shelf micro-controller (e.g, 8051,or i486), program an FPGA, design a custom ASIC (chip). The architect will then proceed to design an algorithm to solve the presented problem (within the constraints placed by the system requirements). This algorithm is then *captured* in an HDL for simulation and synthesis. Simulation is yet another area of VLSI CAD which is very important, but we have side-stepped, as writing a correct simulator for a modern HDL is non-trivial. By simulating the HDL of the elevator-controller module, its behavior under various conditions can be tested, and its properties measured. CAD tools to calculate expected power dissipation are also available, and expected wait-times can be checked with random initial conditions. The controller's response to emergency inputs under various states and its guarantee of a maximum service time per request can be checked formally using a technique known as *model checking*, which is derived from finite-state-machine state exploration.

15.2 HDL Capture

Common HDLs are Verilog and VHDL. Both offer similar features in terms of specifying the structure of the logic module, and its behavior using (i) combinational, and (ii) sequential processing elements. Combinational elements are logic circuits which have no state, and their outputs depend on their current inputs *only*. There may still be a logic-delay between the time an input changes, and the time this change is reflected in a change to the output. Combinational logic circuits are manipulated using Boolean algebra, and their properties have been extensively studied and understood. Sequential processing elements have *state*, and thus, their current output depends not only on their input, but also the current state of the element. This make sequential analysis more complex, but it also allows for a significant reduction in the size of the Boolean circuit needed to implement a given algorithm. For example, 16-bit multiplication would need thousands of logic elements to produce a valid output given only combinational element, but a sequential process combined with combinational logic can perform the same computation (albeit in multiple clock cycles) in only few hundred elements. Sequential elements are often modeled as *state-machines*, and are represented in Verilog/VHDL as encoded state-registers. Consider the following Verilog HDL fragment:

```
module ElevatorControl( clk, rst, swButton,
                        elvButton, emgButton,
                        sensor, opControl);
input    clk,rst;
input [31:0] swButton; //switches in floor, control room
input [63:0] elvButton;// switches inside  4 elevators
...
output [31:0] opControl;// motor control switch lights
endmodule;
```

Although automated tools to convert a given algorithmic description (which may have been done using another language) to HDLs exist, most of the time, the HDL is written by hand in conjunction with the system architect. Key system parameters, such as, interface registers, IO register locations, interface pin-diagrams, need to be created and frozen. Many of these physical parameters have no equivalence in the algorithmic description (what is the weight of an algorithm ?, the controller once manufactured will have weight and dimensions which need to be specified so that the rest of the system can be built around the controller).

As we mentioned earlier, HDL descriptions are simulated and analyzed for their properties. Once the designers are satisfied that the design meets the functional requirements, the design is *synthesized*. Synthesis is the process of taking HDL descriptions, converting them to Boolean logic expressions (using combinational and sequential elements), and finally writing out an equivalent Boolean description of the circuit with each logic elements *mapped* to a given primitive logic element. The set of primitives is also given to the synthesis tool, and this element set is called a *library*. Each library element has been designed to have a physical representation on the chip, and once the circuit is mapped onto the library elements, it can be manufactured purely as a connected graph of these library elements. Each library element provides the synthesis tool the *single* (for single-output standard cells) Boolean function it can implement. It is the task of the synthesis tool to *map* the given Boolean circuit using these primitives. Obviously, an universal primitive like two-input NAND can be used to map the complete combinational portion of the circuit, but it has been known that providing higher levels of Boolean primitives (for example, primitives which compute AND-OR-INV, or AOI, combinations of their input, not only reduce the number of total primitives used, but also reduce the maximum depth of the circuit). The maximum depth of the combinational circuit determines the clock cycle of the chip, as each clock cycle must leave enough time for the output of every combinational block in the chip to be computed.

Synthesis tools have a rich history, and UC Berkeley provided many of the fundamental results which are now part of every synthesis CAD tool. Thus it is only fitting that we present BLIF (Berkeley Logic Interchange Format) syntax before continuing on to the next steps of the VLSI flow.

15.3 BLIF Format in a nutshell

The BLIF format describes a circuit in terms of models. A model is an arbitrary combinational or sequential network of logic functions. A circuit is built up of modules connected together to form a connected, directed graph. Each cyclic edge in this network must go through a *latch* element. Each net has a single driver, which must be named without ambiguity. A model is declared in BLIF as

```
.model <name>
.inputs <input-list>
.outputs <output-list>
<command>
...
<command>
.end
```

The *command* is one of:

1. logic gate
2. generic latch
3. library-gate: library format is described below,
4. model reference:
5. subfile reference:
6. fsm-description:
7. clock constraint
8. delay constraint

A # begins a comment line which continues to the end of the line. A *logic gate* associates a logic function with a signal in the model, which can be used as an input to another logic function. A logic gate is declared as follows:

```
.names <in-1> <in-2> ... <in-n> <output>
<single-output-cover>
```

An example of a logic-gate is shown below:

```
.names x y z
1- 1
01 1
```

The single-output cover may include – to represent a don't care. Library gates are specified using *.gate* construct, where the syntax is as follows:

```
.gate <name> <formal-argument>
```

The argument list of the gate is looked up from the library file, and is matched by name.

Consider a very small model represented as a BLIF file:

```
.model small
.inputs ck  i0 i1 i2 i3
.outputs  o0 o1
.names  i0 i1 i2 i3 o0
0001 1
0000 1
0010 1
.names  i0 i1 i2 i3 o1
0001 1
0000 1
0010 1
.end
```

This model has 4 inputs (i0,i1,i2 and i3), and 2 outputs (o0 and o1). The truth table (or single-output cover) for o0 and o1 is given in the BLIF file.

Next we write a library description file containing the four gates of INV, BUF, 2-AND and 2-OR. This is shown below:

```
GATE inv_x1 1  q=!i;
     PIN    i  INV 1 999 1.00 0.00 1.00 0.00
GATE buf_x2 1  q=i;
     PIN    i  INV 1 999 1.00 0.00 1.00 0.00
GATE a2_x2  2  q=(i0*i1);
     PIN    i0 INV 1 999 1.00 0.00 1.00 0.00
     PIN    i1 INV 1 999 1.00 0.00 1.00 0.00
GATE o2_x2  2  q=(i0+i1);
     PIN    i0 INV 1 999 1.00 0.00 1.00 0.00
     PIN    i1 INV 1 999 1.00 0.00 1.00 0.00
```

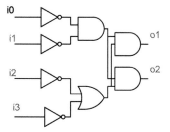

Fig. 15.1 Gate level netlist, schematic is shown after mapping onto library.

Using ABC Synthesis tool we map the BLIF file using the library to produce a gate-level mapped netlist as shown below. The ABC commands necessary to do the mapping were: (i) read_blif (ii) read_library, (iii) map, and (iv) write_blif. We get the following BLIF file.

```
.model small
.inputs ck i0 i1 i2 i3
.outputs o0 o1
.gate inv_x1 i=i2 q=n7
.gate inv_x1 i=i3 q=n8
.gate o2_x2  i0=n8 i1=n7 q=n9
.gate inv_x1 i=i0 q=n10
.gate inv_x1 i=i1 q=n11
.gate a2_x2  i0=n11 i1=n10 q=n12
.gate a2_x2  i0=n12 i1=n9 q=o0
.gate a2_x2  i0=n12 i1=n9 q=o1
.end
```

Once the following gate is introduced in the library, the number of elements and the critical path get reduced.

```
GATE   noa22_x1  3    nq=((!i0+!i1)*!i2);
       PIN       i0 INV 1 999 1.00 0.00 1.00 0.00
       PIN       i1 INV 1 999 1.00 0.00 1.00 0.00
       PIN       i2 INV 1 999 1.00 0.00 1.00 0.00
```

We again map the original input BLIF using the augmented library to get the following BLIF file.

```
.model small
.inputs ck i0 i1 i2 i3
.outputs o0 o1
.gate o2_x2    i0=i1 i1=i0 q=n7
.gate noa22_x1 i0=i3 i1=i2 i2=n7 nq=o0
.gate noa22_x1 i0=i3 i1=i2 i2=n7 nq=o1
.end
```

Thus, it is obvious, how large an impact, choosing a good set of primitive elements in the library can have on the size of the chip and its critical path.

This gate level netlist is the traditional *hand-off* from front-end design tools to the *back-end* processing, also known as the *physical design flow*. In the physical design flow, the netlist is first divided into block regions at the chip level using an automated tool called the *floorplanner*. The input to this tool is the bounding box of each module of the netlist, and it calculates relative positions of the modules to minimize the overall size of this chip.

Once the floorplan is complete, we perform *cell-placement*. Standard cell placement has been an extensively researched topic, with recent publications (as late as Jan. 2009) reporting break-throughs in technology. We present simple, but effective and time-tested placement solutions based on simulated annealing, and spectral partitioning. Both methods are fully described earlier in the context of graph partitioning (Breuer had shown in the 1970s an interesting relationship between placement and partitioning).

Placement is followed by *global routing*, a process, in which we assign nets to *channels*, which are defined as pathways on the grid of the chip with exclusive reservation of a limited number of nets (to avoid congestion when we perform final detailed routing). An obvious optimization criterion is the minimization of wirelength used to connect any net, but secondary objectives, such as minimizing via-counts (a *via* occurs whenever a wire changes direction on a net), minimizing coupling-length between long and *crosstalk sensitive* wires, and minimizing skew. In our formulation of global routing we have used wirelength as the primary objective, with via-minimization and coupling analysis and minimization as the secondary objectives. Our formulation of global routing is based on *flows* in the grid-graph.

Global routing is followed by detailed routing. Detailed routing has developed into its own industry as it needs to cater to foundry specific design rules, timing calculations, and yield analysis. Thus, we shall not present any detailed router in this text (it is fit to be the topic of a book by itself).

A completely routed design database is processed at the *mask shop* where image processing of the data is done to verify printability on the wafer. This is done using a combination of image processing tasks such as optical proximity correction, resolution enhancement and sub-resolution assist features. Like detailed routing, mask optimization is also an industry within its own right, and we defer on that topic as well.

15.4 Schematic capture

15.4.1 Xcircuit

The program xcircuit is a generic drawing program tailored especially for making publication-quality renderings of circuit diagrams (hence the name). The output is pure PostScript, and the graphical interface attempts to maintain as much consistency as possible between the X11 window rendering and the final printer output. An example of a simple circuit being drawn in Xcircuit is shown in Figure 15.2.

Xcircuit has five drawing elements:

1. polygon: (multiple lines which may or may not be closed and filled)
2. arc: (ellipse segment which may be closed and/or filled as above)
3. label: any text placed on the sheet,
4. curve: based on PostScript *curveto* command,
5. instance: object instance of another block.

Schematic representation (as produced from Xcircuit) of the Verilog design as shown in Listing 15.1 is shown in Figure 15.3(b).

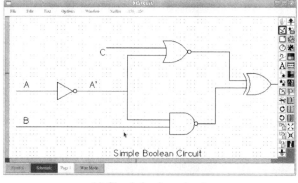

(a) Simple boolean circuit

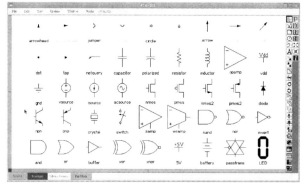

(b) Library elements

Fig. 15.2 Xcircuit : publication quality rendering of circuit schematics

15.4.2 GNU gschem

GNU gschem is part of gEDA suite of CAD tools. It is a full featured schematic
editor. An example of gschem is shown in Figure 15.4. GNU gschem internally
saves the schematic in its own format, but using the gnetlist command, SPICE
netlist can be exported:

```
$gnetlist -g spice-sdb simple_circuit.sch
```

More details on SPICE processing is presented in Section 15.9.

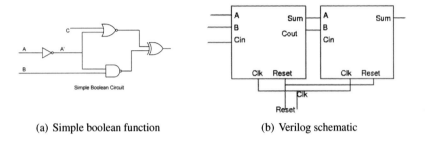

(a) Simple boolean function   (b) Verilog schematic

Fig. 15.3 Xcircuit used for drawing schematic

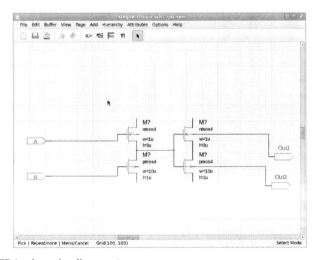

Fig. 15.4 gEDA schematic editor `gschem`

15.5 Verilog Processing

Verilog is a high-level hardware description language (HDL). Although it was originally designed for digital design simulation and verification, it is now also used

for design realization using *logic synthesis*. Alongwith, VHDL, Verilog is the dominant language for expressing digital logic. Structurally, Verilog is similar to C (by design), and VHDL has superficial resemblance to ADA. An example of a Verilog design (for a full adder) is shown in Listing 15.1.

```
    /* \file adder.v
       \author Sandeep Koranne (C) 2010
       \description Example of Verilog file for adder
    */
5   module adder( A, B, Cin, Cout, Sum, Clk, Reset );
       input A,B,Cin,Clk,Reset;
       output Cout, Sum;
       reg    Cout, Sum;
       always @ (posedge Clk or negedge Reset) begin
10       if( Reset == 1'b0 ) begin
         {Cout,Sum} <= 0;
           end
           else begin
         {Cout,Sum} <= A + B + Cin;
15         end
       end
    endmodule // adder
```

Listing 15.1 Verilog design of full adder

Verilog is *timed*, whereas conventional C language has no notion of synchronization. Verilog (and indeed most other hardware description languages) describe *data flow* from *input* to *output*. The bit-width of expressions, and the operators which process these signals have to be carefully defined in Verilog as they have a strong impact on the *area* of the circuit produced. In C or other high level computer languages, the underlying execution machine is the CPU, which already has a predefined bus and register width (e.g., 32-bit or 64-bit). Thus, the cost of a 16-bit operation can be assumed to be the same as a 24-bit operation. In Verilog (and other HDLs) this is not the case, as the area of a 24-bit width data-path is significantly larger than a 16-bit datapath. Hardware designers use *design space exploration* to choose the smallest area producing circuit which meets the design goals.

Once a design has been created as a Verilog file, it can be simulated, synthesized and also tested for resilence against single *stuck-at* faults and delay faults. The specification of the design intent and its implementation should be carefully matched. In C or C++, the design intent is the design documentation, and its implementation if the executable. These are compared using tests. Since, HDLs undergo various transformations (logic synthesis, boolean optimizations, rewriting, retiming), the process of logic verification is important. Moreover, unlike C/C++ executables, where the fix to a detected problem may simply involve a patch, hardware circuits once implemented in chips, require expensive design *respins* costing millions of dollars. This provides additional impetus to the design verification step.

Open-source tools for HDL simulation and synthesis include the Verilog simulators Icarus and GPL `cver`, VHDL synthesis tools from the Alliance CAD suite. These are discussed below.

```
/* \file adder_tb.v
   \author Sandeep Koranne, (C) 2010
```

```
    \description Test bench for adder
 */

module adder_tb;
 reg A,B,Cin,Clk,Reset;
 wire temp;
 wire Cout,Sum;

 initial begin
  $dumpfile("adder_tb.vcd");
  $dumpvars(0,adder_tb);
  $monitor("Time = %d Clk = %b Sum = %d Cout = %d",
           $time, Clk, Sum, Cout );
  Clk <= 0;      A <= 0;
  B <= 0;       Cin <= 0;
  #20 Reset <= 0;
  #5  Reset <= 1;
  #10 A <= 1;
  #30 B <= 1;
 end

 always #10 Clk = ~Clk;
 always #100 $finish;
 adder U1(A,B,Cin,temp,,Clk,Reset);
 adder U2(temp, A, 0, Cout,Sum, Clk, Reset);
endmodule // adder_tb
```

15.5.1 Icarus Verilog Simulator

Icarus Verilog is a Verilog simulation and synthesis tool. It is a native code compiled Verilog simulator which generates machine code for Verilog models, the intermediate machine code is generated in a form called *vvp* assembly. This assembly code is executed on a virtual machine called *vvp*. The synthesis engine generates netlists in diverse formats including XNF (Xilinx Netlist Format). The compiler proper (or the compiler driver to be exact) is called *iverilog*. To prepare the above listed design for simulation we can run iverilog adder.v.

```
$ iverilog -h
Usage: iverilog [-ESvV] [-B base] [-c cmdfile|-f cmdfile]
               [-g1995|-g2001|-g2005] [-g<feature>]
               [-D macro[=defn]] [-I includedir]
               [-M depfile] [-m module][-N file]
               [-o filename] [-p flag=value][-s topmodule]
               [-t target] [-T min|typ|max] [-W class]
               [-y dir] [-Y suf] source_file(s)
```

```
$iverilog adder.v
```

A part of the generated VVP assembly is shown below for reference:

```
#! /usr/bin/vvp
:ivl_version "0.9.2 " "(v0_9_2)";
:vpi_time_precision + 0;
:vpi_module "system";
:vpi_module "v2005_math";
:vpi_module "va_math";
S_0x8ac00d8 .scope module, "adder" "adder" 2 5;
 .timescale 0 0;
v0x8ac0168_0 .net "A", 0 0, C4<z>; 0 drivers
v0x8ae0140_0 .net "B", 0 0, C4<z>; 0 drivers
v0x8ae01a0_0 .net "Cin", 0 0, C4<z>; 0 drivers
v0x8ae0200_0 .net "Clk", 0 0, C4<z>; 0 drivers
v0x8ae0268_0 .var "Cout", 0 0;
v0x8ae02c8_0 .net "Reset", 0 0, C4<z>; 0 drivers
v0x8ae0348_0 .var "Sum", 0 0;
E_0x8ae3da8/0 .event negedge, v0x8ae02c8_0;
E_0x8ae3da8/1 .event posedge, v0x8ae0200_0;
E_0x8ae3da8 .event/or E_0x8ae3da8/0, E_0x8ae3da8/1;
    .scope S_0x8ac00d8;
T_0 ;
    %wait E_0x8ae3da8;
    %load/v 8, v0x8ae02c8_0, 1;
    %cmpi/u 8, 0, 1;
    %jmp/0xz  T_0.0, 4;
    %ix/load 0, 1, 0;
    %assign/v0 v0x8ae0348_0, 0, 0;
    %ix/load 0, 1, 0;
    %assign/v0 v0x8ae0268_0, 0, 0;
    %jmp T_0.1;
T_0.0 ;
    %load/v 8, v0x8ac0168_0, 1;
    %mov 9, 0, 1;
    %load/v 10, v0x8ae0140_0, 1;
    %mov 11, 0, 1;
    %add 8, 10, 2;
    %load/v 10, v0x8ae01a0_0, 1;
    %mov 11, 0, 1;
    %add 8, 10, 2;
    %ix/load 0, 1, 0;
    %assign/v0 v0x8ae0348_0, 0, 8;
    %ix/load 0, 1, 0;
    %assign/v0 v0x8ae0268_0, 0, 9;
T_0.1 ;
    %jmp T_0;
    .thread T_0;
# The file index is used to find the file
# name in the following table.
:file_names 3;
    "N/A";
```

```
"<interactive>";
"adder.v";
```

The form of the genrated VVP assembly is instructive to understand the internal working of Icarus Verilog.

15.5.2 Pragmatic GPL cver

GPLCVER 2.11a is the name of a Verilog simulator written by Pragmatic C Software Corp and placed under GNU GPL. Simulating the Verilog design shown in Listing 15.1 gives us:

```
Compiling source file "adder.v"
Compiling source file "adder_tb.v"
Time =        0 Clk = 0 Sum = x Cout = x
Time =       10 Clk = 1 Sum = x Cout = x
Time =       20 Clk = 0 Sum = 0 Cout = 0
Time =       30 Clk = 1 Sum = 0 Cout = 0
Time =       40 Clk = 0 Sum = 0 Cout = 0
Time =       50 Clk = 1 Sum = 1 Cout = 0
Time =       60 Clk = 0 Sum = 1 Cout = 0
Time =       70 Clk = 1 Sum = 1 Cout = 0
Time =       80 Clk = 0 Sum = 1 Cout = 0
Time =       90 Clk = 1 Sum = 0 Cout = 1
Halted at location **adder_tb.v(26)
                time 100 from call to $finish.
  There were 0 error(s),
   2 warning(s), and 8 inform(s).
```

15.5.3 GTKWave: Waveform Viewer

The VCD (value change dump) file generated during the simulation can be viewed graphically using the GTKWave waveform viewer. Running GTKWave on the VCD file gtkwave adder_tb.vcd and adding the signals from instance U1 is shown in Figure 15.5.

15.6 VHDL Processing

Very high-speed integrated circuits hardware design language (VHDL) is a standardized language for describing the behavior and implementation of digital cir-

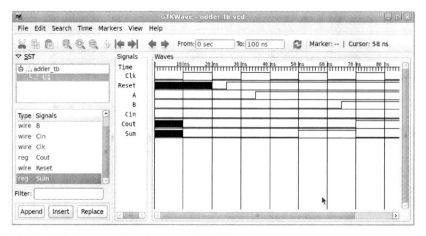

Fig. 15.5 GTKWave, waveform viewer from VCD files

cuits. It has syntactical resemblance to the ADA computer programming language.
An example of an adder-accumulator is shown in Listing 15.2.

```
    -- Adder-accumulator example in VHDL, from Alliance
    library IEEE;
    use IEEE.STD_LOGIC_1164.ALL;
    use IEEE.STD_LOGIC_arith.ALL;
5   use IEEE.STD_LOGIC_unsigned.ALL;


    entity AddAccu is

10      port ( CLK : in Std_Logic;
                CLR : in Std_Logic;
                LD  : in Std_Logic;
                A   : in Std_Logic_Vector(15 downto 0) ;
                RESULT : out Std_Logic_Vector(15 downto 0) );
15
    end AddAccu;


    architecture DataFlow OF AddAccu is
20  signal resultint : Std_Logic_Vector(15 downto 0) ;
    begin
    process (CLK)
    begin
      if CLK'event and CLK='1' then
25        if CLR = '1' then resultint  <= ( others => '0' );
          elsif LD = '1' then resultint <= resultint + A;
          end if;
     end if;
    end process;
30  RESULT <= resultint;
    end DataFlow;
```

Listing 15.2 VHDL example of adder-accumulator

We shall use the Alliance CAD system's VHDL simulator *asimut* to simulate this
design.

15.7 Alliance CAD System

Alliance CAD is a complete VLSI CAD system. It includes tools for VHDL processing, schematic capture, logic synthesis, test pattern generation, layout editing, standard cell placement, routing, design rule checking (DRC), circuit extraction, layout versus schematic (LVS) comparison, and mask generation. We use the design shown in Listing 15.2 to examine the above mentioned CAD tools.

15.7.1 Alliance CAD VHDL processing

The input VHDL can include both structural (component interconnection) and behavioral statements. The Alliance CAD tool `vasy` (VHDL Analyzer for Synthesis) is used to process a given VHDL file. `vasy` accepts a subset of the complete VHDL language. VASY has a number of command line arguments which control its behavior. To write a '.vbe' behavioral file we add the '-a' argument, and the '-p' argument adds the power (VDD and VSS connectors). A complete list of the commandline arguments can be found by running `man vasy`.

```
vasy -a -B -o -p -I vhdl addaccu
```

The output of this tool is a '.vbe' file which expands any operators which are present in the VHDL file. This processed '.vbe' file can then be simulated using the VHDL simulator `asimut`.

15.7.2 Alliance CAD tool `asimut`

In Verilog, we had written an explicit test-bench module which contained the stimulus for the CLK, RESET, and the input signals. In Alliance CAD, the simulation *patterns* are listed in an external file. An example is shown below:

```
-- input / output list :
in        clk B;;
in        clr B;;
in        ld B;;
in        a (15 downto 0) X;;;
out       result (15 downto 0) X;;;

begin

<   0ns>  : 0 1 0 0000   ?****   ;
< +5ns>   : 0 1 0 0000   ?****   ;
< +5ns>   : 0 1 0 0000   ?****   ;
```

```
..
< +5ns>  : 1 0 1 0004  ?0004  ;
< +5ns>  : 0 0 1 0004  ?0004  ;
end;
```

The format of the pattern file is shown above. The list of input and output signals is followed by a sequence of time steps at which the input signals change values. In the above example, the clk, clr, and ld signal are 1 bit wide, whereas the input A signal is 15-bit. The advantage of using an external pattern file is that it can be generated from a program and can contain *expected* values for outputs. When simulated, any mismatch between the computed value and the value stored in the pattern file is treated as a simulation mismatch error and printed on the console.

   The VHDL simulator in Alliance is asimut. It compiles and loads a complete hardware description written in VHDL . The hard- ware description may be structural (a hierarchy of instances) or behavioral. asimut also expects a pattern file as the input; the format of the pattern file was discussed above. The pattern file is compiled, loaded and linked with the hardware description, and the simulation is started.

```
$asimut -b addaccu addaccu res_vasy_1
..linking ...
executing ...
###----- processing pattern 0 : 0 ps -----###
###----- processing pattern 1 : 5000 ps -----###
###----- processing pattern 2 : 10000 ps -----###
..
```

The speed of the simulation can be increased by using BDD (binary decision diagrams), by specifying the -bdd commandline argument to asimut. Another advantage of using external pattern files is that post-synthesis gate level netlist can be simulated with the same pattern file. This provides a check for synthesis mismatch. The logic synthesis step of converting high-level VHDL circuit description to elementary gates is described next.

15.7.3 VHDL Logic Synthesis using Alliance CAD tool Boom

The Alliance CAD tool Boom is used for logic synthesis. The commandline for using boom is shown below:

```
$boom [options] [Algorithm] Input Output
```

The options available are:

```
-V      Sets verbose mode on
-T      Sets trace and verbose modes on
-O      Reverses initial Bdd variables order
```

```
-A       Keeps all auxiliary variables
-P       Uses a parameter file (Input_file_name.boom)
-L       Uses literal's number for surface estimation
-l num Optimization level [0-3] (default 0, low level)
-d num Delay optimization percent (default 0 %)
-i num Iteration count
-a num Amplitude
```

And the algorithm available to Boom are:

```
-s       Simulated annealing (default)
-j       Just do it algorithm
-b       Burgun algorithm
-g       Gradient algorithm
-p       Procrastination algorithm
-w       Window bdd reorder
-t       Top bdd reorder
-m       Simple bdd reorder
-o       One pass (faster algorithm)
-r       Random bdd reorder
-n       No optimization algorithm
```

We use the `addaccu` example to show the logic synthesis procedure. Using `vasy` we had produced the '.vbe' file which is given as input to `boom`:

```
$boom -V -T -j addaccu.vbe
--> Parse BEH file addaccu.vbe

--> Check figure addaccu

--> Optimization parameters
    Algorithm : just do it
    Keep aux  : no
    Area      : 100 %
    Delay     :   0 %
    Level     :   0

--> Initial cost
    Surface  : 349750
    Depth    : 49
    Literals : 266

--> Translate Abl to Bdd
    Keep register signal resultint 14
    Keep signal rtlcarry_0 1

--> Optimization % 100
```

```
--> Final cost
    Surface  : 289500
    Depth    : 37
    Literals : 230

--> Post treat figure addaccu

--> Drive BEH file addaccu_o
```

The resulting '.vbe' file (addaccu_.vbe) consists of a gate level netlist. The *depth* of the circuit is 7, which means that there is atleast 1 path of 7 components from the input to output. Using different algorithms, the area cost of the circuit can be reduced, as well as a tradeoff between area and speed of the circuit performed. While the produced gate-level file is logically equivalent to the given input, it can be further optimized using Boolean transformations and mapped to a standard cell library. In Section 15.3 we had discussed the BLIF format and its specification of standard cell libraries. The impact of a good standard cell library function in reducing area and delay was shown in Figure 15.1, where the same logical function was implemented in smaller area by using a more complex standard cell. In Alliance, standard cell libraries are defined using Catalog files, and the library mapper tool is called boog. boog takes as input the gate-level '.vbe' file, and the target library. It produces a gate-level netlist.

```
boog <input_file> [-o <output_file>] [-l <lax_file>]
                  [-x <xsch_mode>] [-m <optim_mode>]
MBK_TARGET_LIB : /usr/lib/alliance/cells/sxlib
```

We use boog to *map* the addaccu_o.vbe gate-level file to a netlist of cells from the Alliance sxlib standard cell library:

```
$boog addaccu_o -o addaccu -x 1 -m 2
```

We specify the optimization mode (using -m 2) as well as generate a highlighted schematic file (the critical path is marked) using the -x 1 commandline arguments. The input and output file names are also given to the tool.

```
Reading default parameter...
50% area - 50% delay optimization
Reading file 'addaccu_o.vbe'...
Controlling file 'addaccu_o.vbe'...
Reading lib '/usr/lib/alliance/cells/sxlib'...
Preparing file 'addaccu_o.vbe'...
Capacitances on file 'addaccu_o.vbe'...
Unflattening file 'addaccu_o.vbe'...
Mapping file 'addaccu_o.vbe'...
Saving file 'addaccu.vst'...
Adding signal 'not_a 3'
```

```
Adding signal 'not_a 7'
Quick estimated area
    (with over-cell routing)...265250 lambda
Details...
inv_x2: 36
xr2_x1: 31
buf_x2: 16
sff1_x4: 16
oa2a22_x2: 16
...
no2_x1: 1
Total: 144
Saving delay gradient in xsch color file
'addaccu.xsc'...
End of boog...
```

The standard cell mapping produces a '.vst' file (addaccu.vst) as well as a schematic tile (addaccu.xsc).

15.7.4 Alliance CAD tool *xsch* schematic viewer

We have seen the XCircuit tool for schematic depiction in Section 15.4. The Alliance CAD tool for schematic viewing is called xsch. Using the generated 'addaccu.xsc' file from boog we can run the schematic viewer as:

```
$xsch -l addaccu
```

The produced schematic plots are shown in Figure 15.6. The critical path is marked. xsch can also generate Xfig '.fig' format files. Xfig is a vector graphics editor (see Section 19.7).

In addition to the schematic file, the main output of boog is the gate-level netlist which represents our adder-accumulator circuit in terms of standard cell library elements. This gate-level netlist is optimized and converted to physical geometry in the next sections.

15.7.5 Gate level processing in Alliance CAD

The gate-level netlist format used in Alliance is structural VHDL. This is a subset of VHDL used for component netlisting. An example is shown below:

```
entity addaccu is
   port (
      clk    : in      bit;
```

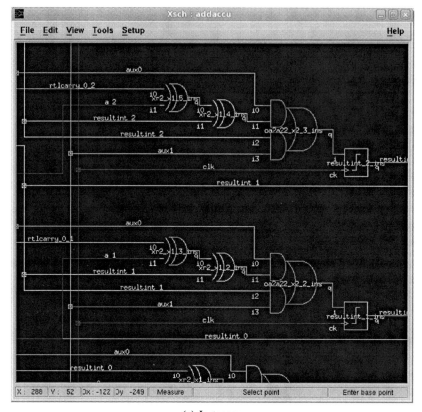

(a) Instances

Fig. 15.6 Schematic viewer tool in Alliance CAD `xsch`

```
        clr    : in       bit;
        ld     : in       bit;
        a      : in       bit_vector(15 downto 0);
        result : out      bit_vector(15 downto 0);
        vdd    : in       bit;
        vss    : in       bit
    );
end addaccu;

architecture structural of addaccu is
Component no2_x1
    port (
        i0 : in       bit;
        i1 : in       bit;
```

```
         nq  : out      bit;
         vdd : in       bit;
         vss : in       bit
  );
end component;
....
not_rtlcarry_0_12_ins : inv_x2
   port map (
       i   => rtlcarry_0_12,
       nq  => not_rtlcarry_0_12,
       vdd => vdd,
       vss => vss
   );

not_resultint_11_ins : inv_x2
   port map (
       i   => resultint(11),
       nq  => not_resultint(11),
       vdd => vdd,
       vss => vss
   );
end structural;
```

The gate-level '.vst' file declares an *entity* which is the top-level design addaccu. Subsequently, every standard cell which is instanced in this design is *declared*. Then a *structural* architecture of the design is defined as the interconnection of instances. The interconnections are made using *signals* which are equivalent to wires connection outputs of components to inputs of other instances. This gate-level netlist is actually a collection of directed acyclic graphs from the *primary inputs* of the design, to *primary outputs* of the design. The *critical path* in this graph is the path from an input to output which has the maximum cumulative delay on it. To increase the speed of the circuit, this path delay should be minimized. In VLSI designs, increasing the area of the driver cells can reduce the path delay (but at a cost of increasing area). This analysis, of which cells to *resize* is complex, and dependent on the graph theoretic analysis of the design, as well as transistor level analysis of the standard cells. The Alliance CAD tool loon (local optimization on nets) performs this optimization. loon takes as input the gate-level .vst file and produces an equivalent (but optimized) gate-level netlist.

```
$loon addaccu addaccu_o
Reading default parameter...
50% area - 50% delay optimization
Reading file 'addaccu.vst'...
Reading lib '/usr/lib/alliance/cells/sxlib'...
Capacitances on file 'addaccu.vst'...
Delays on file 'addaccu.vst'...13297 ps
Area on file 'addaccu.vst'...
  265250 lamda (with over-cell routing)
```

```
Details...
inv_x2: 36 (10%)
xr2_x1: 31 (26%)
buf_x2: 16 (6%)
sff1_x4: 16 (27%)
oa2a22_x2: 16 (13%)
a2_x2: 13 (6%)
nao2o22_x1: 12 (7%)
oa2ao222_x2: 2 (1%)
an12_x1: 1 (0%)
no2_x1: 1 (0%)
Total: 144
Worst RC on file 'addaccu.vst'...438 ps
Inserting buffers on critical path
...15 buffers inserted -> 12698 ps
Improving RC on critical path for file
  'addaccu_o.vst'...12558 ps
Improving all RC for file 'addaccu_o.vst'...
Worst RC on file 'addaccu_o.vst'...438 ps
Area on file 'addaccu_o.vst'...
  281250 lamda (with over-cell routing)
Details...
inv_x2: 33 (8%)
xr2_x1: 31 (24%)
buf_x2: 30 (10%)
sff1_x4: 16 (25%)
oa2a22_x2: 16 (12%)
a2_x2: 13 (5%)
nao2o22_x1: 12 (7%)
inv_x4: 3 (1%)
oa2ao222_x2: 2 (1%)
buf_x4: 1 (0%)
an12_x1: 1 (0%)
no2_x1: 1 (0%)
Total: 159
Critical path (no warranty)...12558 ps
  from 'resultint 0' to 'resultint_15_ins'
Saving file 'addaccu_o.vst'...
```

The loon tool has resized some of the cells on the critical path to use larger drive strength variants. It can also add inverter-pairs as buffers on some paths. This optimized gate-level netlist is the handoff from the front-end design team to the back-end team which converts this textual information to physical mask geometry which is then sent off to the VLSI fab house for manufacturing as computer chips.

15.7.6 Physical design with Alliance CAD

The first step in converting a gate-level standard cell netlist to manufacturable geometry is *standard cell placement*. To aid in automatic placement, all standard cells have the same height (thats why they are called standard). A standard cell placer arranges these cells in *rows* while minimizing the wire length. Placement is a hard combinatorial problem, and therefore advanced techniques have been developed to approximate a good solution. The Alliance CAD tool's standard cell placer is called `ocp` and implements a simulated-annealing based placer. In addition to placing the cells, `ocp` can also perform IO ring placement, from a given IO constraint file:

```
TOP ( # IOs are ordered from left to right
    (IOPIN clk.0 );
    (IOPIN clr.0 );
    (IOPIN ld.0 );
    (IOPIN a(15).0 );
)
BOTTOM ( # IOs are ordered from left to right
    (IOPIN result(15).0 );
)
IGNORE ( # IOs are ignored(not placed) by IO Placer
)
```

The number of rows can be input to the placer program. We run `ocp` on our design:

```
$ocp -v -gnuplot -ioc addaccu_o addaccu addaccu_p

 o Number total of instances is ....  144
 o Number of instances to place is ....  144
 o Number of instances already placed is ....  0
 o Number of nets is .... 170
 o Sum of instances to place widths is ... 1061
 o Computing Initial Placement ...
 o User Margin : 20%
 o Number of Rows : 11
 o Real Margin : 16.8495%
 o Width of the abutment box : 116
 o Height of the abutment box : 110
 o conspace : 6.10526 1st connector : 3.05263
 o adding connector   : clk x : 3 y : 110
 o adding connector   : clr x : 9 y : 110
 o adding connector   : ld x : 15 y : 110
 o adding connector   : a 15 x : 21 y : 110
 o Initial Placement Computing ... done.
 o Beginning global placement ....
Loop = 1, Temperature = 0.245678, Cost = 0.959008
```

```
 RowCost = 129.455, BinCost = 260.597, NetCost = 15557.5
 Success Ratio = 99.5688%, Dist = 1, Delta = 0.5
o Total impossible movements = 650
o 3.84615 % suroccupied target
o Final Optimization succeeded ...
o Final Net Cost ..... 5094.5
o Final Net Cost Optimization ..... 23.7407%
o Total Net Optimization .... 68.5961%
```

```
Ocp : placement finished
```

The produced placement is a coordinate assignment of (X,Y) for every instance of
the netlist:

```
V ALLIANCE : 6
H addaccu_p,P,13/ 7/2010,100
A 0,0,58000,55000
I 46000,0,tie_x0,tiex0_93,SYM_Y
I 44000,45000,tie_x0,tiex0_92,NOSYM
I 43000,0,tie_x0,tiex0_91,SYM_Y
I 42000,50000,tie_x0,tiex0_90,SYM_Y
...
```

The placement can be viewed using the graphical symbolic viewer tool in Alliance
CAD graal as shown in Figure 15.7.

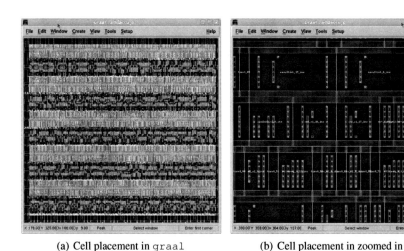

(a) Cell placement in graal   (b) Cell placement in zoomed in

Fig. 15.7 Alliance CAD tool graal showing OCP placement

A more detailed view of the placement coordinates is shown in Figure 15.8(a), and the metal interconnect layer which lines up in the standard cell rows is shown in Figure 15.8(b).

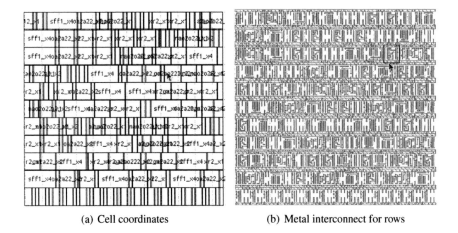

(a) Cell coordinates   (b) Metal interconnect for rows

Fig. 15.8 Standard cell placement showing cell blocks and metal interconnect

15.7.7 Alliance CAD tool for standard-cell routing: `nero`

The standard-cell router in Alliance CAD tools is called `nero`, for *Negotiating Router*. It takes as input the number of metal tracks to use for routing, the design file containing cell coordinates (placement output) and the gate level netlist. It produces a symbolic placement file containing valid track assignments for metal wires which connect the various components of the netlist together. VLSI routing is another hard combinatoric problem, and the router implemented in Alliance CAD uses *negotiation*, and priority ordering to converge to a valid routing solution. The output of the router is shown in Figure 15.9(a) and Figure 15.9(b).

Till this point, the physical geometries are maintained in symbolic, or *lambda* form. The advantage of symbolic geometry representation is the ease of technology migration, scaling, design rule verification and circuit extraction. The tool for circuit extraction is `cougar` which takes as input the post routed layout, and extracts the netlist of interconnections (including parasitic resistance and capacitance). The extracted netlist can be in hierarchical SPICE format:

```
* Spice description of addaccu_e
* Spice driver version 11952808
* Date ( dd/mm/yyyy hh:mm:ss ): 13/07/2010 at 16:12:02
```

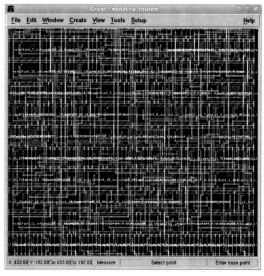

(a) Routing output in `graal`

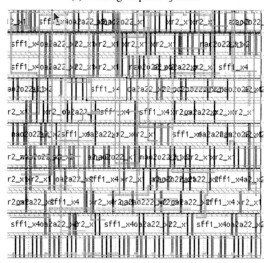

(b) Metal track for routing

Fig. 15.9 Output from `nero` negotiating router

```
* INTERF a[0] a[1] a[2] .. a[11] a[15]
* INTERF clk clr ld
* INTERF result[0] ... result[15]
* INTERF vdd vss
```

```
.INCLUDE rowend_x0.spi
.INCLUDE tie_x0.spi
...
C102 rtlcarry_0_3 vss 8.90499e-15
C101 a[3] vss 2.31066e-14
C100 xr2_x1_7_sig vss 6.83829e-15
C99 mbk_buf_rtlcarry_0_3 vss 6.7001e-15
C1 result[15] vss 3.77352e-15
.ends addaccu_e
```

The extracted SPICE netlist can be analyzed using `ngspice` SPICE simulator. See Section 15.9 for more details on using SPICE simulators.

Using the Alliance CAD tool `s2r` (symbolic to real), symbolic geometries are converted to actual mask data. The tool can moreover perform post translation treatment of the layout.

15.7.8 QUCS : Universal Circuit Simulator

QUCS is an integrated electronic circuit simulator. An example of a simple digital circuit designed in QUCS is shown in Figure 15.10.

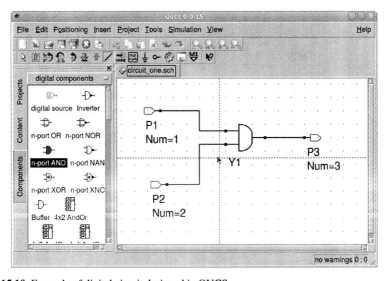

Fig. 15.10 Example of digital circuit designed in QUCS

The circuit can be simulated within QUCS, and even DC, AC and S-parameter analysis can be performed. The schematic design format of QUCS is XML:

```
<Qucs Schematic 0.0.15>
```

```
<Properties>
  <View=13,12,396,265,1.5,0,0>
  <Grid=10,10,1>
..
</Properties>
<Components>
  <Port P3 1 300 100 4 12 1 2 "3" 1 "out" 0>
  <Port P1 1 80 60 -23 12 0 0 "1" 1 "in" 0>
  <Port P2 1 90 170 -23 12 0 0 "2" 1 "in" 0>
  <AND Y1 1 210 100 -26 27 0 0 "2" 0 "1 V" 0 "0" 0 "10" 0 "old" 0>
</Components>
```

15.8 Magic VLSI Editor

Magic is a full custom VLSI layout editor.

It can represent Manhattan mask geometry for a wide variety of process nodes, and uses a clever *corner stitched* data representation alongwith layer *tiling* which allows Magic to perform design-rule checking (DRC) and layout extraction very efficiently. Another advantage of Magic is that it represents the design geometry as simple TEXT files. Consider a small layout example shown in Figure 15.11.

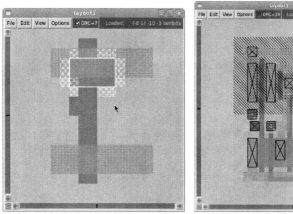

(a) Partial layout in Magic   (b) Example of Standard cell

Fig. 15.11 Magic VLSI layout editor

The same layout in Magic's layout file is shown below:

```
magic
tech scmos
timestamp 1279089463
<< error_s >>
```

```
rect 2 7 3 8
rect 5 7 7 8
rect 1 5 2 7
rect 1 4 3 5
rect 7 4 8 7
rect 2 3 3 4
rect 5 3 7 4
<< polysilicon >>
rect 3 7 5 9
rect 3 3 5 4
rect 2 1 5 3
rect 3 -6 5 1
<< metal1 >>
rect 0 5 11 8
rect 0 -5 11 -2
<< ntransistor >>
rect 2 4 7 7
<< end >>
```

15.9 NGSpice SPICE Engine

SPICE is an acronym for simulation program with intergraged circuit emphasis. `ngspice` is a general purpose circuit simulator for linear and non-linear circuit analysis. It is an update of the Berkeley Spice 3f5 program. `ngspice` can perform: (i) analog simulation, (ii) mixed-mode simulation as well as (iii) digital circuit simulation. The digital simulation model takes advantage of *event propagation* to optimize the computation. `ngspice` supports the following types of analysis:

1. DC analysis: operating point and DC sweep, with inductors shorted and capacitors opened. A DC analysis is automatically performed before transient analysis,
2. AC small signal analysis:limited to analog nodes, and represents the steady-state behavior at a particular set of stimulus frequencies,
3. Transient analysis: extension of DC analysis to the time domain,
4. Pole-zero analysis: computes poles and zeros in the small-signal AC transfer function,
5. Small-signal distortion analysis: computes steady-state harmonics,
6. Sensitivity analysis: calculates either the DC operating-point sensitivity, or AC small-signal sensitivity,
7. Noise analysis: analyzes device generated noise.

Circuits in SPICE are described using topological conventions: (i) circuit cannot contain a loop of voltage sources, (ii) each node in the circuit must have a dc path to ground, and (iii) every node must have at least two connections. A simple SPICE circuit is shown in Figure 15.12.

The circuit shown in Figure 15.12 can be defined in SPICE as:

```
* Example of a simple circuit in Spice
VS 1 0 DC 10V
```

Fig. 15.12 Example circuit
for SPICE analysis

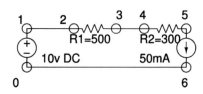

```
IS 5 0 DC 50MA
R1 1 3 500
R2 3 5 300
.END
```

The circuit has a voltage source between nodes 1 and 0 of 10 volts, and a current source of 50 mA between nodes 5 and 0. The two resistors are placed in series. The SPICe file contains a title, and the last line contains .end to depict end of circuit. Hierarchical SPICE netlists can be generated:

```
.subckt VMODULE 1 2 3 ; defines a subcircuit
r1 1 2 500
r2 2 3 100
.ends

U1 5 6 9 VMODULE ; refers to an instance of VMODULE
```

15.9.1 Elementary devices in SPICE

SPICE already knows about elementary devices such as resistors, capacitors, voltage sources, current sources and MOSFETS. Resistors are introduced using the R<n> notation, capacitors with the C<n> notation, inductors with L<n>, coupled inductors with K<n>, switches with S<n>, voltage sources with VS<n>, current sources with IS<n> and MOSFETS with M<n>. Diodes are represented with D<n> and BJT models with Q<n>.

A MOSFET model is defined as:

```
M<n> nd ng ns nb mname <m=val> <l=val> <w=val>
```

where *nd* denotes the connection to *drain*, *ng* connection to *gate*, *ns* connection to *source*. The model name is given in mname, and device parameters *m* denotes multiplicity, while *l* and *w* denote the *length* and *width* of the MOSFET channel, respectively. ngspice supports advanced device models in MOSFETS, including the BSIM1, and BSIM3 models.

15.9.2 Performing TRANSIENT analysis

An example of performing transient analysis on the above circuit is shown below.

```
ngspice 1 -> source ex1.cir
Circuit: * example of a simple circuit in spice
ngspice 2 -> tran 1ns 10ns
Doing analysis at TEMP = 27.000000 and TNOM = 27.000000
Initial Transient Solution
--------------------------
Node                                           Voltage
----                                           -------
1                                                   10
5                                                  -30
3                                                  -15
vs#branch                                        -0.05
%100.00
No. of Data Rows : 60
```

ngspice also has the ability to plot signal waveforms directly from the interactive command terminal.

15.9.3 Conclusion

In this section we have discussed the many open-source tools we have for doing VLSI CAD. Verilog, VHDL simulation, logic synthesis, placement, routing, schematic generation, physical layout editing and SPICE simulation were discussed in this section. In addition to VLSI CAD, some of the software libraries developed for VLSI have also found use in othe domains. For example, Tcl/Tk were initially developed for glue logic between CAD tools, binary-decision diagram libraries for logic synthesis and verification are also used in combinatoric search, and satisfiability testing is also used as a search space exploration tool in integer linear programming.

Chapter 16
Math libraries

Abstract Most of the open-source mathematics software is built upon libraries which implement mathematical functions and computations. Since mathematics is an integral part of most scientific computing and research, it is to be expected that these libraries can also often be used in application software written by the reader. In this chapter we discuss some of the most popular mathematical libraries which are used in open-source applications. The libraries cover almost all aspects of mathematics including linear algebra, number theory, graph theory, FFT, multi-precision computation and linear programming.

Contents

In this chapter we discuss several interesting and useful tools available for performing mathematics using computers. They include symbolic algebra, matrices, linear algebra, group theory, polynomials, statistics. Specifically we shall cover LAPACK, BLAS and ATLAS foundation libraries. NTL, number theory library, GNU Scientific library, GNU Multiprecision bignum library. For sparse solvers, TAUCS, FFT using FFTW library. Linear programming with glpk. Complete open sourced mathematical systems can be found in GNU Octave, Maxima, R, and Sage. Group theoretic software pari and its front-end 'gp'. A formalized mathematical system AXIOM and the REDUCE computer algebra software package. A front end to many of these software alongwith publication quality typesetting is 'TeXMacs'. We round-off this section with the discussion of the Computer Geometry Algorithms Library (CGAL).

S. Koranne, *Handbook of Open Source Tools*,
DOI 10.1007/978-1-4419-7719-9_16, © Springer Science+Business Media, LLC 2011

Linear algebra is significant component of almost all engineering and scientific work loads. In this section we discuss some of the software libraries which are available for linear algebra computation.

16.1 BLAS

Basic Linear Algebra Subprograms (BLAS) is an application programming interface (API) standard for publishing libraries to perform basic linear algebra operations such as vector and matrix multiplication. They are most commonly used in high-performance computing and are used to develop larger packages such as LAPACK. Optimized implementations of the BLAS interface have been developed by hardware vendors such as Intel and AMD as part of their MKL and ACML libraries respectively. BLAS is divided into 3 levels:

1. Level-1: vector only operations of the type $y \leftarrow \alpha x + y$,
2. Level-2: matrix-vector operations, $y \leftarrow \alpha A x + \beta y$,
3. Level-3: matrix-matrix operations, such as: $C \leftarrow \alpha A B + \beta C$. This level contains the GMM (or general matrix multiply) operation.

The type of the vector or matrix is defined as part of the name of the operation; the type is one of single-float, double-float, or complex. As mentioned above many optimized implementation of BLAS are available, and since it is the building block of more complex libraries, performance improvement in BLAS are often critical to improving system performance. An example of a BLAS function to compute matrix-vector product is shown in Listing 16.1.

```
   // \file blas_example.cpp
   // \author Sandeep Koranne (C) 2010
   // \description Example of using BLAS in C++
   #include <iostream>          // Program IO
5  extern "C" {
   #include <cblas.h>           // BLAS header
   }
   #if 0
   cblas_dgemv(const enum CBLAS_ORDER Order,
10       const enum CBLAS_TRANSPOSE TransA,
         const int M, const int N,
         const double alpha,
         const double *A, const int lda,
         const double *X, const int incX,
15       const double beta,
         double *Y, const int incY);
   #endif
   static double X[] = { 1.0, 2.0, -2.0, -1.0 };
   static double A[] = {
20    4.0, 3.0, 1.0, 2.0,
      2.0, 1.0, 5.0, 3.0,
      6.0, 3.0, 1.0, 2.0,
      1.0, 4.0, 6.0, 2.0
   };
25
   int main( int argc, char* argv[] ) {
      std::cout << "A = ";
      for( int i=0; i < 4; ++i ) {
```

```
            for( int j=0; j < 4; ++j ) {
30            std::cout << A[i*4+j] << " ";
            }
            std::cout << "\n";
          }
          double *y = new double[4];
35        // compute Ax
          cblas_dgemv( CblasRowMajor, CblasNoTrans, 4, 4, 1.0,
                A, 4, X, 1, 0.0, y, 1 );
          for(int i=0; i < 4; ++i ) std::cout << y[i] << " ";
          std::cout << std::endl;
40        return(0);
        }
```

Listing 16.1 Example of using BLAS library

The use of the `cblas_dgemv` function demonstrates the flexibility of the underlying API.

16.2 ATLAS

ATLAS is an acronym for Automatically Tuned Linear Algebra Software. It provides an implementation of BLAS, but it is unique in the sense that it provides a completely automatic method of generating optimized BLAS libraries from the same code on any machine or architecture. Even though the performance of ATLAS cannot compare to hand-optimized machine specific software, ATLAS is often used to build a quick and optimized version of BLAS while the customized functions are implemented.

ATLAS uses automatic tuning of code parameters based on experimental runs of standardized software on the target machine as part of the compilation. Thus it is critical to use a representative machine (under nominal workload) to compile ATLAS. Given the speculative nature of modern architecture's instruction pipeline, and large dependency of data movement on cache, ATLAS uses data layout and cache line optimization to derive its performance.

When compiling code which uses the ATLAS libraries, the linking step should be performed in the following order:

```
-llapack -lcblas -lf77blas -latlas
```

This ordering is important for the linker to find the ATLAS library functions.

16.3 LAPACK

LAPACK (Linear Algebra PACKage) is a software library for numerical linear algebra. It provides routines for solving systems of linear equations and linear least squares, eigenvalue problems, and singular value decomposition. It also includes routines to implement the associated matrix factorizations such as LU, QR, Cholesky and Schur decomposition.

16.4 NTL

Number Theory Library (NTL) is a C++ library for performing experiments and computations in *number theory*. NTL supports arbitrary length integer computation, computation over finite fields, vectors, matrices, polynomials and lattices.

```
#include <NTL/ZZ.h>
NTL_CLIENT
int main() {
    ZZ a, b, c;
5   cin >> a;
    cin >> b;
    c = (a+1)*(b+1);
    cout << c << "\n";
}
```

Listing 16.2 Example of using NTL

Running this example produces:

```
123456789
123456789
15241578997104100
```

16.5 GSL

GNU Scientific Library (GSL) is a software library for numerical computations in science and mathematics. The library implements the calculation of numerically evaluated functions such as Bessel functions. Consider the program shown in Listing 16.3.

```
#include <stdio.h>
#include <gsl/gsl_sf_bessel.h>
int main(void) {
    double x = 5.0;
5   double y = gsl_sf_bessel_J0(x);
    printf("J0(%g) = %.18e\n", x, y);
    return 0;
}
```

Listing 16.3 Example of using GNU Scientific Library (GSL)

Running this program requires:

```
gcc $(gsl-config --cflags) gsl_example.c
    $(gsl-config --libs)
J0(5) = -1.775967713143382920e-01
```

The GSL library includes functions for:

- Physical constants and Basic mathematical functions
- Complex numbers, Polynomials, Vectors and matrices
- Special functions
- Permutations, Combinations and Multisets
- Linear algebra, Eigensystems
- FFT, Discrete Hankel transform and Discrete wavelet transform
- Numerical differentiation and integration
- Random number generation and Quasi-random sequences
- Statistics, Histograms, N-tuples
- Simulated annealing
- Ordinary differential equations
- Interpolation, Chebyshev approximations
- Root-finding in one and multiple dimensions
- Minimization in one and multiple dimensions
- Least-squares fitting, Nonlinear least-squares fitting
- IEEE floating-point arithmetic

16.6 GMP

GNU multi-precision library (GMP) is a library for arbitrary precision arithmetic on signed integers, rational numbers and floating point numbers. The only limit to the precision is imposed by the available memory of the machine. The applications of GMP include cryptography, computational geometry and computer algebra.

16.7 MPFR

GNU multi-precision float rounding (MPFR) is a library for arbitrary precision floating point computation with *correct rounding*. The computation is efficient and has correct mathematical semantics. It is used for computational geometry applications, and like GNU GMP, is now required to build GCC.

16.8 FFTW

FFTW is an acronym for "fastest Fourier transform in the west". It is a software library for computing discrete FFT and was developed at MIT. Using FFTW, transforms of real and complex values vectors can be carried out. It is one of the fastest FFT libraries. Like ATLAS, FFTW is based on a *plan* based computation, except in this case the user can define a plan at run time depending upon the type and size of the vector. For a sufficiently large number of repeated transforms it is advantageous to use FFTW's ability to choose the fastest algorithm by actually measuring the performance of (some or all of) the supported algorithms on the given array size and platform. These measurements, which the authors call wisdom can be stored in a file or string for later use. A good example of the FFT can be shown using the GIMP fft filter:

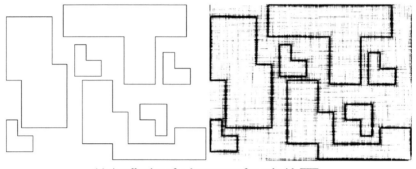

(a) A collection of polygons transformed with FFT

Fig. 16.1 FFT transformations

The particular call to FFTW in this filter is shown below:

```
fftw_plan p;
p = fftw_plan_dft_r2c_2d(w, h, fft_real,
    (fftw_complex *) fft_real, FFTW_ESTIMATE);
fftw_execute(p);
```

This performs a real-to-complex in-place FFT on the data (assuming that the vector has memory for complex sized data).

16.9 GLPK

The GNU Linear Programming Kit (GLPK) is a software system designed to solve large-scale linear programming (LP), mixed integer programming (MIP) problems. The software comprises of a C language library which implements a revised simplex method, primal-dual method and Gomory's mixed integer cuts alongwith branch-and-bound algorithms. GLPK also supports the GNU MathProg modeling language which is a subset of AMPL. Consider the following GLPK sample code (which is available from the examples directory):

```
..
s1:    lp = glp_create_prob();
s2:    glp_set_prob_name(lp, "sample");
s3:    glp_set_obj_dir(lp, GLP_MAX);
```

```
 5  s4:   glp_add_rows(lp, 3);
    ..
    s11:  glp_add_cols(lp, 3);
    s12:  glp_set_col_name(lp, 1, "x1");
    s13:  glp_set_col_bnds(lp, 1, GLP_LO, 0.0, 0.0);
10  s14:  glp_set_obj_coef(lp, 1, 10.0);
    ..
    s29:  ia[9] = 3, ja[9] = 3, ar[9] =  6.0; /* a[3,3] =  6 */
    s30:  glp_load_matrix(lp, 9, ia, ja, ar);
    s31:  glp_simplex(lp, NULL);
15  s32:  z = glp_get_obj_val(lp);
    s33:  x1 = glp_get_col_prim(lp, 1);
```

The GLPK library has functions for defining the problem, specifying whether the objective function should be maximized or minimized, adding constraints and the objective function. We compile this program and run it as:

```
gcc sample.c -o sample -lglpk
* 0: obj =   0.000e+00 infeas = 0.0e+0 (0)
* 2: obj =   7.333e+02 infeas = 0.0e+0 (0)
OPTIMAL SOLUTION FOUND

z = 733.33; x1 = 33.33; x2 = 66.667; x3 = 0
```

16.10 COIN-OR: Comp. Infrastructure for OR

Another software system for linear programming, and much more, is the Computational Infrastructure for Operations Research (COIN-OR). It contains the following tools:

1. Developer tools: UNIX development tools, mainly developed for COIN-OR internal development,

   - BuildTools: CoinBazaar, CoinBinary, CoinWeb, and Coopr,

2. PFunc: Parallel Functions, API for task parallelism,
3. Graphs: tools and libraries for operating on graphs:

   - CGC: COIN-OR Graph Classes:
   - LEMON: Library of Efficient Models and Optimization in Networks, a C++ template library for combinatorial optimization,

4. Interfaces

   - CoinMP: API and library for CLP, CBC and CGL,
   - OSI: Open Solver Interface (see also Section 16.10.1,
   - PuLP: Python library for modeling LPs, and IPs.

5. Modeling Systems: Coopr, FlopC++, PuLP,
6. Optimizers

   - CLP: COIN-OR LP, a simplex solver,

- CoinMP: API and library for CLP, CBC and CGL,
- BCP: Branch-Cut-price framework,
- CBC: COIN-OR branch-and-cut, LP based branch-and-cut,
- CGL: Cut Generator Library,
- CHiPPS: COIN-OR High Performance Parallel Search framework,
- DIP: Decomposition in IP,
- SYMPHONY: callable library for MIPs,

16.10.1 Open Solver Interface

The COIN-OR Open Solver Interface is a standardized application programming interface (API) for calling solver libraries and external functions. It supports open-source as well as commercial solvers, such as COIN-OR LP, BCP, and GLPK.

16.11 Conclusion

A large fraction of engineering tools are based on a fixed set of mathematical libraries. The most important libraries which implement mathematical features are described in this chapter. In particular the mathematical libraries of BLAS, ATLAS, LAPACK, NTL, GSL, GMP and MPFR were discussed. The FFTW system for computing FFTs was also described with the help of an image-processing example. Linear programming has become an important solution method for engineering disciplines (especially financial applications). The GNU GLPK and COIN-OR systems for linear programming was discussed in Section 16.10.

Chapter 17
Mathematics Software

Abstract In this chapter we describe open-source software for mathematics. We discuss Maxima, a general purpose symbolic math software system. GNU Octave which operates on matrices and can be used for signal processing functions. Statistical computing with R and PSPP is described, as well as number theory, group theory and graph isomorphism checking using PARI, and Nauty. The open-source math software Axiom, REDUCE, Singular, CoCoA, and Macaulay are discussed in their context of algebraic geometry and commutative algebra. The Polytope analysis software 'polymake' is shown, and with the help of examples, its use in calculating polytope properties is discussed. The TeXmacs front-end and editing platform is described as well. To end the chapter we present, Sage, which is a Python interface to many of the other software mentioned in this chapter. The Computational Geometry Algorithms Library (CGAL) is described for its use in solving computational geometry and discrete geometry problems.

Contents

S. Koranne, *Handbook of Open Source Tools*,   357
DOI 10.1007/978-1-4419-7719-9_17, © Springer Science+Business Media, LLC 2011

17.1 Maxima

Maxima is an open-source symbolic math software written in Common Lisp. It is
based upon Macsyma (developed in 1982 at MIT). It supports symbolic differentiation, integration and operations on polynomials.

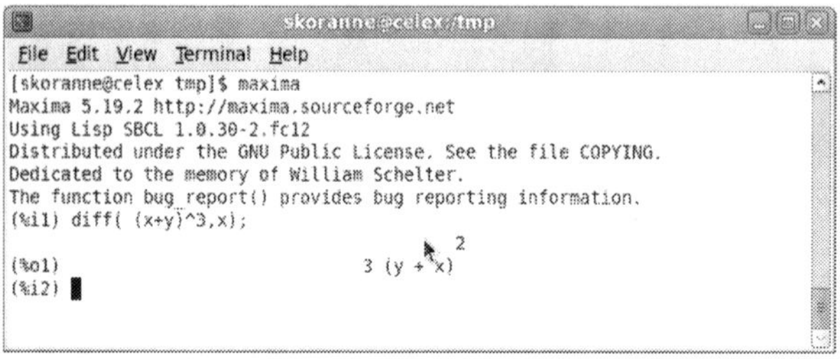

Fig. 17.1 Maxima text console

```
(%i1) u : expand((a+b)^4);
         4        3      2 2       3      4
(%o1) b  + 4 a b  + 6 a  b  + 4 a  b + a
(%i3) factor(10!);
              8  4  2
(%o3)        2  3  5  7

(%i5) integrate( (1/(1-x^2)),x);
      log(x + 1)    log(x - 1)
(%o5) ---------- - ----------
           2             2
```

17.2 GNU Octave

GNU Octave is a high-level language for manipulating matrices and performing
numerical computations. The command-line interpreter is called `octave` and the
software provides an interface for solving linear and non-linear problems. GNU
Octave in turn is built upon many of the mathematical libraries described above (including FFTW, LAPACK and BLAS), and as such provides an uniform syntax and
mechanism to use these libraries. The Octave language is mostly compatible with
Matlab and indeed matrix manipulation and computation is one of the mainstays of
GNU Octave's usage.

GNU Octave also has interfaces to plot graphs, load audio sample files, perform
image processing (using the ImageMagick library to perform image IO). A graphical
front end for Octave using Qt (called `qtoctave`) is also available (see Figure 19.6

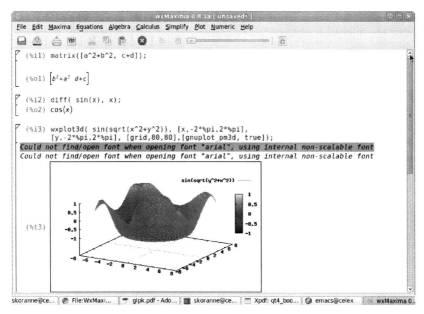

Fig. 17.2 Maxima software running in wxMaxima

for a screen-shot). In this section we describe some useful functions available in GNU Octave. To launch GNU Octave on the command-line we run `octave`:

```
octave:1> a=rand(3,3)
a =

   0.108517   0.813629   0.677108
   0.464601   0.260034   0.506348
   0.077527   0.382319   0.129084

octave:2> b=rand(3,3)
b =

   0.441143   0.096095   0.551832
   0.907029   0.146410   0.290589
   0.264928   0.842573   0.647098

octave:3> a+b
ans =

   0.54966   0.90972   1.22894
   1.37163   0.40644   0.79694
   0.34245   1.22489   0.77618

octave:4>
```

The GNU Octave command to make a matrix is shown above. A random matrix can be generated using the `rand` function. Octave supports built-in datatypes as well as user-defined data types. The built-in datatypes supported by Octave are:

1. int8,
2. uint8,
3. int16,
4. uint16,
5. int32,
6. uint32,
7. int64,
8. uint64,
9. double.

In addition, GNU Octave also supports complex and integer scalars (and matrices of the above types). String objects consist of sequence of characters. The type of the variable can be inspected by using the `typeinfo` function.

User-defined data structures can be defined as:

```
octave:1> p = struct();
octave:2> p.x = 10;
octave:3> p.y = 20;
octave:4> p
p =
{
  x =    10
  y =    20
}
```

17.2.0.1 Index expressions

A *scalar*, *row-vector* and *column-vector* can be defined as:

- `a(2)` : defines a scalar,
- `a(1:2)` : defines a row-vector,
- `a([1;2])` : defines a column vector

To create a row vector using an index expression:

```
for i = 1:4
    mat(i) = i * i
endfor
mat = 1 4 9 16
```

GNU Octave can return multiple values from functions. Octave supports arithmetic operators, comparison operators, and Boolean operators. In addition to using the interpreter, Octave programs can be stored in a file. Consider a program as shown in Listing 17.1 which defines a function `hello_world`.

```
# GNU Octave file

function hello_world(name)
  printf (mfilename("fullpath") );
5 printf ("\nHello from %s\n", name );
endfunction
```

Listing 17.1 GNU Octave program script

When the path in which this file is placed located in the Octave search path (a path can be added to the Octave search path using the `addpath` function), the function `hello_world` is available to use in the interpreter. One thing to keep in mind is that the function should not be called inside the file (otherwise it results in infinite recursion). To prevent this, place an expression which not a function as the first non-comment token in the file. Using the `mfilename` we can print the name of the current file (using the "fullpath") argument to `mfilename`, we can print the full path to the file. Any function which is not named the same as the file is a function which is only visible inside the file and cannot be called from outside. Other important functionality of Octave are listed below:

1. Input/output functions,
2. Plotting,
3. Matrices:

   • Diagonal matrix
   • Sparse matrix

4. Linear algebra,
5. Non-linear equation,
6. Numerical integration,
7. Differential equations,
8. Optimization,
9. Statistics,
10. Polynomial manipulations,
11. Signal processing:

    • Audio processing,
    • Image processing.

17.3 R : A Programming Environment for Data Analysis and Graphics

R is a programming language and software environment for statistical computing and graphics. It is an implementation of the S programming language with lexical scoping semantics inspired by Scheme (see Section 1.6.9). The R language has become standard among statisticians for the development of statistical software, and is widely used for statistical software development and data analysis. Amongst its many features are:

1. Operators for calculation on arrays,
2. Simple but effective programming language 'S',
3. Tools for data analysis,
4. Effective data I/O and storage,
5. Integrated graphical display and plotting.

R provides a wide variety of statistical (including linear and nonlinear modeling, classical statistical tests, time-series analysis, classification, and clustering, etc.) and graphical techniques. Consider a small transcript of an R session below:

```
> x <- c(1,3,5,7,11,13)
> y <- x^2
> mean(y)
[1] 62.33333
> var(y)
[1] 4599.467
> summary(lm(y ~ x))

Call:
lm(formula = y ~ x)

Residuals:
        1        2        3        4        5        6
   19.478   -1.043  -13.565  -18.087   -3.130   16.348

Coefficients:
             Estimate Std. Error t value Pr(>|t|)
(Intercept)    -32.74      13.03  -2.513 0.065821 .
x               14.26       1.65   8.644 0.000985 ***
---
Signif. codes:  0 *** 0.001 ** 0.01 * 0.05 . 0.1   1

Residual standard error: 17.09 on 4 degrees of freedom
Multiple R-squared: 0.9492,Adjusted R-squared: 0.9365
F-statistic: 74.71 on 1 and 4 DF,  p-value: 0.0009853
```

In the next section we shall describe another software system for statistical analysis, PSPP see Section 17.4. R, although used for statistical analysis is designed to be a flexible environment which supports statistical analysis. Moreover, in R the computed results of operations are stored in objects on which subsequent analysis can be performed. PSPP is designed to produce output immediately upon computation. Some of the useful command in R are given below:

```
> q() # quit R
> ?solve # give help on solve function
> example(solve) # show an example of the solve function
> plot( x <- rnorm(64), type="s", main="Title")
> log10(2)
[1] 0.30103
> sqrt(100)
```

```
[1] 10
> exp(1.01)
[1] 2.745601
> source( "file.R")  # read command from file
> sink( "file.log") # write output to file
> objects() # data persistence
> ls()
"A" h8" "hilbert"
> ls
<environment: namespace:base>
> c(1,1,2,3)  # construct a vector of 1,1,2,3
[1] 1 1 2 3
> x <- c(1,1,2,3) # assign this vector to var. x
> x
[1] 1 1 2 3
> c(x,0,x) -> y # assignment in the other direction
> x[1] = 8 # vectors in x start at 1
> x        # x is now modified
[1] 8 1 2 3
> y        # but y is not, vectors are copied in assignment
[1] 1 1 2 3 0 1 1 2 3
> sum(x)
[1] 14
> mean(x)
[1] 3.5
> x <- c(1:10)
[1]   1  2  3  4  5  6  7  8  9 10
> x <- c(1,2,...,8)
seq(from=1,to=10, length=3) # functions in R can have named params
[1]   1.0  5.5 10.0
> logical_vector <- x > 2 # vector of booleans if x > 2
m <- array(23:43, dim=c(4,4))
> m
      [,1] [,2] [,3] [,4]
[1,]   23   27   31   35
[2,]   24   28   32   36
[3,]   25   29   33   37
[4,]   26   30   34   38
> eigen(m)
$values
[1]  1.226523e+02+0.000000e+00i -6.522506e-01+0.000000e+00i
[3]  2.949706e-15+1.401517e-15i  2.949706e-15-1.401517e-15i

$vectors
                [,1]           [,2] ...
[1,] -0.4752248+0i  0.7359579+0i ...
[2,] -0.4915202+0i  0.2934262+0i ...
[3,] -0.5078157+0i -0.1491055+0i ...
[4,] -0.5241111+0i -0.5916372+0i ...

m_transpose = aperm(m, c(2,1)) # transpose is a permutation
> m_transpose
      [,1] [,2] [,3] [,4]
[1,]   23   24   25   26
```

```
[2,]    27    28    29    30
[3,]    31    32    33    34
[4,]    35    36    37    38
>
```

As can be seen above, there is a difference between calling a function `ls()` versus the object representing the function `ls`. The former calls the `ls` function, while the latter returns the definition of the function. Vectors can be assigned to other vectors using the `c()` function. Vectors are copied by value and not by reference. Matrix examples are shown as well; also see Section 17.2 for GNU Octave for matrix computations.

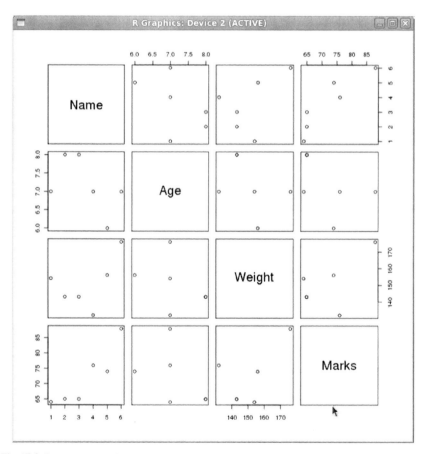

Fig. 17.3 Data summary plot using R

Data present in external file can be read using the `read.table` function. The data file should have a header line defining the name of the variable present in each subsequent row. The data is read into a *data-frame*.

```
roll_table <- read.table("rollcall.data", header=TRUE)
> roll_table
    Name Age Weight Marks
1   Jack   8    143     65
2   Jill   8    143     65
...
> sum(roll_table["Age"])
[1] 43
mean(roll_table["Age"])
     Age
7.166667
> mean( roll_table["Marks"])
Marks
   72
>summary(roll_table["Marks"])
     Marks
 Min.   :64.0
 1st Qu.:65.0
 Median :69.5
 Mean   :72.0
 3rd Qu.:75.5
 Max.   :88.0
> plot( roll_table)
```

A complete summary of this data can be generated using the `plot` function as shown in Figure 17.3. In addition to the built-in functions, R also has a complete programming language in which users can define their own functions. The programming language supports the usual paradigms of looping constructs, **if**-then-**else** and object-oriented programming.

```
> my_sum <- function(a,b) { a+b }
> my_sum(10,20)
[1] 30
```

17.4 PSPP

PSPP is a program for statistical analysis of sampled data. It is designed to be an open-source alternative of other proprietary statistical analysis programs. PSPP can be used to perform descriptive statistics, T-tests, non-parametric tests, linear regressions alongwith the basic analysis of frequency, mean, median mode on variable data. Some of the distinctive features of PSPP are:

- Portability: available as open-source software on many operating systems including GNU/Linux,
- Has both command-line and GUI interface,
- High performance: can support billion cases and variables,
- Compatibility: data and syntax are compatible with other statistical analysis programs,

- Plotting: can produce output directly in Postscript,
- Easy import of data: from comma-separated files, spreadsheets, and databases.

A sample dataset can be managed by creating a data view containing columns of data types as shown in Figure 17.4.

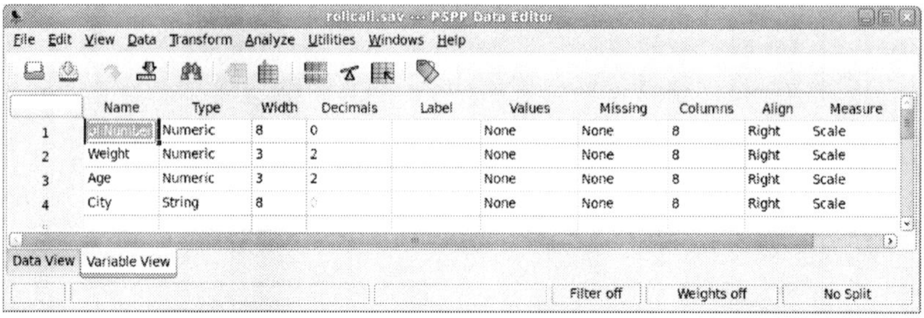

Fig. 17.4 Variable view in PSPP

PSPP (in batch mode) reads a syntax file, and a given data file, performs the analysis and writes the output to a listing file (or standard output). The output tables and charts can be in ASCII or Postscript and HTML. In batch mode `pspp` accepts various command-line arguments specifying the input file, output file, and other options. For graphical user interface the command `psppire`. An example session with PSPP to analyze data is shown in Figure 17.5. PSPP is designed for statistical anal-

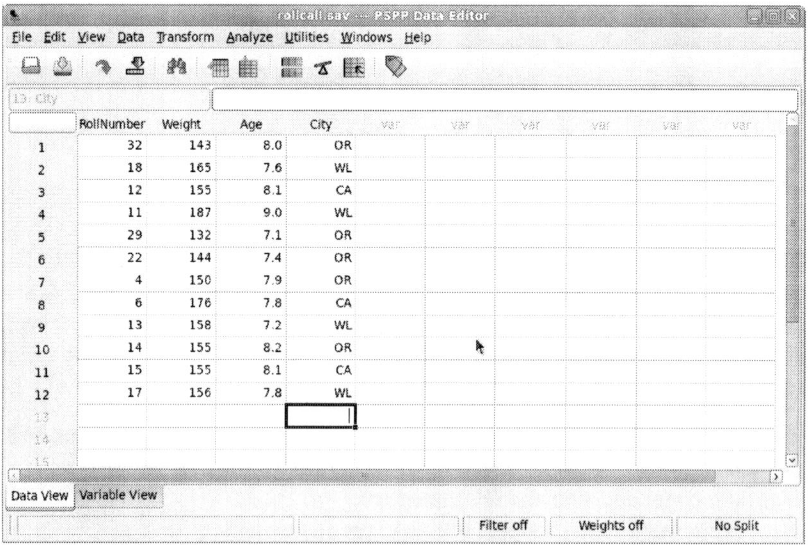

Fig. 17.5 PSPP software to analyze data

ysis of data, thus data input/output is central to the program. Each datum represents (or belongs to) a *case*, also called an observation. In Figure 17.5 12 observations are listed. The data can also be read from TEXT files, PostgreSQL databases, and GNUmeric spreadsheets. Consider an example to read the UNIX /etc/passwd file:

```
GET DATA /TYPE=TXT /FILE='/etc/passwd' /DELIMITERS=':'
         /VARIABLES=username A20
                    password A40
                    uid F10
                    gid F10
                    ...
                    shell A40
```

Additionally the data can be included as part of the program for simple analysis. Consider the example given below:

```
DATA LIST /X 1-3
BEGIN DATA.
109
2
65
4
432
10
15
20
198
24
12
END DATA.
COMPUTE X=X/2.
DESCRIPTIVES X.
```

This PSPP program performs a division of 2 on the sample data before computing the statistical properties. Running pspp <file> gives the analyzed listing as:

```
1.1 DATA LIST.  Reading 1 record from INLINE.
+--------+------+-------+------+
|Variable|Record|Columns|Format|
#========#======#=======#======#
|X       |    1|  1-  3|F3.0  |
+--------+------+-------+------+

2.1 DESCRIPTIVES.  Valid cases = 11;
        cases with missing value(s) = 0.
+--------#--+-----+-------+-------+-------+
|Variable# N| Mean|Std Dev|Minimum|Maximum|
#========#==#=====#=======#=======#=======#
|X       #11|40.50|  65.43|   1.00| 216.00|
+--------#--+-----+-------+-------+-------+
```

Another example using statistical regression with formatted input is shown below:

```
echo "Rollcall in PSPP"
DATA LIST /Name 1-10 (A) Age 11-12 (0) Weight 13-15 (0)
                        Marks 16-17 (0)
BEGIN DATA.
Jack        814365
Jill        814365
John        713276
Mark        615674
Daniel M    715464
Zack, Paul  717688
END DATA.
LIST.
REGRESSION /VARIABLES=Age Marks /STATISTICS DEFAULTS
           /DEPENDENT=Marks.
REGRESSION /VARIABLES=Age Weight /STATISTICS DEFAULTS
           /DEPENDENT=Weight.
REGRESSION /VARIABLES=Weight Marks /STATISTICS DEFAULTS
           /DEPENDENT=Marks.
```

The output of this program is:

```
Rollcall in PSPP

1.1 DATA LIST.  Reading 1 record from INLINE.
+--------+------+-------+------+
|Variable|Record|Columns|Format|
#========#======#=======#======#
|Name    |    1|  1- 10|A10   |
|Age     |    1| 11- 12|F2.0  |
|Weight  |    1| 13- 15|F3.0  |
|Marks   |    1| 16- 17|F2.0  |
+--------+------+-------+------+

      Name Age Weight Marks
---------- --- ------ -----
Jack         8    143    65
Jill         8    143    65
John         7    132    76
Mark         6    156    74
Daniel N     7    154    64
Zack, Paul   7    176    88

2.1 REGRESSION.  Model Summary
#==============#===#=========#=================#========================#
#          R #R Square|Adjusted R Square|Std. Error of the Estimate#
#==============#===#=========#=================#========================#
#    |.45#    .21|            .21|                     9.32#
#==============#===#=========#=================#========================#

2.2 REGRESSION.  ANOVA
#==============#=================#==#============#====#============#
#              #Sum of Squares|df|Mean Square| F |Significance#
#==============#=================#==#============#====#============#
#   |Regression#       90.35| 1|     90.35|1.04|       .37#
#   |Residual  #      347.65| 4|     86.91|    |         #
#   |Total     #      438.00| 5|          |    |         #
#==============#=================#==#============#====#============#

2.3 REGRESSION.  Coefficients
#==============#======#==========#=====#======#============#
#        #   B  |Std. Error|Beta| t  |Significance#
#==============#======#==========#=====#======#============#
#   |(Constant)#112.47|    39.87|  .00| 2.82|       .22#
#   |      Age # -5.65|     5.54|-.45|-1.02|       .35#
#==============#======#==========#=====#======#============#

3.1 REGRESSION.  Model Summary
#==============#===#=========#=================#========================#
#          R #R Square|Adjusted R Square|Std. Error of the Estimate#
#==============#===#=========#=================#========================#
#    |.36#    .13|            .13|                    15.78#
#==============#===#=========#=================#========================#
```

Using PSPP we can easily compute the dependence of variables on others and compute their statistical relationships. PSPP includes several statistical analysis functions, as shown in Figure 17.6.

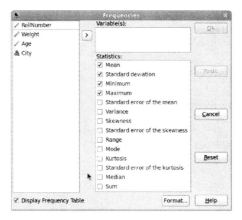

Fig. 17.6 Statistical analysis function in PSPP

When the data is analyzed PSPP produces a detailed report as given below:

```
3.1 FREQUENCIES.  Weight
+-----+--------+---------+--------+--------+--------+
|     |        |         |        | Valid  |  Cum   |
|Label|  Value |Frequency| Percent| Percent| Percent|
#=====#========#=========#========#========#========#
|     |    132 |      1 |   8.33 |   8.33 |   8.33 |
|     |    143 |      1 |   8.33 |   8.33 |  16.67 |
|     |    144 |      1 |   8.33 |   8.33 |  25.00 |
|     |    150 |      1 |   8.33 |   8.33 |  33.33 |
|     |    155 |      3 |  25.00 |  25.00 |  58.33 |
|     |    156 |      1 |   8.33 |   8.33 |  66.67 |
|     |    158 |      1 |   8.33 |   8.33 |  75.00 |
|     |    165 |      1 |   8.33 |   8.33 |  83.33 |
|     |    176 |      1 |   8.33 |   8.33 |  91.67 |
|     |    187 |      1 |   8.33 |   8.33 | 100.00 |
#=====#========#=========#========#========#========#
|       Total|     12 |  100.0 |  100.0 |        |
+------------+--------+--------+--------+--------+

+--------------------+------+   +------------+------+
|N          Valid    |   12 |   |Range       | 55.00|
|           Missing  |    0 |   |Minimum     |132.00|
|Mean                |156.33|   |Maximum     |187.00|
|Mode                |155.00|   | 50 (Median)|  156 |
|Std Dev             | 14.69|   +------------+------+
|S.E. Kurt           | 1.23 |
|S.E. Skew           |  .64 |
+--------------------+------+
```

17.5 Pari

PARI/GP is a mathematical software system for performing research in number theory and group theory. Running `gp` on GNU/Linux gives:

```
[skoranne@celex tmp]$ gp
            GP/PARI CALCULATOR Version 2.3.4 (released)
     i686 running linux (ix86/GMP-4.3.1 kernel) 32-bit version
compiled: Oct 20 2009, gcc-4.4.2 20091018 (Red Hat 4.4.2-4) (GCC)
          (readline v6.0 enabled, extended help available)

               Copyright (C) 2000-2006 The PARI Group

PARI/GP is free software, covered by the GNU General
Public License, and comes WITHOUT ANY WARRANTY WHATSOEVER.

Type ? for help, \q to quit.
Type ?12 for how to get moral (and possibly technical) support.

parisize = 4000000, primelimit = 500000
?
```

The software comprises of two components:

1. PARI: is the underlying C library which implements the required functions for operating on groups,
2. gp: is a command line interface to the above library. It is structured as a calculator which interfaces to the user.

PARI/GP supports arbitrary precision computation on integers and thus can easily perform factorial and prime number computations.

17.6 Nauty

Nauty is a computer program and library for computing automorphism group of graphs and digraphs. It can also be used to produce a canonical labeling of the vertices of the graph. Nauty is written in C language and is thus easily portable. For graphs which can fit in the wordsize of the machine (32-bit or 64-bit) it has excellent performance. Nauty can be used as a library, or it can be used as a graph isomorphism checker. The command-line program `dreadnaut` which ships with Nauty provides orbit (automorphism group of the vertices of the graph) calculation. Consider the 3-dimensioned simplex with 4 vertices. This is a complete graph (every vertex is connected to every other vertex). See Figure 17.7.

```
Dreadnaut version 2.4 (32 bits).
> n=4
> g
 0 : 1 2 3;
```

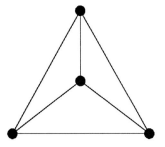

Fig. 17.7 Graph of 3-simplex

```
1 : 2 3;
2 : 3;
3 : .
> x
(2 3)
level 3:   3 orbits; 2 fixed; index 2
(1 2)
level 2:   2 orbits; 1 fixed; index 3
(0 1)
level 1:   1 orbit; 0 fixed; index 4
1 orbit; grpsize=24; 3 gens; 10 nodes; maxlev=4
tctotal=16; cpu time = 0.00 seconds
```

Using dreadnaut we first specify that the graph has 4 vertices, then we input the adjacency of the graph. The 'x' command runs the canonical labeling computation which prints the automorphism groups (orbits). The graph is completely symmetrical and thus there is only 1 orbit.

17.7 Axiom

OpenAxiom is an open-source symbolic math software. However, unlike Maxima and other symbolic math software Axiom has been designed for conducting research in mathematics. It has an integrated environment, compiler, large set of mathematical libraries. In text mode Axiom returns:

```
        AXIOM Computer Algebra System
          Version: Axiom (March 2008)
   Timestamp: Friday April 25, 2008 at 17:38:07

   Issue )copyright to view copyright notices.
   Issue )summary for a summary of useful system commands.
   Issue )quit to leave AXIOM and return to shell.


     Re-reading compress.daase   Re-reading interp.daase
```

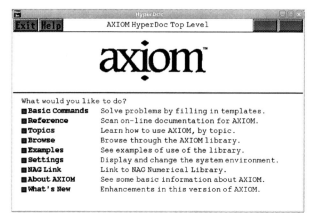

Fig. 17.8 AXIOM software

```
    Re-reading operation.daase
    Re-reading category.daase
    Re-reading browse.daase
(1) -> integrate(1/(x**3 * (a+b*x))),x)
          2 2                 2 2                         2
      - 2b x log(b x + a) + 2b x log(x) + 2a b x - a
(1)   ------------------------------------------------
                              3 2
                          2a x
```

The above example shows the symbolic math capabilities of Axiom. However, the
most powerful feature of Axiom lies in its design and programming languages
(SPAD and Aldor). Axiom is type aware, not only in the programming language
sense, but also mathematically. Using *Categories* which define algebraic properties
mathematical correctness is ensured. Categories allow the algorithms implemented
in Axiom's programming language to be defined in their natural settings; Axiom's
interpreter constructs the domain based on user input as shown below:

```
  (5)   5
            Type: PositiveInteger
(6)  -> (x+y)**2
(6)  ->
            2               2
  (6)   y  + 2x y + x
            Type: Polynomial Integer

-> M = [ [ x**2+1, 0], [0, x]]::Matrix( POLY(FRAC(INT)))
      +1  0+ 2    +0  0+    +1  0+
  M= |    |x  + |    |x + |    |
      +0  0+      +0  1+    +0  0+
Type: Equation Polynomial SquareMatrix(2,Fraction Integer)
```

Axiom is aware of the type (mathematical type) of the input entered into the interpreter. We can ask Axiom for *type conversion*:

```
(10) -> (22/7)::Float
(10) ->

   (10)   3.1428571428 571428571
       Type: Float

(11) -> %::Fraction Integer

(11) ->
          22
   (11)   --
           7
       Type: Fraction Integer
```

Axiom also supports complex numbers as well as arbitrary long integers. To quit the interpreter, enter:

```
-> )quit
```

17.8 Reduce

Reduce is an open-source interactive math software for general algebraic computations. It is written in Lisp (can be compiled using Codemist Standard Lisp (csl) or SLisp). It supports the following functionality:

1. Expansion and ordering of polynomials,
2. Automatic simplification,
3. Calculation on symbolic matrices,
4. Ability to add new functions using Lisp,
5. Polynomial factorization,
6. Analytic differentiation and integration.

Reduce is available as open-source and is portable, being able to run on many computer systems from GNU/Linux personal computers to supercomputers. Moreover, Reduce has built up a number of useful packages. Many of these add significant functionality such as: (i) tensor manipulation, (ii) Boolean algebra, (iii) Groebner basis and commutative algebra, (iv) overdetermined systems of partial differential equations, and (v) polynomial ideals. Invoking Reduce we get:

```
REDUCE, 15-Sep-08 ...

1: (a+b)^3;
```

```
 3     2           2     3
a  + 3*a *b + 3*a*b  + b
2: m := mat((a,b),(c,d));

      [a  b]
m := [    ]
      [c  d]



3: m*m;

[ 2                    ]
[a  + b*c   b*(a + d)]
[                    ]
[              2     ]
[c*(a + d)   b*c + d  ]
```

Interaction with Reduce can be done completely through the interpreter, or programs can be written in Reduce which can use the underlying Lisp system for efficiency, eg:

```
7: first {a,b,c};
a
8: rest {a,b,c};
{b,c}
9: reverse {a,b,c};
{c,b,a}
```

In this sense, Reduce can be used as a special purpose Lisp system which has significant mathematical library support built into it. Reduce has a number of mathematical functions already built into the system and new ones defined by the user can be compiled and brought into the running image.

```
11: solve( x^2-8*x+1, x);
{x=sqrt(15) + 4,
 x= - sqrt(15) + 4}
```

Reduce can also generate FORTRAN statements which implement and evaluate the computation:

```
COMMENT Switch on Fortran mode
on fort;
x := (a+b+c)^2;

      x=a**2+2.0*a*b+2.0*a*c+b**2+2.0*b*c+c**2
off fort;
```

In this example, Reduce has printed the FORTRAN statement to calculate value of x from a, b, c using the formula we entered in the interpreter. This facility can be used to generate FORTRAN code for simulations, and can be compiled and run on HPC machines.

17.9 Singular Computer Algebra System

Singular is an open-source computer algebra system for polynomial computation. It was designed to solve problems in commutative algebra, algebraic geometry and singularity theory. Singular algebra system features one of the fastest and most general polynomial implementation and has been used to implement Buchberger's algorithm for Groebner basis. Its main computational objects are ideals and modules over baserings (which are polynomial rings or localization of polynomial rings over fields including finite fields, rationals, floats, and algebraic extensions).

Functions included in Singular include polynomial operations, factorization, resultant, characteristic set and GCD computation. Singular has an interactive shell and a C-like programming language. Using the programming language, users have contributed extensions and application software which is now available with the Singular distribution. This is also an intended consequence of the open-source methodology.

Singular is invoked on GNU/Linux system as:

```
$./Singular/3-1-1/ix86-Linux/Singular
> 12 + 32;
44
> int x = _;
> x;
44
> int y = 12; // this is a comment
> help factorial; // launches the help in browser
```

In the above example, we can see the Singular facilities for arithmetic, variable assignment, and text comments. The last displayed result (non-assigned) is stored in an internal variable _, and variable assignment is done using the = operator. All variables in Singular must be typed, in the above assignment we have used the integer type. To use a function which is defined in a library we have to use:

```
LIB "general.lib";
// Singular proceeds to load general.lib and dependencies
> factorial(10);
3628800
> binomial(5,3);
10> kmemory;
// libname   : general.lib
// procname  : kmemory
// type      : singular
```

```
> kmemory();
206
```

The "general.lib" library has many useful functions including binomial, fibonacci, exponential, prime number, sorting, factorization and more. As is shown in the above example, typing the name of the function kmemory returns information about the type of the variable, and we can indeed call the function using kmemory() which returns the number of bytes used by Singular thus far (in KB).

To perform polynomial operations in Singular we first have to define a Ring, then create a polynomial in that ring:

```
> ring R = 0, (x,y),dp;
> poly Q = (2x+y);
> Q;
2x+y
> Q*Q;
4x2+4xy+y2
> poly P = ((x+y)^2)*(x2+y5);
> P
. ;
y7+24y6+144y5+x2y2+24x2y+144x2
> factorize(P);
[1]:
   _[1]=1
   _[2]=y5+x2
   _[3]=y+12
[2]:
   1,1,2
```

The definition of a ring in Singular involves setting the ground field (0 in the above example), the names of the ring variables (x,y), and the monomial ordering to be used (dp stands for degree reverse lexicographical ordering). The notation for polynomials in Singular is coefficient, variable, power. Factorization of polynomials can also be computed with a time constraint using timeFactorize which accepts a second argument (number of seconds); if the factorization does not complete in the stipulated time the polynomial is considered irreducible. A field on a characteristic can be defined as:

```
> ring PZ7 = 7,(x(1..10)),ds;
> poly ZQ = (x(1)+x(2))^5 + (x(6)+x(4))^3;
> ZQ;
x(4)^3+3*x(4)^2*x(6)+3*x(4)*x(6)^2+x(6)^3+
x(1)^5-2*x(1)^4*x(2)+
3*x(1)^3*x(2)^2+3*x(1)^2*x(2)^3-2*x(1)*x(2)^4+x(2)^5
> factorize(ZQ);
[1]:
   _[1]=1
```

```
    _[2]=x(4)^3+3*x(4)^2*x(6)+3*x(4)*x(6)^2+
         x(6)^3+x(1)^5-2*x(1)^4*x(2)+
         3*x(1)^3*x(2)^2+3*x(1)^2*x(2)^3-2*x(1)*x(2)^4+
         x(2)^5
[2]:
   1,1
> poly ZQ = (x(1)+x(2))^5;
// ** redefining ZQ **
> factorize(ZQ);
[1]:
   _[1]=1
   _[2]=x(1)+x(2)
[2]:
   1,5
```

Rings can also be defined on complex numbers:

```
> ring cpz = (complex,2,i),(a,b),lp;
> cpz;
//   characteristic : 0 (complex:6 digits, additional 6 digits)
//   1 parameter    : i
//   minpoly        : (i^2+1)
//   number of vars : 2
//        block   1 : ordering lp
//                  : names     a b
//        block   2 : ordering C
> poly CQ = a2 + 3*i;
> CQ;
a2+(i*3)
> CQ*CQ;
a4+(i*6)*a2-9
// going back to a previous ring
> setring PZ7;
> PZ7;
//   characteristic : 7
//   number of vars : 10
//        block   1 : ordering ds
//                  : names     x(1).. x(10)
//        block   2 : ordering C
>
```

Matrices can be constructed in row-major order:

```
> intmat M0[3][3] = 1,2,3,4,5,6,7,8,9;
> M0;
1,2,3,
4,5,6,
7,8,9
> int j;
> int M0_trace;
> for( j=1; j <= 3; j++ )
```

```
  { M0_trace = M0_trace + M0[j,j]; }
> M0_trace;
15
```

One of the most common use of Singular is performing computations on *ideals* and *varieties* generated by a system of polynomial equations. Consider the following example:

```
> ring r = 0,(x,y,z),dp;
> poly f =   x3+y3+(x-y)*x2y2+z2;
> poly g = f^2 * (2x-y);
> ideal I = f,g;
> ideal J = subst(I, var(1),1);
> J;
J[1]=y2+z2+1
J[2]=-y5-2y3z2-yz4+2y4+4y2z2+2z4-2y3-2yz2+4y2+4z2-y+2
> J = subst(J,var(2),2);
> J;
J[1]=z2+5
J[2]=0
// Compyute ideal using Groebner basis
> ideal sI = groebner(f);
> sI;
sI[1]=x3y2-x2y3+x3+y3+z2
> reduce(g,sI);
0
// Compute the Jacobian ideal
> ideal J = jacob(f);
// ** redefining J **
> J;
J[1]=3x2y2-2xy3+3x2
J[2]=2x3y-3x2y2+3y2
J[3]=2z
```

Groebner basis computations are very useful in many scientific disciplines are the subject of active research (many using Singular). New functions (procedures) can be defined in Singular as:

```
> proc mysize (poly Q) { return (size(Q)); }
> mysize(f);
5
>
```

Singular's procedure syntax is similar to other programming languages such as C.

17.10 polymake: software to analyze Polytopes

Polymake is an open-source math software to study, generate and operate on the combinatorics and geometry of convex polytopes and polyhedra. Polymake includes a number of tools and methods which can be used to perform experiments on polytopes. Consider the following example:

```
$ cube
usage: cube <file> <dimension> [<scale>]
$ simplex
usage: simplex <file> <dimension> [<scale>]
$ simplex 3simplex.poly 3
```

The file '3simplex.poly' is shown below:

```
_application polytope
_version 2.3
_type RationalPolytope

AMBIENT_DIM
3

DIM
3

VERTICES
1 0 0 0
1 1 0 0
1 0 1 0
1 0 0 1

N_VERTICES
4

SIMPLICIALITY
3

BOUNDED
1

CENTERED
0
```

The above example has generated the 3-simplex as shown in Figure 17.7. At this point the polytope is represented by its coordinates as shown in the Polymake section titled VERTICES. Polymake (upto version 2.3) operates on polytopes which are present in files. Polymake version 2.9.7 has moved to a Perl based frontend. Polytope properties can be queried using Polymake as shown below:

```
$ polymake 3simplex.poly DIM
DIM
3
$ polymake 3simplex.poly DIAMETER
DIAMETER
1
$ polymake 3simplex.poly F_VECTOR
F_VECTOR
4 6 4
$ polymake 3simplex.poly G_VECTOR
G_VECTOR
1 0
$ polymake 3simplex.poly H_VECTOR
H_VECTOR
1 1 1 1
$ polymake 3simplex.poly CD_INDEX
CD_INDEX
c^3 + 2cd + 2dc
$ truncation -h
usage: truncation <out_file> <in_file>
       { <vertex> [ <vertex> ... ] | all }
       [ -cutoff <cf> | -noc ] [ -relabel ]
$ truncation 3simplex_0.poly 3simplex.poly 0
$ polymake 3simplex_0.poly GRAPH
GRAPH
{1 2 3}
{0 2 4}
{0 1 5}
{0 4 5}
{1 3 5}
{2 3 4}
```

The examples shown above compute certain properties of the 3-simplex. The F_VECTOR, G_VECTOR and H_VECTOR are combinatorial flag vectors computed from the face-lattice of the polytope. The CD-index encodes the face inclusion of the lattice. The last operation *truncation* produces a new polytope from the old one by truncating (removing) the specified vertex. Polymake can also be used for interactive viewing of polytopes. Version 2.9.7 and onwards have an integrated Perl frontend which can be used to perform computations on the polytope. Consider the following example:

```
$polymake
polytope > $cube = load("3cube.poly");

polytope>print"cube has dimension=",$cube->DIM,"\n";
cube has dimension = 3
polytope > print "cube graph = ", $cube->GRAPH,"\n";
cube graph = Polymake::graph::
        Graph__Undirected::__as__Polytope__GRAPH=
```

```
      ARRAY(0xacdb34c
polytope >
```

The Perl language's feature can be used alongwith Polymake's polytope data structures and functions.

17.11 Other Math Systems

In addition to the math software we have described above, there are a number of other specialized mathematics software which we briefly mention below:

17.11.1 Macaulay 2

Macaulay 2 is an open-source mathematics computer system for computation on commutative algebra and algebraic geometry. On GNU/Linux, Macaulay2 can be started using the command-line M2:

```
$M2
Macaulay2, version 1.3.1
with packages: ConwayPolynomials, Elimination,
               IntegralClosure, LLLBases,
               PrimaryDecomposition, ReesAlgebra,
               SchurRings, TangentCone

i1 : 2+3;
i2 : o1
o2 = 5
i3 : 10!
o3 = 3628800
i4 : 22/7 + 3/11

      263
o4 = ---
       77
o4 : QQ
```

The last example shows that rational (QQ) numbers are supported in Macaulay2. A ring can be created in Macaulay2 as follows:

```
i1 : R = QQ[x,y,z]
o1 = R
o1 : PolynomialRing
i2 : (x+y+z)^3
```

```
          3      2        2      3      2
o2 = x   + 3x y + 3x*y   + y   + 3x z +

                    2        2        2     3
      6x*y*z + 3y z + 3x*z   + 3y*z   + z
o2 : R
i6 : c = matrix { { x^2 + 1, y^2, z^2+1}}
o6 = | x2+1 y2 z2+1 |

                    1       3
o6 : Matrix R   <--- R
i7 : M = coker b
o7 = cokernel | x y z |
                              1
o7 : R-module, quotient of R
```

17.11.1.1 Algebraic Geometry with Macaulay2

The most common use of Macaulay2 is studying geometric objects in (complex) affine spaces. We can define a Ring as above and construct ideals from polynomial equations as:

```
i3 : curve = ideal( x^2+y^2, x-2*y )

             2   2
o3 = ideal (x  + y , x - 2y)

o3 : Ideal of R

i4 : gb curve

o4 = GroebnerBasis[status: done;
         S-pairs encountered up to degree 1]

o4 : GroebnerBasis

i5 : o4

i6 : dim curve

o6 = 1

i7 : codim curve

o7 = 2

i8 : degree curve

o8 = 2
```

Above we construct the ideal of the curve defined by the polynomials $x^2 + y^2$ and $x - 2y$, the dimension of this curve is 1, while the co-dim and degree are 2. Recall, that the degree of a curve is the number of intersections of the curve with a general plane. We can calculate the intersection of this curve with a *surface* as:

```
i9 : surface = ideal(x+y-2)

o9 = ideal(x + y - 2)

o9 : Ideal of R

i12 : find_point = intersect(curve,surface)

            2              2
o12 = ideal (3x  - 3x*y - 6y  - 6x + 12y,
      --------------------------------
          2          2     3      2
        - 18x y - 27x*y - 9y  + 12x  + 24x*y
      --------------------------------
         2                 3       2
       - 6y  - 24x + 48y, - 18x  - 27x y -
      --------------------------------
         2      2           2
       9x*y  + 30x  + 24x*y + 12y  + 12x - 24y)
      --------------------------------
o12 : Ideal of R
```

The ring defined can have variable ordering based on degree, lexicographical ordering, or reverse lexicographical or a combination of the above.

17.11.2 CoCoA

The CoCoA System (Computations on Commutative Algebra) is an open-source mathematical system for performing research and computation in commutative algebra. CoCoA is written around a C++ library CoCoALib. The text console for CoCoA is shown below, and the Qt GUI frontend is shown in Figure 17.9.

```
--------------------------------------------------------------
---          __/     __/         \              ---
--         /   _ \  /    _ \    , \             --
--         \  |  | \   |  |  | ___  \           --
---         ____, __/ ____, __/ _/   _\         ---
--------------------------------------------------------------
--    Version     : 4.7.4                            --
--    Online Help : type  ?   or   ?keyword          --
--    Web site    : http://cocoa.dima.unige.it       --
--------------------------------------------------------------


--------------------------------
-- The current ring is R ::= QQ[x,y,z];
```

Fig. 17.9 xcocoa : GUI frontend for CoCoA

Interacting with CoCoA using the text console is similar to the other math software described earlier in this chapter. Variables in CoCoA have to be defined starting with uppercase letters as:

```
A := 5;
5A;
25
-------------------------------
Use S ::= Z/(5)[a,b,c];
F := a-b;
I := Ideal( F^2, c );
I;
Ideal(a^2 - 2ab + b^2, c)
```

Like Singular, CoCoA also needs a ring to operate in. In the above example we have a created a ring S which is defined as modulo 5 with variables a, b, c. In another example with modulo-5 ring:

```
Use S ::= Z/(5)[a,b,c];
```

```
F := (3*a + 2*b);
F;
-2a + 2b
```

```
F*F
;
-a^2 + 2ab - b^2
```

```
F^5;
-2a^5 + 2b^5
```

We define a polynomial, and then compute the ideal of another polynomial. Groebner basis computations can also be carried out in such rings. Consider the following example of elimination:

```
Use R ::= Q[t,x,y];
GBasis( Ideal( t^3+3*t-x, 2*t^2+y));
[
    2t^2 + y,
    1/2ty - 3t + x,
    -2tx + 1/2y^2 - 3y,
    1/4y^3 + 2x^2 - 3y^2 + 9y]
Elim(t, Ideal(t^3+3*t-x, 2*t^2+y));
Ideal(1/4y^3 + 2x^2 - 3y^2 + 9y)
```

Variable substitution can be performed as:

```
Use R ::= Q[x,y,z];
F := (x+y+z)*(x+1);
Eval(F,[1]);
2y + 2z + 2
```

```
Subst(F,z,0);
x^2 + xy + x + y
```

17.12 CGAL (Computer Geometry Algorithms and Library)

CGAL is an open-source C++ library which aims to provide data structures and algorithms to solve computational geometry problem such as like triangulations, Voronoi diagrams, Boolean operations on polygons and on polyhedra, arrangements of curves, mesh generation, geometry processing, and convex hull algorithms.

Consider the following example using CGAL as shown in Listing 17.2.

```
   // \file chull_example.cpp
   // \author Sandeep Koranne (C) 2010
   // \description Example use of CGAL library
   #include <iostream>          // for Program IO
5  #include <cassert>           // assertion checking
   #include <vector>            // std::vector of points
   #include <CGAL/Exact_predicates_inexact_constructions_kernel.h>
   #include <CGAL/convex_hull_2.h>

10 typedef CGAL::Exact_predicates_inexact_constructions_kernel K;
   typedef K::Point_2 Point_2;
   typedef std::vector<Point_2> Point_Vector;

   int main() {
15   Point_Vector inputP, outputP;
     inputP.push_back( Point_2(0,0 ) );
     inputP.push_back( Point_2(0,5 ) );
     inputP.push_back( Point_2(2,5 ) );
     inputP.push_back( Point_2(2,7 ) );
20   inputP.push_back( Point_2(7,7 ) );
     inputP.push_back( Point_2(7,0 ) );
     inputP.push_back( Point_2(3,0 ) );

     CGAL::convex_hull_2( inputP.begin(), inputP.end(),
25           std::back_inserter( outputP ) );
     std::cout << "Convex hull has " << outputP.size() << " points.\n";
     std::copy( outputP.begin(), outputP.end(),
         std::ostream_iterator<Point_2>( std::cout, " " ) );
     std::cout << std::endl;
30   return(0);
   }
```

Listing 17.2 Example of using CGAl to compute convex hull

We can compile and run this program as follows:

```
$g++ -I$CGAL/include chull_example.cpp
    -l CGAL_Core -l CGAL
$ ./a.out
Convex hull has 4 points.
0 0 7 0 7 7 2 7
```

CGAL Library has the following functionality:

1. Geometry Kernels: as we saw in the above example, CGAL has inexact and exact computation kernels. Many algorithms need exact computation to arrive at the correct answer in the presence of degenerate input. CGAL's design allows the use of inexact kernels where the algorithm can work around the loss of precision,

2. Combinatoric algorithms: CGAL supports monotone and sorted matrix search, linear and quadratic programming; these operations are available as standalone, but usually they are used in a geometric setting to implement a higher level algorithm,

3. Convex hull: above we have seen an example of 2d convex hull. CGAL has convex hull algorithms for 2d, 3d and dD, including Delaunay Triangulations. Voronoi Diagrams are also included.

4. Polygons and Nef Polygons: CGAL supports operations, including generalized Booleans on polygons, collections of polygons and Nef polygons. Moreover, CGAL also supportes operations on Nef polygons embedded on the sphere. It

also has functions for polygon partitioning, straight skeleton and polygon offsetting. Minkowski sums of polygons can also be computed,

5. Operations on Polyhedra: CGAL includes the *half-edge* data structure for polyhedra representation and supports 3d boolean operations on Nef polyhedra,

6. 2d Arrangements: arrangements are combinatoric structures representing a collection of intersecting lines in the plane (for 2d, generalized to any dimension). 2d intersection and arrangement of curves is also supported. CGAL also includes an implementation of *snap rounding* to represent arrangements in finite precision,

7. Mesh Generation: CGAL includes mesh generation tools and algorithms to generate conforming triangulations, and surface reconstruction from points,

8. Spatial Searching and Sorting: CGAL includes the interval skip list data structure, and implements a 2d range search and neighbor search algorithm. It also includes an implementation of dD range trees, segment trees, and AABB trees. Rectangle intersection (and its generalization to dD iso-oriented box intersection) is also implemented.

17.13 TeXMacs

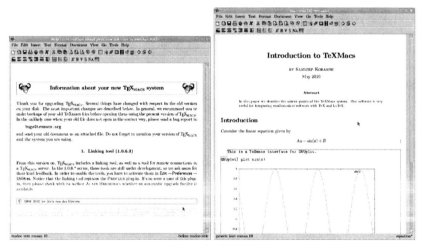

(a) Example of TeXMacs

Fig. 17.10 TeXMacs software

GNU TeXmacs is an open-source WYSIWYG (what you see is what you get) style editing platform with special features for scientists. It can render equations and mathematics formulas, an example is shown in Figure 17.10(a) and (b).

(a) More example of TeXMacs

Fig. 17.11 TeXMacs software calling external math programs

Texmacs is also integrated with mathematics software such as GNU Octave and Maxima. Example is shown in Figure 17.11 (a) and (b).

17.14 Sage

In this chapter we have seen many mathematics software, each having its own syntax for defining variables, polynomials, matrices and solving equations. Sage is an open-source mathematics software system which combines many of the above software system using a common Python based interface. Sage includes a GUI which executes in a browser, as well as a text interface. These are shown in Figure 17.12 and Figure 17.13.

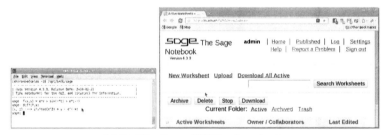

(a) SAGE software TEXT and notebook

Fig. 17.12 SAGE mathematical software

It is also possible to try out Sage online before installing it, however, we greatly recommend the software if you need access to multiple math software systems using a common interface. Consider the simple example of solving a quadratic equation in x:

```
$/opt/SAGE/sage
```

```
-------------------------------------------------------
| Sage Version 4.3.3, Release Date: 2010-02-21
| Type notebook() for GUI, and license() for information.
-------------------------------------------------------
sage: x = var('x')
sage: solve(x^2-6*x+8,x)
[x == 4, x == 2]
sage: diff(cos(x^2),x,4)
16*x^4*cos(x^2) + 48*x^2*sin(x^2) - 12*cos(x^2)
```

The second example computes the 4th differential of the expression $\cos x^2$. Sage can also perform integration, and solving partial differential equations. Internally, Sage is connected to many of the mathematical software we have discussed in this chapter. For example, Sage can connect to PARI to compute the 'chebyshev_U' function:

```
sage: y = polygen(QQ,'y')
sage: che
chebyshev_T    chebyshev_U    checkbox
sage: cheby
chebyshev_T    chebyshev_U
sage: chebyshev_U(2,y)
4*y^2 - 1
sage: bessel_I( 1,1, "pari", 100)
0.56515910399248502720769602761
```

Sage supports TAB-completion, to complete names of functions as shown in the above example.

(a) SAGE notebook and help screen

Fig. 17.13 SAGE mathematical software

As shown in Figure 17.13, Sage can be executed on a more powerful computer and then multiple remote users can connect to the Sage software using a browser. The *notebooks* which represent the computation in Sage are persisted on the Sage server and can be used to perform computation. This greatly reduces the administrative overhead with maintaining Sage for a group of scientists. To invoke a specific math software that is integrated with Sage, we can use the following form:

```
sage: maxima.eval("diff(sin(x),x)")
'cos(x)'
```

Thus, Sage can be used a single, homogeneous software system which can be used to perform computations in any one of the software, and also to use a combination to solve a problem.

17.15 Conclusion

In addition to mathematical libraries, there exists a number of open-source mathematical software systems which implement complete mathematical framework. These are described in this chapter, and include Maxima, GNU Octave, R, PSPP, Pari/GP, Nauty, OpenAxiom, Reduce, Singular, and polymake. Other specialized math software for algebraic geometry include Macaulay2 and CoCoA. The CGAL (computer geometry algorithms library) implements many computational geometry algorithms and data-structures. Mathematical front-ends to the above mentioned software include TeXMacs and SAGE. SAGE, in particular combines many of the features of the previously described software with uniform notation and the ability to pass data from one tool to another.

Chapter 18
Artificial Intelligence and Optimization

Abstract Artificial Intelligence (AI), and automated problem solving have been the epitome of computer science for many decades. While some of the lofty goals of completely automated reasoning have not come to pass, AI has nevertheless cemented its place as one of a number of useful techniques to bear on a hard problem. In this chapter we discuss tried and proven open-source tools and libraries which implement AI concepts. The tools include CLIPS (expert system) and ACL2 (theorem prover). The libraries include simulated annealing, genetic algorithm, machine learning using SVM and general backtracking.

Contents

In this chapter we discuss open-source tools and libraries for performing artificial intelligence tasks, and problem solving.

18.1 Introduction to AI Problems

Artificial Intelligence is defined as the ability of computer systems to solve ill-defined problems using methods which are inspired by the human brain. Many techniques in AI, nevertheless, are derived from algorithms in graph theory and discrete mathematics (search, backtracking), thermodynamics (simulated annealing), biology (genetic algorithms and ant-colony optimization, neural networks). During the 1950s to 1980s, AI was thought to be a panacea, with which to solve all the complex problems of the day. But due to a confluence of factors, including lack of compute power and memory, over promised visions and other non-computer related factors, use of AI declined (this was termed the AI winter).

In the recent decade, AI has made a resurgence; even though researchers are still loath to actively claim a method as being AI. They refer to the method as machine learning, or guided optimization (to name some examples). Artificial Intelligence based computer programs are also find use in robotics where the persona of a human has to be adopted by the robot. In this chapter, however, we stick to *traditional* AI problems and present some of the advances which have been made. We present open-source solutions to many of the problems including developing an expert system, machine learning, general unconstrained optimization and the use of neural networks. Complete applications which solve an AI problem such as automatic theorem proving are also described.

18.2 CLIPS: C Language Integrated Production System

CLIPS is an expert system software, its name is an acronym for C Language Integrated Production System. On GNU/Linux system CLIPS can be started on the command line as:

```
$ ./source/clips/clips
            CLIPS (V6.0 05/12/93)
CLIPS>
```

Some useful functions with CLIPS are:

1. (exit): quits CLIPS and returns back to the operating-system,
2. (clear): removes all rules and facts from memory; equivalent to shutting down CLIPS and restarting,
3. (reset): removes only facts (not rules) from memory,
4. (run): starts executing the expert system.

Expert systems are comprised of *facts* about the universe in which they are expected to be experts, and *rules* which govern the relationships between the facts. CLIPS internally maintains a list of facts and rules. A fact can be *asserted* to CLIPS using the (assert) function. A fact is simply a piece of information deemed to be true:

```
CLIPS> (assert (direction east))
<Fact-0>
CLIPS> (assert (direction west))
<Fact-1>
CLIPS> (assert (direction north))
<Fact-2>
CLIPS> (assert (direction south))
<Fact-3>
```

Like facts in real life, sometimes they turn out to be false, in **CLIPS** a fact can be *retracted* as:

```
CLIPS> (assert (direction market))
<Fact-4>
CLIPS> (retract 4)
CLIPS> (facts)
f-0      (direction east)
f-1      (direction west)
f-2      (direction north)
f-3      (direction south)
For a total of 4 facts.
```

In the above example, we had added "market" as a direction (in some applications, a direction such as "head to the market" may make sense), but later we wished to remove this fact from CLIPS memory, and we used the `retract` function with the fact index to remove this fact. The function `(facts)` lists all facts known to **CLIPS** at this point of time. In addition to facts, an expert system also needs rules which combine facts. Consider the following example:

```
(defrule print-directions
   (direction ?dir)
=>
   (printout t ?dir " is a known direction" crlf))
(defrule street-name
   (direction street-name)
   =>
   (assert  (stname-direction name))
   (printout t "Street name has direction in it" crlf))
(run)
south is a known direction
north is a known direction
west is a known direction
east is a known direction
CLIPS>
```

CLIPS can also prompt the user for information, which is useful when writing an expert system to diagnose problems. The user can respond to the questions:

```
$ ~/EXPERT/source/clips/clips
          CLIPS (V6.0 05/12/93)
CLIPS> (defrule are-you-awake
(sleep ?)
=>
(printout t "Are you awake (yes or no)?")
(assert (awake (read))))
CLIPS> (assert (sleep sandeep))
<Fact-0>
```

```
CLIPS> (facts)
f-0     (sleep sandeep)
For a total of 1 fact.
CLIPS> (rules)
are-you-awake
For a total of 1 defrule.
CLIPS> (run)
Are you awake (yes or no)?Jack
CLIPS> (facts)
f-0     (sleep sandeep)
f-1     (awake Jack)
For a total of 2 facts.
```

Using CLIPS, it is simple to build up a chain of facts and rules and then the expert system can perform pattern matching and rule-chaining to arrive at the goal. Facts relevant to the domain can be stored in a database (or external file) and then loaded into CLIPS at run-time. Similarly, rules can be stored in a file and loaded into the memory. Expert systems in general, and CLIPS in particular, are examples of *data-driven* programming. It is instructive to see the program use data (facts) as the axis on which to guide the computation, as opposed to procedural programming where the functions manipulate the data.

18.3 ACL2: automatic theorem proving

ACL2 is an acronym for "A Computational Logic for Applicative Common Lisp." ACL2 is an interactive mechanical automated theorem prover which was designed for proving theorems modeling the behavior of hardware and software systems. In particular, ACL2 has been used to model and verify properties about the design of floating-point units of processors. ACL2 was written by Kaufmann and Moore as a successor to the Boyer-Moore theorem prover. ACL2 can be launched using a Common Lisp system which is loaded with the ACL2 generated core memory image as shown below:

```
$ ./saved_acl2
CMU Common Lisp 19f Fedora
    release 2.fc12 (19F), running on celex
With core: acl2-sources/saved_acl2.core
Dumped on:  on celex
Loaded subsystems:
    Python 1.1, target Intel x86
    CLOS based on Gerd's PCL 2008-11-12 16:36:41
Sat, 2010-06-12 03:10:36-07:00
ACL2 Version 3.6.  Level 1.
Cbd "/home/skoranne/acl2-sources/".
Distributed books directory "acl2-sources/books/".
Type :help for help.
```

```
Type (good-bye) to quit completely out of ACL2.

ACL2 !>
```

ACL2 is built on top of Common Lisp (see Section 1.6.8), thus all the features of Common Lisp are directly available inside ACL2. The "books" ACL2 is refering to in the above transcript refer to the collected knowledge that has been communicated to ACL2.

```
ACL2 !>(thm (equal (car (cons x y)) x ))

But we reduce the conjecture to T, by the
simple :rewrite rule CAR-CONS.

Q.E.D.

Summary
Form:  ( THM ...)
Rules: ((:REWRITE CAR-CONS))
Warnings:  None
Time:  0.00 seconds (prove: 0.00,
       print: 0.00, other: 0.00)

Proof succeeded.
```

In the above example we ask ACL2 to prove the identity which is latent in the definition of the Lisp function `car`. ACL2 uses the rewrite rule from its internal definitions to arrive at a proof. We can ask it to prove de'Morgan's law about Boolean algebra.

```
ACL2 !>(thm (implies (and a b) (not (or (not a) (not b)))))

But we reduce the conjecture to T, by case analysis.

Q.E.D.

Summary
Form:  ( THM ...)
Rules: NIL
Warnings:  None
Time:  0.01 seconds (prove: 0.00,
       print: 0.00, other: 0.01)

Proof succeeded.
```

The advantage of using Common Lisp as the underlying language to describe the theorem constituent is shown in the next example, where we define a new predicate `is-perm`:

```
     (defun is-perm (x y)
     "Is X a permutation of Y"
     (if (consp x)
       (and (member (car x ) y )
5      (is-perm (cdr x) (remove (car x) y)))
     (not (consp y))))
```

This is a simple recursive function which checks where (car x) is a member of Y and if so, deletes that element and checks the remainder of the list. We can use the above defined predicate to check whether a sorting function is producing a permutation of the input. Consider the following definition of a insert function:

```
(defun insert (e x)
  (cond ((endp x) (cons e x))
    ((< e (car x)) (cons e x))
    (t (cons (car x) (insert e (cdr x))))))
```

We can test this function as:

```
ACL2 !>(defun insert (e x)
        (cond ((endp x) (cons e x))
              ((< e (car x)) (cons e x))
              (t (cons (car x)
                  (insert e (cdr x))))))

The admission of INSERT is trivial, using the
relation O< (which is known to be well-founded on
the domain recognized by O-P) and the measure
(ACL2-COUNT X).  We observe that the type of
INSERT is described by the theorem
(CONSP (INSERT E X)).
We used primitive type reasoning.

Summary
Form:   ( DEFUN INSERT ...)
Rules: ((:FAKE-RUNE-FOR-TYPE-SET NIL))
Warnings:  None
Time:  0.00 seconds (prove: 0.00,
        print: 0.00, other: 0.00)
 INSERT
ACL2 !>(insert 3 '(1 2 5 6 7 8))
(1 2 3 5 6 7 8)
ACL2 !>
```

Using the insert function we can write a simple sorting function as:

```
(defun isort (x)
  (if (endp x) nil
    (insert (car x)
      (isort (cdr x)))))
```

And test it as:

```
Form:   ( DEFUN ISORT ...)
Rules: ((:TYPE-PRESCRIPTION INSERT))
Warnings:  None
Time:  0.00 seconds (prove: 0.00,
```

```
        print: 0.00, other: 0.00)
  ISORT
ACL2 !>(isort '(2 3 4 1 ))
(1 2 3 4)
```

To prove that indeed this sorting is producing an ordered output, as well as a permutation of the input we ask ACL2 to prove the latter automatically.

```
CL2 !>(defthm perm-isort
          (is-perm (isort x) x))
```

We try to assert the fact that a sorted output is basically a permutation of the input. This theorem is automatically proven by ACL2 (the output of the rules used is too large to include here). Above we have seen a simple example of using the Common Lisp sub-language within ACL2 to define functions and predicates over which ACL2 can perform automated reasoning. ACL2 can also be used deduce new theorems over knowledge fields, by systematic generation and refutation of new theorems. The theorems ACL2 can prove can then be added to its repertoire of knowledge for further sessions.

18.4 GAUL : Genetic Algorithms Utility Library

Genetic algorithms are an optimization method which use an iterative search technique based on evolution. Problem variables are modeled as *individuals*, and by combining traits of parents, the offsprings are expected to have a better objective function. The GAUL library implements several variations of genetic algorithms. The `entity` data-structure stores individuals, while `population` data-structure stores multiple entities. The population genesis per iteration is provided by the following function:

```
    *ga_genesis_integer( const int            population_size,
                         const int            num_chromo,
                         const int            len_chromo,
                         GAgeneration_hook    generation_hook,
 5                       GAiteration_hook     iteration_hook,
                         GAdata_destructor    data_destructor,
                         GAdata_ref_incrementor data_ref_incrementor,
                         GAevaluate           evaluate,
                         GAseed               seed,
10                       GAadapt              adapt,
                         GAselect_one         select_one,
                         GAselect_two         select_two,
                         GAmutate             mutate,
                         GAcrossover          crossover,
15                       GAreplace            replace,
          vpointer    userdata );
```

We model a floor-planning problem using genetic-algorithms.

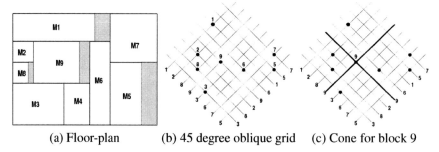

(a) Floor-plan   (b) 45 degree oblique grid   (c) Cone for block 9

Fig. 18.1 Sequence-pair representation

18.5 Representing floor-plans by $k-tuples$

We introduce the concept of $k-tuples$, which is an efficient and compact representation of floor-plans. The idea of $k-tuples$ is based upon the seminal work of Murata *et al.* in solving the rectangle placement problem using a notation called *sequence-pair*. Since $k-tuples$ are based on sequence-pair we give a brief introduction of sequence-pair notation below.

18.5.1 Sequence-pair Notation

A sequence-pair is a pair of sequences of n elements representing a list of n blocks. Given a block placement (like the one given in Figure 18.1(a)) Murata *et al.* describe a procedure for creating a linear order amongst the blocks using a 45-degree oblique grid notation as shown in Fig. 18.1(b). For every block, the plane is divided by the two crossing slope lines into four cones as shown in Figure 18.1 (c). Given a sequence-pair representation for a block placement (test schedule, equivalently), the horizontal topological relationship (the linear order amongst the block on the x axis) amongst the blocks can be represented as a horizontal constraint graph $G_h(V,E)$, which can be constructed as follows:

1. $V = \{s_h\} \cup \{t_h\} \cup \{v_i | i = 1, \ldots, n\}$, where v_i corresponds to a block, s_h is the source node representing the left boundary and t_h is the sink node, representing the right boundary;
2. $E = \{(s_h, v_i) | i = 1, \ldots, n\} \cup \{(v_i, t_h) | i = 1, \ldots, n\} \cup \{(v_i, v_j) |$ block v_i is to the left of block $v_j \}$.

The vertical constraint graph $G_v(V,E)$ can be similarly constructed. The corresponding constraint graphs $G_h(V,E)$ and $G_v(V,E)$ for the example schedule shown in Figure 18.1(a) are shown in Figure 18.2(a) and (b), respectively.

A small example of using GAUL to optimize a floor-planning problem is shown in Listing 18.1. We model each sequence-pair by a chromosome of 16-bit, and we define a cost function which returns true when the sequence-pair (as represented by

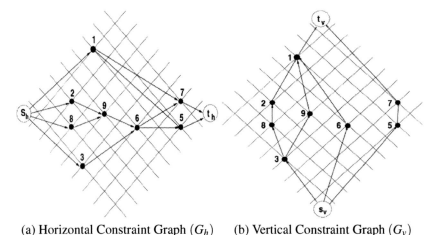

(a) Horizontal Constraint Graph (G_h) (b) Vertical Constraint Graph (G_v)

Fig. 18.2 Graph representation of floor-plans (transitive edges have been omitted.)

the chromosome) is a feasible solution. GAUL library then implements the actual genetic algorithm iteration.

```
    // \file gaul_example.cpp
    // \author Sandeep Koranne, (C) 2010
    // \description Example of genetic algorithm optimization
    #include <iostream>           // Program IO
5   #include <cassert>            // assertion
    #include <cstdlib>            // exit
    extern "C" {
    #include <gaul.h>             // GAUL library
    }
10  int main(int argc, char* argv[] ) {
       random_init(); // RNG init for good mix
       unsigned int num_blocks = 10; // there are 10 blocks in floorplan
       const unsigned int NUM_GEN = 100; // number of generations
       population *pop = NULL; // collection of solutions
15     for(unsigned int i=0; i < NUM_GEN; ++i ) {
         if( pop ) ga_extinction( pop );
         random_seed( i );
         pop = ga_genesis_integer( num_blocks, // population size
                 1,              // number chromosomes
20               16,             // length chromosome
                 NULL, NULL, NULL, NULL,
                 cost_function,
                 ga_seed_integer_random,
                 NULL,
25               ga_select_one_bestof2,
                 ga_select_two_random,
                 ga_mutate_integer_singlepoint_drift,
                 ga_crossover_integer_singlepoints,
                 NULL,
30               NULL );


         ga_population_set_parameters(pop,
                 GA_SCHEME_DARWIN,
35               GA_ELITISM_PARENTS_SURVIVE,
                 0.5,0.05,0.0 );
         ga_evolution( pop, NUM_GEN );
       }
       ga_extinction( pop );
```

```
40    std::cout << std::endl;
      return(0);
   }
```

Listing 18.1 Example of using GAUL for floorplanning

This use of genetic algorithm to solve floor-planning is one of the classic examples of genetic algorithm (since this was one of the early problems for which GA was used). It can be seen that genetic algorithms can offer an innovative solution, and with the advent of parallel processing, their use is expected to rise.

18.6 ASA : Adaptive Simulated Annealing Library

Simulated annealing (SA) is a stochastic optimization technique, which although slower than other analytical techniques, often provides better quality of solution. It can be used in general unconstrained combinatorial search applications where the objective function is not amenable to analytic implementation. However, there exist functions (in particular linear and quadric forms) which can be optimized more efficiently using linear programming, or quadratic programming. Thus, the reader should always consider an analytic approach before deciding to use simulated annealing (as it can be many times slower).

To use SA with constrained problem, a technique called SUMS (scalable unconstrained minimization solver) can be used which converts the non-feasible region (lying outside the constraints to a very large cost region), thus allowing the annealer to avoid it. Simulated Annealing itself is inspired by thermodynamics, where crystalline forms were heated and allowed to cool slowly, thus forming perfectly aligned shapes. Similarly, SA is organized and implemented so that the search space is explored iteratively; in the starting the *temperature* is set to a high value. The objective function is evaluated at configurations, if the configuration has a lower cost of the objective function (for minimization) it is accepted, else, a random number generator is consulted, and a probability function is used to accept the configuration (even when the cost is more than the current configuration). This feature of accepting increasing cost configuration (albeit with a decreasing probability) allows SA to escape *local minima* and differentiates it from *greedy descent*.

The probability function is architected so that it depends on the temperature. When the temperature is high, more configurations are accepted. Iteratively, the temperature is reduced (*cooling*), and finally only configurations which reduce the cost are accepted. Choosing the initial starting temperature, number of trial configurations per temperature, iteration count, acceptance criteria, temperature schedule, random number generation and roll-back (reverting back to a known good configuration, if the cost becomes too high) are all known SA implementation criteria. There is significant theory involved in choosing the above correctly, but mostly they are experimentally chosen.

Adaptive Simulated Annealing (ASA) is a variation of SA where the cooling schedule and other factors listed above are automatically controlled by the anneal-

ing software. Once such library is the ASA library. To use the ASA library, the
user has to define a cost function, acceptance criteria and configuration space exploration function. ASA library then automatically configures (adaptively), the cooling
schedule and other metrics for efficient simulated annealing. The cost function is the
called in the inner-loop of the SA algorithm, so care must be taken that implementation of the cost function is efficient. The user cost function and acceptance criteria
can be defined as:

```
   /* user-defined */
   double USER_COST_FUNCTION (double *cost_parameters,
                              double *parameter_lower_bound,
                              double *parameter_upper_bound,
5                             double *cost_tangents,
                              double *cost_curvature,
                              ALLOC_INT * parameter_dimension,
                              int *parameter_int_real,
                              int *cost_flag,
10                            int *exit_code, USER_DEFINES * USER_OPTIONS);

   // Acceptance criteria test function
     void user_acceptance_test (double current_cost,
                              double *parameter_lower_bound,
15                            double *parameter_upper_bound,
                              ALLOC_INT * parameter_dimension,
                              const void *OPTIONS_TMP);

   // Search space exploration, called generating function
20   double user_generating_distrib (LONG_INT * seed,
                                     ALLOC_INT * parameter_dimension,
                                     ALLOC_INT index_v,
                                     double temperature_v,
                                     double init_param_temp_v,
25                                   double temp_scale_params_v,
                                     double parameter_v,
                                     double parameter_range_v,
                                     double *last_saved_parameter,
                                     const void *OPTIONS_TMP);
```

To optimize the performance, the user can also specify a *re-anneal* cost, which is
used by ASA to decide when to call annealing on a configuration space.

```
   int user_reanneal_cost (double *cost_best,
                           double *cost_last,
                           double *initial_cost_temperature,
                           double *current_cost_temperature,
5                          const void *OPTIONS_TMP);
```

For self-optimization, the cost function maybe called recursively, the declaration of
the recursive cost function is:

```
   double RECUR_USER_COST_FUNCTION (double *recur_cost_parameters,
                                    double *recur_parameter_lower_bound,
                                    double *recur_parameter_upper_bound,
                                    double *recur_cost_tangents,
5                                   double *recur_cost_curvature,
                                    ALLOC_INT * recur_parameter_dimension,
                                    int *recur_parameter_int_real,
                                    int *recur_cost_flag,
                                    int *recur_exit_code,
10                                  USER_DEFINES * RECUR_USER_OPTIONS);
```

ASA library has been used in many application where multi-dimensional search space optimization using simulated annealing was required, and is stated to be much faster than standard Boltzman annealing. Since a lot of the asa_usr code is used in the final application (only the cost function, and generating function templates in the code have to be modified), this library is not used as a conventional API library. In fact the code is modified and compiled to create the user application. We compiled and ran the code as:

```
gcc -O3 -o asa_test asa_usr.c asa_usr_cst.c asa.c -lm
```

where we had modified the templates in asa_usr.c and asa_usr_cst.c to reflect our application.

18.7 Artificial Neural Networks : FANN

Artificial Neural Networks (ANN) are biologically inspired graphs which exhibit *solution learning* behavior. By *training* an ANN on a set of problem-solution pairs, the network adapts its internal weights to seemingly become intelligent at solving new problem of the same nature. This description also contains the inherent limitation of ANNs, that the problem class has to be sufficiently narrow, or the learning time sufficiently large (alongwith memory to store the network) to solve a wider variety of problems. Even with this caveat, ANNs prove to be an useful tool in applications such as speech and hand writing recognition, where classical methods perform poorly.

ANNs have inputs, a computation network and a set of outputs. The computation network is defined using activation functions which define when a particular neuron should *fire* based on the strength of its inputs (and a function on them).

Implementing ANNs requires a good deal of good deal of knowledge on the workings of neural networks, however, the Fast ANN library already implements most of the commonly used networks. Using FANN in an application is straightforward as shown in Listing 18.2.

```cpp
   // \file fann_example.cpp
   // \author Sandeep Koranne (C) 2010
   // \description Example of using FANN
   // Fast Artificial Neural Network library
5  #include <iostream>          // program IO
   #include <cassert>           // assertion checking
   #include <fann.h>            // FANN library

   int main(int argc, char* argv [] ) {
10   float connection_rate = 1.0;
     float learning_rate = 0.85;
     unsigned int number_layers = 3;
     unsigned int number_inputs = 16;
     unsigned int number_outputs = 4;
15   unsigned int number_internals = 10;
     struct fann *nw =
       fann_create_standard( number_layers, number_inputs,
          number_internals, number_outputs );
```

```
20     assert( nw && "Unable to construct ANN" );
       fann_train_on_file( nw, "classify.dat", 1000,50,0.001 );
       fann_save( nw, "bucket_classify.net" );
       fann_destroy( nw );
       std::cout << std::endl;
25     return (0);
    }
```

Listing 18.2 Example of using FANN neural network library

In Listing 18.2 we show the creation and training of an artificial neural network (ANN) using the FANN library. In the code we create a network using the `fann_create_standard` which takes as input the number of layers of the network, number of inputs, internal layers and number of outputs. Consider a data grouping or classification application which classifies a given input set of 16 data values into one of four categories; thus number of inputs is 16 and number of outputs is 4 (binary). The network has to be trained on a sample set. In this example we generate a sample data file. The header of this file states that there are 5 patterns of 16 inputs and 4 outputs. Compiling the program in Listing 18.2 and running it we get:

```
$ cat classify.dat
2 16 4
0.1 0.4 0.6 0.3 0.8 0.1 0.1 0.0 0.2 0.1 0.4 0.6 0.3 0.8 0.1 0.1
1 0 0 0
0.7 0.64 0.66 0.3 0.58 0.2 0.1 0.0 0.2 0.2 0.4 0.6 0.3 0.8 0.1 0.1
0 0 0 1
0.1 0.4 0.6 0.3 0.8 0.1 0.1 0.0 0.2 0.1 0.4 0.6 0.3 0.8 0.1 0.1
0 1 0 0
0.1 0.14 0.6 0.3 0.8 0.1 0.1 1.0 1.0 0.1 0.4 0.6 0.3 0.8 0.1 0.1
0 0 1 0

$ ./fann_example
Max epochs      1000. Desired error: 0.0010000000.
Epochs             1. Current error: 0.9598405957. Bit fail 8.
Epochs            50. Current error: 0.2475839257. Bit fail 1.
Epochs            88. Current error: 0.0007690609. Bit fail 0.
```

The result of this training session is the file 'bucket_classify.net' which contains a representation of the ANN:

```
FANN_FLO_2.1
num_layers=3
learning_rate=0.700000
connection_rate=1.000000
network_type=0
learning_momentum=0.000000
training_algorithm=2
train_error_function=1
train_stop_function=0
cascade_output_change_fraction=0.010000
quickprop_decay=-0.000100
quickprop_mu=1.750000
rprop_increase_factor=1.200000
rprop_decrease_factor=0.500000
```

```
rprop_delta_min=0.000000
rprop_delta_max=50.000000
rprop_delta_zero=0.500000
cascade_output_stagnation_epochs=12
... // more properties
neurons (num_inputs, activation_function,
        activation_steepness)=(0, 0,
0.00000000000000000000e+00)
```

The network file contains the network property and the activation function for
the neurons. Once the network has been created and stored in a file it can be used
for future classification as shown in Listing 18.3.

```cpp
// \file fann_use.cpp
// \author Sandeep Koranne (C) 2010
// \description Use of FANN produced network for classification
#include <iostream>        // program IO
#include <cassert>         // assertion checking
#include <fstream>         // for file IO
#include <fann.h>          // FANN library

int main( int argc, char *argv [] ) {
  if( argc != 3 ) {
    std::cerr << "Usage : ./fann_use <nw> <data>...\n";
    exit(1);
  }
  struct fann *nw = fann_create_from_file( argv[1] );
  assert( nw && "Unable to create network from file" );
  float data[16];          // remember 16 inputs
  std::ifstream ifs( argv[2] );
  unsigned int count = 0;
  while( ifs ) {
    ifs >> data[count++];
  }
  float *classification; // remember 4 outputs
  classification = fann_run( nw, data );
  std::cout << "Classified data set = "
      << "\nData[0] probability = " << classification[0]
      << "\nData[1] probability = " << classification[1]
      << "\nData[2] probability = " << classification[2]
      << "\nData[3] probability = " << classification[3];

  std::cout << std::endl;
  return (0);
}
```

Listing 18.3 Using FANN produced ANN for classification

Compiling and running the program on a new sample data (of 16 inputs) we get:

```
$ ./fann_use bucket_classify.net sample.dat
Classified data set =
Data[0] probability = -0.001551
Data[1] probability = 1.85513
Data[2] probability = 3.20333e-32
Data[3] probability = 0.23337
```

Thus, we conclude the sample data can be classified in the second category. The
above examples are deliberately simple, but FANN is used in this manner for complex classification tasks where network training is comprised of many millions of

trials on valid data. The simplicity of the API is well demonstrated using these small examples. Some of the other useful functions from FANN are given below in Table 18.1.

Table 18.1 FANN API Functions

Name
fann * fann_create_standard(**unsigned int** num_layers, ...);
fann * fann_create_sparse(**float** connection_rate)
fann * fann_create_sparse_array(**float** connection_rate)
fann * fann_create_shortcut(**unsigned int** num_layers, ...);
fann * fann_create_shortcut_array(**unsigned int** num_layers)
void fann_destroy(fann *ann);
fann_type * fann_run(fann *ann, fann_type * input);
void fann_randomize_weights(fann *ann, fann_type min_weight)
void fann_init_weights(fann *ann, fann_train_data *train_data);
void fann_print_connections(fann *ann);
void fann_print_parameters(fann *ann);
unsigned int fann_get_num_input(fann *ann);
unsigned int fann_get_num_output(fann *ann);
unsigned int fann_get_total_neurons(fann *ann);
unsigned int fann_get_total_connections(fann *ann);
enum fann_nettype_enum fann_get_network_type(fann *ann);
float fann_get_connection_rate(fann *ann);
unsigned int fann_get_num_layers(fann *ann);
void fann_get_layer_array(fann *ann, **unsigned int** *layers);
void fann_get_bias_array(fann *ann, **unsigned int** *bias);
void fann_get_connection_array(fann *ann)
void fann_set_weight_array(fann *ann
void fann_set_weight(fann *ann
void fann_set_user_data(fann *ann, **void** *user_data);
void * fann_get_user_data(fann *ann);
unsigned int fann_get_decimal_point(fann *ann);
unsigned int fann_get_multiplier(fann *ann);

18.8 LIBSVM : Support Vector Machines

Support Vector Machines are a class of AI tools and techniques used for automatic data classification. LIBSVM is an integrated software for support vector classification (C-SVC), regression estimation, and distribution estimation. It implements an SMO-type algorithm and is available as an open-source library.

Like Artificial Neural Networks (ANN) (see Section 18.7) tasks involved in SVM classification can be divided into (i) training, and (ii) actual classification. The goal of the training data sets is to construct a model (based on the training data) which predicts the target value of the test data (given only the test data attributes). The mathematical principles behind SVM are based on the mapping of the training vectors into a higher (maybe infinite) dimensional space. SVM then calculates a linear separating hyperplane with the maximal margin in this higher dimensional space.

Due to the principle of dimensional mapping, SVM requires that each data instance be represented as a vector of real numbers. Hence, if the data consists of categorical attributes, these have to be converted into numeric data. A simple *one-hot* encoding can be used to encode attribute data, using Boolean 0 and 1 to represent exclusion and inclusion in the category set. For example, the pieces of a chess game can be represented by the tuple:

$$CP = rook, bishop, knight \qquad rook = 1,0,0 \qquad bishop = 0,1,0 \qquad knight = 0,0,1$$

18.8.1 SVM Tools

LIBSVM consists of a number of tools which are described below:

1. svm-train: The command-line usage of this tool is shown below:

```
$ ./svm-train
Usage: svm-train [options]
       training_set_file [model_file]
options:
-s svm_type : set type of SVM (default 0)
-t kernel_type : type kernel function
-d egree : degree in kernel function (default 3)
-g gamma : gamma in kernel function
-r coef0 : coef0 in kernel function (default 0)
-c cost : C of C-SVC, epsilon-SVR (default 1)
-n nu : nu of nu-SVC, (default 0.5)
-p epsilon : epsilon-SVR (default 0.1)
-m cachesize : cache size in MB (default 100)
-e epsilon : tolerance of termination
-h shrinking : use shrinking heuristics
-b probability_estimates :
   whether to train a SVC or SVR model
   for probability estimates, 0 or 1
-wi weight : C of class i to weight*C
-v n: n-fold cross validation mode
-q : quiet mode (no outputs)
```

2. svm-predict: The command-line usage for this tool is shown below:

```
$ ./svm-predict
Usage: svm-predict [options] test_file
       model_file output_file
```

```
options:
-b probability_estimates:
  whether to predict probability estimates,
  0 or 1 (default 0);
  for one-class SVM only 0 is supported
```

3. svm-scale: performs scaling of the data prior to classification. Scaling is an important pre-processing check which can improve the quality of the classification. The example usage of this program is shown below:

```
$ ./svm-scale
Usage: svm-scale [options] data_filename
options:
-l lower : x scaling lower limit (default -1)
-u upper : x scaling upper limit (default +1)
-y y_lower y_upper :
  y scaling limits (default: no y scaling)
-s save_filename :
  save scaling parameters to save_filename
-r restore_filename :
  restore scaling parameters from restore_filename
```

4. checkdata: a simple Python program which checks the training data for syntax correctness,

The usual method of classifying data with LIBSVM is to train the SVM model on pre-defined training data (using svm_train), and then perform actual classification using svm_predict. The format of the data (both training and classification) is shown below:

```
<label> <index1>:<value1> <index2>:<value2> ..
```

where each line denotes an instance and is ended by the newline character. For classification, <label> is an integer indicating the class label. The <index>:<value> are index, value where indices start from 1 and value is a real number. Consider the same classification data which we used for the artificial neural network (except the attributes have been reduced to 4):

```
+1 1:0.1 2:0.4 3:0.6 4:0.3
-1 1:0.7 2:0.64 3:0.66 4:0.3
+1 1:0.1 2:0.4 3:0.6 4:0.3
+1 1:0.1 2:0.14 3:0.6 4:0.3
```

The data file should not have blank trailing lines. The Python program checkdata can be used to check the data for syntax. We now train our SVM model on this data using:

```
$ ~/libsvm-2.91/svm-train data.dat
*
optimization finished, #iter = 2
nu = 0.500000
obj = -1.900054, rho = -0.900054
```

```
nSV = 2, nBSV = 2
Total nSV = 2
```

The produced SVM model is shown below:

```
svm_type c_svc
kernel_type rbf
gamma 0.25
nr_class 2
total_sv 2
rho -0.900054
label 1 -1
nr_sv 1 1
SV
1 1:0.1 2:0.4 3:0.6 4:0.3
-1 1:0.7 2:0.64 3:0.66 4:0.3
data.dat.model (END)
```

We can now use this model to perform data classification on real data as shown below:

```
$./cat test.dat
+1 1:0.3 2:0.4 3:0.6 4:0.31
-1 1:0.67 2:0.64 3:0.46 4:0.23
+1 1:0.1 2:0.4 3:0.61 4:0.33
+1 1:0.1 2:0.14 3:0.16 4:0.28
$./svm-predict -b 0 test.dat data.dat.model output
Accuracy = 75% (3/4) (classification)
```

It is also possible to use LIBSVM within an application by using the classification facility as an API. The model creation and training can be carried out, and the model loaded into another application which performs the classification.

18.9 Conclusion

Artificial Intelligence (AI) has been a promised goal of computer science research for more than half a century. Nevertheless, recent years has seen many of the tools and techniques which were invented during AI's inception, find use in diverse application domains. In this chapter we have presented the CLIPS expert system, ACL2 theorem prover and a number of API libraries which implement search and optimization techniques such as simulated annealing, genetic algorithms, support vector machines and artificial neural networks. Since AI is often used to solve optimization problems with poorly defined, or ill-conditioned constraints, it is important to know of current state-of-art in AI tools and techniques when faced with difficult optimization problems which may be amenable to AI techniques.

Part V
Scientific Visualization

Chapter 19
Information Visualization

Abstract In this chapter we describe the many GUI libraries on GNU/Linux systems. These include, GTK, Qt, as well as wxWidget and Fox Toolkit. We present OpenGL through many examples, which also present GLUT, GLUI and show example of using OpenGL from within Python. Graphics rendering engines (OGRE) and OpenGL helper libraries are also discussed. In addition to 3d graphics, graphics layout are also available using the Graphviz `dot` tool. Plotting software Gnuplot, and vector drawing tools Xfig and Inkscape are also discussed. Raytracing with PovRay is shown with the help of examples in Section 19.9. Programmatic creation of graphics is shown with the help of `gd` library, and the Asymptote library. Graphics visualization with GeomView, HippoDraw is described in Section 19.15. Multi-dimensional data visualization with GGobi is discussed. High-performance scientific data visualization with ParaView and OpenDX are discussed.

Contents

S. Koranne, *Handbook of Open Source Tools*,
DOI 10.1007/978-1-4419-7719-9_19, © Springer Science+Business Media, LLC 2011

In Chapter 19 we discuss the various software tools and libraries available for information visualization. We discuss both creation of images and pictures for presentation as well rendering infrastructure and graphical user interfaces (GUI). Specifically we discuss, Qt, OpenGL, 'dot', gnuplot, `xfig`, `gd`, `asymptote`. Complete applications for graphics creation include Inkscape, ParaView, Geomview, Hippo-Draw and OpenDX for information presentation.

19.1 Graphical User Interfaces

We discuss graphical user interface systems that are available. In Section 19.1.2 we present the Gimp Tool Kit (gtk), Nokia/Trolltech's Qt is discussed in Chapter 19.1.3. WxWidgets is briefly discussed in Chapter 19.1.5.1 and the Fox toolkit is presented in Chapter 19.1.5. OpenGL front-end GUIs are presented for completeness towards the end of this part in Chapter 19.2.

19.1.1 X Window System

The X Window System is a graphical computer system and network protocol which is pervasive on GNU/Linux systems. X Window System provides the lowest level of hardware abstraction layer, and provides infrastructure for graphical user interfaces across networked computers. It allows for device independence, as well as remote display of screen contents over networks. Since it was designed for thin-clients the client side processing for X is very efficient, and places little demands on the computer (both CPU as well as memory). As a hardware abstraction layer, it provides windowing (including primitives to draw points, lines, polygons, and rasterization) as well as abstraction for user input (including keyboard and mouse). A singular feature of X is its operating system independence across the network. Since X only provides the lowest level primitives, other software such as window managers, or desktop environments such as GNOME and KDE usually is used on top of X11. GUI toolkits, when running on the X Window System make heavy use of the Xlib API. We present a number of such GUI libraries below, and although low level programming in Xlib is not only possible, but also considered to be more efficient, the convenience offered by using a GUI more than offsets the small gain in performance.

It should also be noted, that there also exist a number of problems with X11, which can be solved using external applications. For example, currently it is not possible to detach a client from one server to another, although by using VNC (see Section 1.7.1) stateful sessions on X11 are possible. Moreover, X protocol traffic is sent unencrypted over the network. This can be solved by using OpenSSH tunneling (see Section 1.5), which encrypts all traffic over a designated port.

19.1.2 GIMP Toolkit: GTK

In Section 14.5 we have presented the GIMP (GNU Image Manipulation Program). GIMP Toolkit (GTK) is a cross-platform widget toolkit for designing and implementing graphical user interfaces (GUI) for the UNIX X-Window system. It was initially created for GIMP, but now has expanded significantly in its usage, most importantly, it is the widget toolkit used for GNOME.

GTK is an object oriented (OO) widget toolkit written in the C programming language. The OO is implemented using the GLib object system (GObject). The underlying drawing primitives used by GTK depend on the platform, on UNIX it uses the X-windows xlib system to draw contents on the screen. Like the other toolkits we discuss in this part of the book, GTK admits many language bindings, including but not limited to C, C++, Python, and Java. Consider the example program shown in Listing 19.1.

19.1.2.1 Hello, World in GTK

Consider the following program in C language.

```
    /* GTK Hello, World! program.(C) Sandeep Koranne, 2010 */
    #include <gtk/gtk.h>
    int main (int argc, char *argv[]) {
      GtkWidget *window;
5     GtkWidget *label;
      char *markup, *str = "Hello, World!";
      gtk_init (&argc, &argv);
      window = gtk_window_new (GTK_WINDOW_TOPLEVEL);
      gtk_window_set_title (GTK_WINDOW (window), str);
10    g_signal_connect (window, "destroy",
          G_CALLBACK (gtk_main_quit), NULL);
      markup = g_markup_printf_escaped("<span size=\"x-large\""
      "style=\"italic\">%s</span>", str);
      label = gtk_label_new (str);
15    gtk_label_set_angle( GTK_LABEL( label ), 45.0 );
      gtk_label_set_markup( GTK_LABEL( label ), markup );
      gtk_container_add (GTK_CONTAINER (window), label);
      gtk_widget_show_all (window);
      gtk_main (); return 0;
20  }
```

Listing 19.1 Using GTK GUI Libraries

We compile this program using the `pkg-config` tool as:

```
gcc gtk_hw.c -o gtk_hw `pkg-config --cflags gtk+-2.0` \
    `pkg-config --libs gtk+-2.0`
```

Running this program gives us the picture as shown in Figure 19.1.

The example above shows the use of Pango Text Attribute Markup Language. Using this mechanism the attributes of the text elements can be changed.

Fig. 19.1 Hello, World with
GTK showing Pango

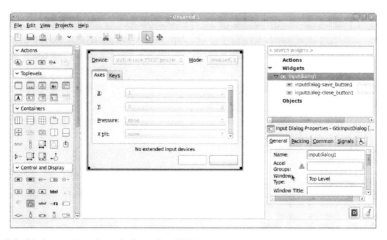

Fig. 19.2 Glade: user interface designer for GTK

19.1.2.2 Using Glade: User Interface Designer

Designing user interfaces by compositing widgets by hand gets tedious very quickly,
thus most widget systems have developed GUI tools to design GUIs. For GTK, the
user interface designer is called Glade. An example is shown in Figure 19.2. Using
Glade a GUI application using GTK can be built very quickly.

19.1.3 Qt: Application development framework

Nokia/Trolltech's Qt application framework library started out as a graphical user
interface widget library, but over the years has grown to include many features which
are not traditionally associated with GUIs. These include threading, internationalization, XML processing and much more. In this chapter we discuss the Qt framework. In the example we have used Qt version 4.6.2. A picture of Qt Creator is
shown in Figure 19.3.

Qt is written in C++, but program bindings exist for many other programming
languages such as Python. Qt is also cross-platform, meaning the same source code

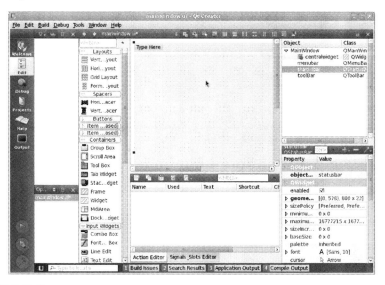

Fig. 19.3 Qt Creator

can be compiled on UNIX, GNU/Linux, Microsoft Windows, Apple Mac OS X, even embedded Linux. In addition to GUI widgets Qt also includes functionality required by application developers to create rich and functional products. Qt is written in an Object Oriented manner (as most GUI toolkits are), and has a unique *signal* and *slot* mechanism which connects user actions to program behavior, see 19.1.3.2.

Starting with Qt 4, the organization of Qt has been changed to consist of several modules, each of which can be used independently (except the Core module). The following modules are most useful to application developers:

19.1.3.1 Qt 4 Module Architecture

- QtCore : Core non-GUI classes,
- QtGUI : GUI interface and widgets,
- QtNetwork : network programming,
- QtOpenGL : OpenGL programming,
- QtSql : database integration using SQL,
- QtSVG : displaying contents of scalable vector graphics (SVG) files,
- QtXML : XML parser and support.

There are other modules which deal with web presentation, multimedia framework, UI testing, multi-threading, Qt 3 support, and dynamic UI creation.

Consider the program presented below:

```
// Hello World from Qt
// (C) Sandeep Koranne, 2010
```

```
    #include <QApplication>
    #include <QLabel>
5
    int main( int argc, char *argv[] )
    {
      QApplication app( argc, argv );
      QLabel *label = new QLabel( "Hello, World from Qt!", 0 );
10    label->show();
      return app.exec();
    }
```

Listing 19.2 Example of using Qt GUI Libraries

To compile this program using Qt 4, we do:

```
$qmake -project
$qmake
$make
```

The constructed project file from Qt is instructive:

```
TEMPLATE = app
TARGET =
SOURCES += qhw.cpp
```

Fig. 19.4 Hello World in Qt

19.1.3.2 Qt Signals and Slots

Before Qt, existing GUI libraries used the concept of *callback* functions (c.f. Motif) or event listeners (main loop) to connect user interaction with program computation. Qt uses an unique C++ pre-processor based solution called signal and slots. Using the `QObject::connect` function, an user can *connect* a *signal* such as button click, menu shortcut to a *slot* such as button-pressed. The Qt framework manages the internal connections between the defined signal and slot connections.

```
    // Qt signal and slot example
    #include <QApplication>
    #include <QLabel>
    #include <QSpinBox>
5   #include <QHBoxLayout>

    int main( int argc, char *argv [] ) {
      QApplication app( argc, argv );
      QWidget window;
10    QHBoxLayout *hbox = new QHBoxLayout( &window );
      QLabel    *label = new QLabel("0");
      QSpinBox *spbox = new QSpinBox;
      hbox->addWidget( spbox );
```

```
        hbox->addWidget( label );
15      QObject::connect( spbox, SIGNAL( valueChanged(int) ),
            label, SLOT( setNum(int) ) );
        window.show();
        return app.exec();
    }
```

Listing 19.3 Signal/slot mechanism in Qt

The output of the program is shown in Figure 19.5.

Fig. 19.5 Signals and slots
example with spinbox

The signal and slots definition in derived classes is not C++ per se, but has to be pre-processed by the `moc` *meta object compiler* from Qt. This compiler generates appropriate code for the class to enable signal and slot connections.

Qt comes complete with instructive examples, and using the Qt Creator and the examples, many complex products have been developed, as an example consider the Qt frontend to Octave as shown in Figure 19.6.

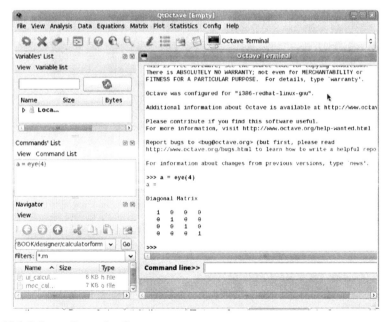

Fig. 19.6 QtOctave frontend to Octave

19.1.4 Qt's application programming API

In addition to providing graphical user interface widgets, Qt API also provides general purpose application development APIs such as database access, inter-process communication, shared memory, multi-threading, Webkits (for rendering multimedia), and much more. We present an example of some of the application development APIs below.

19.1.4.1 Shared Memory

Consider the procedure for shared memory transactions between related processes (which know the *key* to the shared memory):

```
#include <qsharedmemory.h>
{
    QSharedMemory sharedMemory;
    sharedMemory.create(size);
5   sharedMemory.lock();
    char *to = (char*)sharedMemory.data();
    const char *from = buffer.data().data();
    memcpy(to, from, qMin(sharedMemory.size(), size));
    sharedMemory.unlock();
10  sharedMemory.detach();
}
```

Another API for shared memory processing is given in Section 5.8, which uses the Apache Portable Runtime (apr) functions.

19.1.4.2 Network programming using Qt API

To use the TCP networking APIs in Qt, the TCP header files:

```
#include <QTcpServer>
#include <QTcpSocket>
```

should be included in the application, and the `network` library should be added to the `QT` section of the 'qmake' file. To create a TCL server, the `QTcpServer` object and an associated `QTcpSocket` should be created. To start *listening* to a connection the server socket's `isListening()` member function should be invoked. Thereafter, the client can be connected to the server using the `connectToHost` member function of the client socket.

19.1.5 Other GUI Toolkits

In addition to GTK and Qt, several other GUI toolkits find niche use in particular domains. These may exist for a number of reasons, including licensing restriction and developer familiarity.

19.1.5.1 WxWidgets

An introductory program in WxWidgets is shown below.

```
     #include "wx/wx.h"
     class MyApp: public wxApp {virtual bool OnInit();};
     class MyFrame: public wxFrame {
     public:
5        MyFrame(const wxString& title, const wxPoint& pos, const wxSize& size);
         void OnQuit(wxCommandEvent& event);
         void OnAbout(wxCommandEvent& event);
         DECLARE_EVENT_TABLE()
     };
10
     enum { ID_Quit = 1, ID_About, };

     BEGIN_EVENT_TABLE(MyFrame, wxFrame)
         EVT_MENU(ID_Quit, MyFrame::OnQuit)
15       EVT_MENU(ID_About, MyFrame::OnAbout)
     END_EVENT_TABLE()

     IMPLEMENT_APP(MyApp)

20   bool MyApp::OnInit() {
         MyFrame *frame = new MyFrame( _("Hello World"), wxPoint(50, 50),
                                       wxSize(450,340) );
         frame->Show(true);
         SetTopWindow(frame);
25       return true;
     }

     MyFrame::MyFrame(const wxString&title,const wxPoint&pos,const wxSize&size)
     : wxFrame( NULL, -1, title, pos, size ) {
30       wxMenu *menuFile = new wxMenu;
         menuFile->Append( ID_About, _("&About...") );
         menuFile->AppendSeparator();
         menuFile->Append( ID_Quit, _("E&xit") );
         wxMenuBar *menuBar = new wxMenuBar;
35       menuBar->Append( menuFile, _("&File") );
         SetMenuBar( menuBar );
         CreateStatusBar();
         SetStatusText( _("Welcome to wxWidgets!") );
     }
40
     void MyFrame::OnQuit(wxCommandEvent& WXUNUSED(event)){Close(TRUE);}

     void MyFrame::OnAbout(wxCommandEvent& WXUNUSED(event)) {
         wxMessageBox( _("This is a wxWidgets Hello world sample"),
45                     _("About Hello World"),
                       wxOK | wxICON_INFORMATION, this);
     }
```

Listing 19.4 Using wxWidgets GUI library

The resulting GUI is shown in Figure 19.7. A more complex example of GUI development with WxWidgets is the wxmaxima program as shown in Figure 17.2.

19.1.5.2 Java SWT

In addition to C/C++ based GUI construction, Java also has extensive graphical facilities as embodied in the Java SWT toolkit.

Fig. 19.7 Hello World with
WxWidgets

19.1.5.3 FOX Toolkit

FOX stands for Free Objects for X. It is a C++ based class library for GUI. Example
of a FOX developed UI application can be seen in Figure 19.8.

Fig. 19.8 glview example for
FOX

19.1.5.4 GUI Development using Tcl/Tk

An example of using Tcl/Tk for GUI development is shown in Figure 3.1, which
implements a complete CVS/SVN front-end using Tk. A simpler example of using
Tcl/Tk is shown in Section 1.6.6 which shows a push-button implemented using
Tcl/Tk.

19.1.5.5 GUI Development using PyQt

Using Qt from Python is straightforward using the Python bindings. Consider the
example shown in Listing 19.5.

```
#!/usr/bin/python
# \file pyqt.py
# \author Sandeep Koranne
```

```
   import sys
5  from PyQt4 import QtGui
   app = QtGui.QApplication(sys.argv)

   widget = QtGui.QWidget()
   widget.resize(400, 200)
10 widget.setWindowTitle('Main Window in PyQT')
   widget.show()
   sys.exit(app.exec_())
```

Listing 19.5 Example of PyQt

A larger example of PyQt using several Qt functions as well as the Networkx graph toolkit is shown in Listing 19.6.

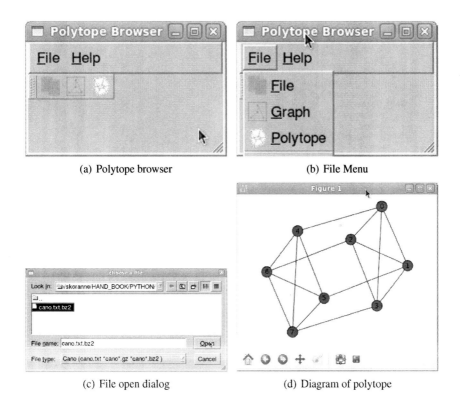

(a) Polytope browser

(b) File Menu

(c) File open dialog

(d) Diagram of polytope

Fig. 19.9 A polytope browser written in PyQt

```
#!/usr/bin/python
# \file main.pyw
# \author Sandeep Koranne

5  try:
       import matplotlib.pyplot as plt
       import sys
       import subprocess
```

```
10   except:
         raise

     import networkx as nx

15   import sys
     import gzip
     import bz2

     import sys
20   from qt import *

     dimension = sys.argv[1]


25   class Polytope:
         """Facets, Fvector, signature and parent information"""
         def __init__(self):
             self.facets = []
             self.fvec = []
30           self.signature = ""
             self.parent = 0
             self.parendId = -1
             self.cut_set = []
         def dump(self):
35           print self.signature
             print " "
             print self.facets
             print " "
             print self.parent
40           print " "
             print self.parentId
         def recreate_format( self, fileName ):
             fs = open( fileName, 'wt' )
             fs.write('VERTICES_IN_FACETS\n')
45           fs.write('\n')
             for i in self.facets:
                 fs.write('{ ')
                 for j in i:
                     fs.write( str(j) )
50                   fs.write(' ')
                 fs.write(' }\n')
             fs.write('\n')
             fs.write('F_VECTOR ( ')
             for i in self.fvec:
55               fs.write( str(i) )
                 fs.write(' ')
             fs.write(' )\n\n ')
             fs.write( self.signature )
             fs.write( ' Parent HC = [ ')
60           fs.write( self.parent )
             fs.write( ' ] ')
             fs.write( ' [ ')
             for i in self.cut_set:
                 fs.write( str(i) )
65               fs.write(' ')
             fs.write( ' ]\n')
             fs.close()


     def IndexPolytopes( polytopes ):
70       """Given a list of polytopes make a dictionary with the index"""
         poly_dict = {}
         i = 0;
         for p in polytopes:
             poly_dict[ p.signature ] = i
75           i=i+1
```

```
            for p in poly_dict:
                me = poly_dict[p] # index of myself
                my_parent_signature = polytopes[me].parent
80              my_parent_id = poly_dict[ my_parent_signature]
                polytopes[me].parentId = my_parent_id

    def ParseFile( fileName ):
            print "Reading file %s" %fileName
85          if fileName.find( '.gz' ) != -1:
                file = gzip.open( fileName, 'rb' )
                lines = file.readlines()
                file.close()
            elif fileName.find( '.bz2' ) != -1:
90              file = bz2.BZ2File( fileName, 'rb' )
                lines = file.readlines()
                file.close()
            else:
                f = open( fileName, 'r' )
95              lines = f.readlines()
                f.close()

            expectFacets = False
            expectFVec = False
100         expectSignature = False
            i = 0
            allPolytopes = []
            for l in lines:
                if l == 'VERTICES_IN_FACETS\n':
105                 if i > 0:
                        allPolytopes.append( next )
                    next = Polytope()
                    i=i+1
                    expectFacets = True
110             elif l == '\n':
                    if expectFacets:
                        expectFacets = False
                        expectFVec = True
                    elif expectFVec:
115                     expectFVec = False
                        expectSignature = True
                    elif expectSignature:
                        expectSignature = False
                        expectFacets = True
120
                else: #Data to be processed
                    if expectFacets:
                        # data is in { 1 2 3 4 }
                        face_str = l.strip()[1:-1]
125                     face = []
                        for vid in face_str.split(' '):
                            try:
                                face.append( int(vid) )
                            except:
130                             pass
                        next.facets.append( face )
                    elif expectFVec:
                        head_string = l[0:8]
                        assert head_string == 'F_VECTOR'
135                     for num in l[8:].split(' '):
                            try:
                                next.fvec.append( int(num) )
                            except:
                                pass
140                 elif expectSignature:
                        # break up the line into components
                        parent_hc = l.find('Parent HC =')
```

```
                        assert parent_hc > 0
                        next.signature = l[0:parent_hc].strip()
145                     parent_hc_end = l[parent_hc+14:].find(']')
                        offset = parent_hc+14+parent_hc_end
                        try:
                            next.parent = l[parent_hc+14:offset].strip()
                        except:
150                         print "Error in parsing parent id"

                        # we have read the parent, now read the cut set
                        for cut_set in l[offset+1:].split(' '):
                            try:
155                             next.cut_set.append( int( cut_set ) )
                            except:
                                pass

            allPolytopes.append( next )
160         print "Collected %d polytopes"%len( allPolytopes )
            IndexPolytopes( allPolytopes )
            return allPolytopes


165 def CalculateHeritage( P, i ):
        """Given the polytope table and the  polytope index find its heritage"""
        answer = []
        P[i].dump()
        parentId = P[i].parentId
170     while i != parentId:
            answer.append( (parentId, P[i].cut_set ) )
            i = parentId
            parentId = P[i].parentId

175
        return answer

    class MainWindow(QMainWindow):

180     def __init__(self, *args):
            apply(QMainWindow.__init__, (self, ) + args)
            self.setCaption("Polytope Browser")

        #load icons from PNG files 60x60 size
185     self.fileOpenIcon = QPixmap("db_icon.png")
            self.graphIcon    = QPixmap("graph_icon.png")
            self.polytopeIcon = QPixmap("poly_icon.png")


190         self.actionLoadPolytope = QAction( self, "Polytope" )
            self.actionLoadPolytope.setText("Load polytope db")
            self.actionLoadPolytope.setMenuText("&File")
            self.actionLoadPolytope.setStatusTip("Load a polytope db file")
            self.actionLoadPolytope.setIconSet( QIconSet( self.fileOpenIcon ) )
195         self.connect( self.actionLoadPolytope, SIGNAL("activated()"),
                          self.loadPolytopeDB )

            self.actionGraph = QAction( self, "Graph" )
            self.actionGraph.setText("Calculate graph")
200         self.actionGraph.setMenuText("&Graph")
            self.actionGraph.setStatusTip("Compute 1-skeleta (graph)")
            self.actionGraph.setIconSet( QIconSet( self.graphIcon ) )
            self.connect( self.actionGraph, SIGNAL("activated()"),
                          self.computeGraph )
205
            self.actionPoly = QAction( self, "Poly" )
            self.actionPoly.setText("Analyze polytope")
            self.actionPoly.setMenuText("&Polytope")
```

```
210          self.actionPoly.setStatusTip("Analyze polytope")
             self.actionPoly.setIconSet( QIconSet( self.polytopeIcon ) )

             self.statusBar = QStatusBar( self )

215          self.fileMenu = QPopupMenu()
             self.actionLoadPolytope.addTo( self.fileMenu )
             self.actionGraph.addTo( self.fileMenu )
             self.actionPoly.addTo( self.fileMenu )

220          self.quitAction = QAction( self, "Quit" )
             self.quitAction.setText("Quit")
             self.quitAction.setMenuText( "&Quit" )
             self.fileMenu.insertSeparator()
             self.quitAction.addTo( self.fileMenu )
225          self.connect( self.quitAction, SIGNAL("activated()"), self.quitApp )

             self.helpMenu = QPopupMenu()
             self.actionAboutQt = QAction( self, "AboutQt" )
             self.actionAboutQt.addTo( self.helpMenu )
230          self.connect( self.actionAboutQt,
                           SIGNAL("activated()"),
                           self.slotAboutQt )

             self.menuBar().insertItem( "&File", self.fileMenu )
235          self.menuBar().insertItem( "&Help", self.helpMenu )

             self.toolBar = QToolBar( self, "Compute" )
             self.actionLoadPolytope.addTo( self.toolBar )
             self.actionGraph.addTo( self.toolBar )
240          self.actionPoly.addTo( self.toolBar )

    def slotAboutQt( self ):
        QMessageBox.aboutQt( self )

245    def loadPolytopeDB( self ):
        self.polytope_db_filename = QFileDialog().getOpenFileName(
            ".",
            "Cano (cano.txt *cano*.gz *cano*.bz2 )",
            self,
250         "open file dialog",
            "choose a file" )
        if self.polytope_db_filename:
            print "Loading polytope file"
            print self.polytope_db_filename
255         self.parsed_db = ParseFile( self.polytope_db_filename )
            self.number_of_polytopes = len( self.parsed_db )

    def gotPolytopeId( self ):
        self.current_polytope_id = int( self.spinBox.value() )
260     self.number_input.hide()
        self.constructGraph()

    def constructGraph( self ):
        print "Calcuating graph of polytope id %d"
265     %(self.current_polytope_id)
        heritage = CalculateHeritage( self.parsed_db,
                                      self.current_polytope_id )
        for h in heritage:
            print h
270     P = self.parsed_db[ self.current_polytope_id ]
        out_file_name = 'g.cano'
        P.recreate_format( out_file_name )
        p0 = subprocess.Popen(['~/cano_proc','-f',out_file_name,
                               '-d',dimension,'-G','-m 1'],shell=False,
275                            stdin=subprocess.PIPE,
                               stdout=subprocess.PIPE)
```

```
          rc = p0.wait()
          if rc != 0:
              print "Graph construction from %s failed." %out_file_name
280       GORIG=nx.read_adjlist("graph.txt", nodetype=int)
          G=GORIG.copy()
          mylabels={}
          for v in G.nodes():
              mylabels[v]=int(v)
285       diam = nx.diameter(G)
          print "Diameter of G = %d"%diam
          #     b=nx.betweenness_centrality(G)
          #     for v in G.nodes():
          #         print "%0.2d %5.3f"%(v,b[v])
290       #p1 = plt.subplot(221)
          nx.draw(GORIG, labels=mylabels,pos=nx.spring_layout(GORIG))
          plt.savefig("san.png") # save as png
          plt.show() # display

295   def computeGraph( self ):
          self.number_input = QDialog( self )
          layout = QHBoxLayout( self.number_input )
          self.number_input.setCaption( "Enter polytope number: ")
          self.spinBox = QSpinBox( self.number_input )
300       layout.addWidget( self.spinBox )
          self.spinBox.setRange( 0, self.number_of_polytopes-1 )
          self.pnum_button = QPushButton( self.number_input, "Accept")
          layout.addWidget( self.pnum_button )
          self.pnum_button.setText("Accept")
305       self.connect( self.pnum_button, SIGNAL("clicked()"),
                        self.gotPolytopeId )
          self.number_input.show()

      def quitApp( self ):
310       # close all files and processes
          exit(0)
      number_of_polytopes = 0
      current_polytope_id = 0
      polytope_db_filename = ""
315   parsed_db = None

  def main(args):
      app=QApplication(args)
      win=MainWindow()
320   win.show()
      app.connect(app, SIGNAL("lastWindowClosed()"), app,
                  SLOT("quit()") )
      app.exec_loop()

325 if __name__=="__main__":
        main(sys.argv)
```

Listing 19.6 Qt integration with Python, polytope browser

The program in Listing 19.6 when run, produces the GUI as shown in Figure 19.9(a). The File menu as well as the file dialog are shown in Figure 19.9(b) and (c), respectively. The Python program uses the `matplot` module which is used to draw the diagram as shown in Figure 19.9(d).

19.2 OpenGL

OpenGL is an industry standard for implementing high-performance graphics rendering programs. It is a high-level API which provides window system and operating system independent software interface to the graphics hardware. An OpenGL compliant program can execute on any platform to produce identical graphics output. Moreover, each graphics hardware provider can implement optimized OpenGL implementations which aid in high-performance.

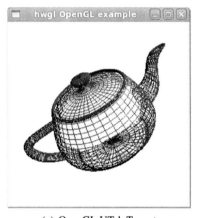
(a) OpenGL UTah Teapot

(b) OpenGL gears example

Fig. 19.10 Classic example of OpenGL

OpenGL provides a number of facilities which can be used by the application program. These include:

1. Graphics primitives: primitives include:

   a. points (GL_POINTS),
   b. lines (GL_LINES),
   c. line-loop (GL_LINE_LOOP),
   d. line-strip (GL_LINE_STRIP),
   e. convex polygon (GL_POLYGON),
   f. quadrilaterals (GL_QUADS),
   g. quad-strip (GL_QUAD_STRIP),
   h. triangles (GL_TRIANGLES),
   i. triangle strip (GL_TRIANGLE_STRIP),
   j. triangle fan (GL_TRIANGLE_FAN).

2. Indexed and RGBA color mode: indexed color mode is generally not used as operations such as *fogging* are complicated, but it is nevertheless used in applications where *object picking* is required,

3. OpenGL pipeline and contexts: OpenGL is a state-machine based drawing system. The OpenGL state-machine state can be inspected and changed by the application. Most of the state pertaining to drawing such as line-width, shading, color is changed by using the API functions,

4. Images and textures: to provide realistic three-dimensional portrayal OpenGL supports application of textures to objects. Images can be transmitted to and from the hardware memory buffer,

5. Extensions provide NURBS and curves support:

Consider the program in Listing 19.7.

```cpp
// \file hwgl.cpp
// \author Sandeep Koranne, (C) 2010
// \description Example of OpenGL primitives
#include <GL/glx.h>
#include <GL/gl.h>
#include <GL/glu.h>
#include <GL/glut.h>

GLfloat light_diffuse[] = {1.0, 0.0, 0.0, 1.0};
GLfloat light_position[]= {1.0, 1.0, 1.0, 0.0};

static int main_window = 0;
static GLuint renderModelList = 1;

static float x,y,z;
static float localScale = 1.0;
static const float MOTION = 1.0;

void processSpecialKeys( int key, int x, int y ) {
  switch( key ) {
  case GLUT_KEY_LEFT: { x += MOTION; break; }
  case GLUT_KEY_RIGHT:{ x -= MOTION; break; }
  case GLUT_KEY_UP:   { y -= MOTION; break; }
  case GLUT_KEY_DOWN: { y += MOTION; break; }
  }
  glutPostRedisplay();
  glFlush();
}


void processNormalKeys( unsigned char key, int x, int y ) {
  if( key == 27 ) { exit( 0 ); }
  if( key == 'b') { localScale /= 1.05f; }
  if( key == 'f') { localScale *= 1.05f; }
  glutPostRedisplay();
  glFlush();
}

void renderScene(void) {
  glClear( GL_COLOR_BUFFER_BIT | GL_DEPTH_BUFFER_BIT );
  glPushMatrix();
  glRotatef( 45.0, 4.0, 2.0, 4.0 );
  glScalef( localScale, localScale, localScale );
  glCallList( renderModelList );
  glPopMatrix();
  glFlush();
}

static void CreateModel(void) {
  GLUquadricObj *obj = gluNewQuadric();
  gluQuadricDrawStyle( obj, GLU_FILL );
  renderModelList = glGenLists( 1 ); // get new display list
  glNewList( renderModelList, GL_COMPILE );
```

```
     glColor3f( 0.4, 0.4,0.6 );
55   glutWireTeapot( 4.0 );
     glEndList();
  }

  void InitGL(void) {
60   x=y=z=0;
     CreateModel();
     glLightfv( GL_LIGHT0, GL_DIFFUSE, light_diffuse );
     glLightfv( GL_LIGHT0, GL_POSITION, light_position );
     glEnable( GL_LIGHTING );
65   glEnable( GL_LIGHT0 );
     glClearColor( 1.0, 1.0, 1.0, 0.5f );
     glEnable( GL_DEPTH_TEST );
     glMatrixMode( GL_PROJECTION ); // setup viewing system
     gluPerspective( 40.0,          // field-of-view
70                   1.0,           // aspect ratio
                     1.0, 10.0 );   // Z-near and far
     glMatrixMode( GL_MODELVIEW ); // scene coordinates
     gluLookAt( 0.0, 0.0, 9,        // eye is at (0,0,9)
                0.0, 0.0, 0.0,      // world center at origin
75              0.0, 1.0, 0.0 );    // Y is up
  }

  int main( int argc, char *argv [] ) {
     glutInit( &argc, argv ); // initialize GLUT sub-system
80   glutInitDisplayMode( GLUT_DEPTH | GLUT_RGBA ); // use RGBA color
     main_window = glutCreateWindow("hwgl OpenGL example");
     InitGL();
     glutKeyboardFunc( processNormalKeys );
     glutSpecialFunc( processSpecialKeys );
85
     glutDisplayFunc( renderScene );
     glutMainLoop();
     return (0);
  }
```

Listing 19.7 Drawing graphics with OpenGL

19.2.1 GLUT : OpenGL Utility Toolkit

GLUT is an utility toolkit for OpenGL which make the interaction between OpenGL and the underlying window system easier to manage as it provides the graphics context (GLX for X-Windows), as well as the initial drawing window where the context is attached. GLUT also supports a wide variety of input interfaces including keyboard, mouse and trackwheel. It also provides high-level graphics primitives such as the Utah teapot (as shown in Figure 19.10(a)) and the OpenGL glxgears program in Figure 19.10(b).

In addition to GLUT, also consider using GLFW (OpenGL contexts and input manager) which aims to provide operating-system independent access to OpenGL, similar to GLUT, but has been developed more recently.

19.2.2 GLUI : GUI for OpenGL

As stated above in Section 19.2, OpenGL by itself is window-system independent, and thus does not provide any graphical user interface as part of the standard. However, all major GUI toolkits have included support for OpenGL. There is also an independent completely OpenGL based user interface library, GLUI, which is described below.

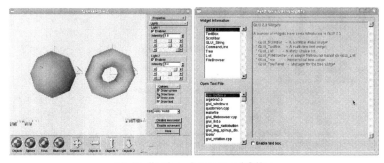

(a) Applications, Internet and Office

Fig. 19.11 OpenGL GLUI examples

Consider the short program in the listing below. It shows a simple example of using OpenGL, GLUT and GLUI within the X-Window system. The libraries are available and ready to be used to create immersive graphics with the high performance coming from OpenGL.

```
    ////////////////////////////////
    // glexample.cpp
    // (C) Sandeep Koranne, 2010
    // Example of OpenGL with GLUT and GLUI
5   ////////////////////////////////
    #include <GL/glx.h>
    #include <GL/gl.h>
    #include <GL/glu.h>
    #include <GL/glut.h>
10  #include <GL/glui.h>

    void InitGL(void);
    void renderScene(void);
    void processSpecialKeys( int key, int x, int y );
15  void processNormalKeys( unsigned char key, int x, int y );
    void changeSize( int w, int h );
    const int WIDTH = 600, HEIGHT = 600;
    GLUI_StaticText *pStaticText = NULL;
    int main_window = 0;
20  int xdim=0, ydim=0;
    void myGlutMouse( int button, int button_state, int x, int y ) {
      //
    }

25  void myGlutMotion( int x, int y ) {
      GLdouble mat[16];
```

```
       GLdouble proj[16];
       GLint viewport[4];
       glGetDoublev(GL_MODELVIEW_MATRIX,mat);
30     glGetDoublev(GL_PROJECTION_MATRIX,proj);
       glGetIntegerv(GL_VIEWPORT,viewport);
       // our viewport is static
       GLdouble dx, dy, dz;
       gluUnProject(x,HEIGHT-y,0,mat,proj,viewport,&dx,&dy,&dz);
35     char ncs[100];
       sprintf(ncs,"   X=%d,Y=%d",(int)dx,(int)dy);
       pStaticText->set_text((char*)ncs);
       glutPostRedisplay();
   }
40
   void myGlutIdle( void ) {
     if ( glutGetWindow() != main_window ) {
       glutSetWindow(main_window);
     }
45   glutPostRedisplay();
   }

   int Initialize( int argc, char *argv[] ) {
     glutInit( &argc, argv );
50   glutInitDisplayMode( GLUT_DEPTH | GLUT_SINGLE | GLUT_RGBA );
     glutInitWindowPosition( 100,100 );
     glutInitWindowSize( WIDTH, HEIGHT );
     main_window = glutCreateWindow("GlExample");
     InitGL();
55   glutDisplayFunc( renderScene );
     glutMotionFunc( myGlutMotion );

     GLUI *gluiTop = GLUI_Master.create_glui_subwindow(
       main_window, GLUI_SUBWINDOW_TOP );
60   GLUI *glui2 = GLUI_Master.create_glui_subwindow(
       main_window, GLUI_SUBWINDOW_LEFT );
     glui2->add_checkbox( "Display WireFrame");
     glui2->add_checkbox( "Black/White");

65   glui2->set_main_gfx_window( main_window );
     pStaticText = glui2->add_statictext("X=0,Y=0");

     glutKeyboardFunc( processNormalKeys );
     glutSpecialFunc( processSpecialKeys );
70   glutReshapeFunc( changeSize );
     glutMainLoop();
     return 0;
   }

75 int main( int argc, char *argv[] ) {
     return Initialize( argc, argv );
   }

   double localScale = 1.0f;
80 void renderScene(void) {
     glClear( GL_COLOR_BUFFER_BIT | GL_DEPTH_BUFFER_BIT );
     glLoadIdentity();
     glTranslatef( xdim, ydim, 0 );
     glScalef( localScale, localScale, localScale );
85   glColor3f(0,0,0);
     glRecti( 0, 0 ,5,1);
     glColor3f( 1.0, 0.0, 0.0 );
     glRecti( 0, 1, 1,2 );
     glFlush();
90 }

   void processSpecialKeys( int key, int x, int y ) {
     switch( key ) {
```

```
     case GLUT_KEY_LEFT: { xdim += 1; break; }
95   case GLUT_KEY_RIGHT:{ xdim -= 1; break; }
     case GLUT_KEY_UP:   { ydim -= 1; break; }
     case GLUT_KEY_DOWN: { ydim += 1; break; }
     }
     glutPostRedisplay();
100  glFlush();
     }

     void processNormalKeys( unsigned char key, int x, int y ) {
     if( key == 27 ) { exit( 0 ); }
105  if( key == 'b') { localScale /= 2.0f; }
     if( key == 'f') { localScale *= 2.0f; }
     glutPostRedisplay();
     glFlush();
     }
110
     void InitGL(void) {
     glClearColor( 1.0, 1.0, 1.0, 0.5f );
     glClearDepth( 1.0 );
     glBlendFunc( GL_SRC_ALPHA, GL_ONE_MINUS_SRC_ALPHA );
115  glEnable( GL_BLEND );
     glEnable( GL_DEPTH_TEST );
     glDepthFunc( GL_LEQUAL );
     glLineWidth( 1.0 );
     }
120
     void changeSize( int w, int h ) {
     int m = ( w < h ? w : h );
     glViewport( 0, 0, m, m );
     glMatrixMode( GL_PROJECTION );
125  glLoadIdentity();
     glOrtho( 0, 0, 50, 50, -1e12, 1e12 );
     glMatrixMode( GL_MODELVIEW );
     glLoadIdentity();
     }
```

Listing 19.8 Example of using OpenGL for graphics

Fig. 19.12 Example of
OpenGL, GLUT and GLUI

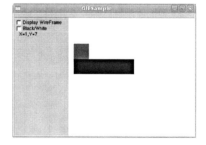

)

19.2.3 Using OpenGL from Python

In addition to calling OpenGL API functions from compiled C and C++ programs
a number of language bindings for OpenGL exist. Python bindings for OpenGL are
available using the PyOpenGL bindings. A short example is shown in Listing 19.9.

```python
# \file cube.py
# \author Sandeep Koranne, (C) 2010
# \description OpenGL in Python to draw a cube

5   from OpenGL.GL import *
    from OpenGL.GLU import *

    import pygame
    from pygame.locals import *
10
    WINDOW_SIZE = (400,400)

    def resize( w, h ):
        print "Resize called"
15      glViewport( 0, 0, w, h )
        glMatrixMode( GL_PROJECTION )
        glLoadIdentity()
        gluPerspective(60.0, float(w)/h, .1, 1.)
        glMatrixMode( GL_MODELVIEW )
20      glLoadIdentity()

    def OpenGL_Init():
        glEnable( GL_DEPTH_TEST )
        glShadeModel(GL_FLAT)
25      glClearColor(1.0, 1.0, 1.0, 0.0)
        glEnable(GL_COLOR_MATERIAL)

        glEnable(GL_LIGHTING)
        glEnable(GL_LIGHT0)
30      glLight(GL_LIGHT0, GL_POSITION,   (0, 1, 1, 0))

    class DOBJ(object):

        def __init__(self, position, color):
35
            self.position = position
            self.color = color


        def render(self):
40          glColor( self.color )
            glBegin(GL_QUADS)
            glVertex3f( 0.0, 0.0, 1.0 )
            glVertex3f( 1.0, 0.0, 1.0 )
45          glVertex3f( 1.0, 1.0, 1.0 )
            glVertex3f( 0.0, 1.0, 1.0 )
            glEnd()

    class Model(object):
50      def __init__(self):
            self.dobj = DOBJ( (0.0,0.0,0.0), (0.5,0.2,0.3) )

        def render(self):
            self.dobj.render()
55
    def run():
        pygame.init()
        screen = pygame.display.set_mode(WINDOW_SIZE, OPENGL)
```

```
60          resize(*WINDOW_SIZE)

            OpenGL_Init()
            model = Model()

65          while True:
                glClear( GL_COLOR_BUFFER_BIT | GL_DEPTH_BUFFER_BIT )
                model.render()
                pygame.display.flip()

70    run()
```

Listing 19.9 OpenGL bindings for Python

19.3 OGRE : OO Graphics Rendering Engine

Ogre is an open-source graphics rendering and scenegraph manager system. Its
name is an acronym for Object Oriented Graphics Rendering Engine, and it aims
to provide a 3D API which includes a scene graph manager, and can call an user
specified 3d rendering library for the actual drawing functions (in our example we
have used OpenGL). The key facilities provided by OGRE can be classified as:

1. Scene management: a scene which is to be rendered (per frame, when using continuous rendering) is described to OGRE as placement of geometric shapes, camera locations and viewing angles. Scenes are thus described at a higher level than
   say point-lists in OpenGL. This also allows application level culling in OGRE.
   Custom scene manager actions can be taken during rendering by installing a
   frame listener class,
2. Resource management: OGRE manages the memory for textures, fonts and other
   resources automatically,
3. Rendering: OGRE implements rudimentary application level culling before calling the low level graphics rendering API to push triangles (or other primitives)
   onto the screen.

OGRE scene graphs are usually constructed from data stored in files. The data
can be directly read in, in a mesh form, or it can be a solid model which is converted
to a mesh form during readin. The typical sequence for OGRE scene graph addition
is:

```
void ReadData( const char* fileName ) {
  Ogre::Entity *crane = mSceneMgr->createEntity(``Crane'', fileName );
  Ogre::SceneNode* head =
    mSceneMgr->getRootSceneNode()->createChildSceneMode();
5 head->attachObject( crane );
}
```

In the above listing, the mesh data for the 'crane' object is read from the file. We in-
quire the rootNode from the scene graph manager, and add the newly created crane
object as a child to the root node. The bounding box calculation for the crane and the
root node establishes a parent-child relationship, wherein, application level culling

(ALC), will use the bounding box (perhaps using a spatial data structure) to reject the display of scene components which are either not visible to the user (shielded) or have low resolution detail (then faster approximate rendering can be performed for those portions). This facility (of performing automatic ALC and resolution drops) , alongwith scene graph maintenance, are some of the most important features of OGRE, and which require significant code and development time, if done without using an API like OGRE. A related software is Coin3d, which is also used for scene-graph representation and developing 3d applications.

19.4 Graphviz: dot

Graphviz ("graph visualization software") is an open-source software written orig-inally by AT&T Research Labs for drawing graphs. Graph drawing is a major area of research in computer science. The input to the graph drawing is specified in the DOT language script. Consider the following description of a polytope graph.

```
graph POLYTOPE {
0 [shape=box,color=red];
0 -- 1 [color=blue];
0 -- 2 [color=blue];
0 -- 3 [color=blue];
0 -- 4 [color=blue];
1 -- 2 ; 1 -- 3 ; 1 -- 4 ;
2 -- 3 ; 2 -- 4 ; 3 -- 4 ;
}
```

By running the dot program on this graph, we can generate the figure as shown in Figure 19.13. The command we used was dot -Tpng a.dot > a.png. dot supports In addition to drawing pictures of graphs, the algorithms in dot have

Fig. 19.13 Graph of a poly-tope generated using dot

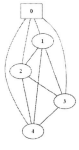

also been used in diverse other projects such as Scribus (see Section 2.3) to layout the page. Doxygen uses the dot binary to draw pictures of class diagrams.

19.4.1 DOT Language

DOT is the input language for graph drawing software. It is a plain text graph description language. DOT language can be used to specify both directed as well as
undirected graphs. An example for DOT to draw an undirected graph of a polytope
was shown above. Directed graphs can also be drawn using:

```
digraph Shape {
 Point -> Line  -> Box;
 Line -> Polygon;
}
```

Nodes in the graph can also have attributes, including the node label, node color.
Edges can also be annotated with attributes.

19.5 gnuplot

GNUplot is an open-source tool which can generate two and three dimensional plots
of data and mathematical functions. Although the name has GNU in it, the software
is not related to the GNU project, although it has its own open-source license.

gnuplot can produce display directly on X-window terminal, but it can also
generate *hardcopy* image files in many formats such as PNG, JPEG, SVG and EPS.
Many mathematical software (such as Octave, Maxima, and SAGE) use gnuplot
for its plotting needs. Examples of plots rendered using gnuplot are shown in Figure 19.14.

Gnuplot has an extensive help system built into it which can be accessed by
typing

```
gnuplot> help
```

The plot command of gnuplot is its most powerful, and can be used to plot the
contents of data files and mathematical functions. The output format can be changed
by using the set term command. One such *terminal* is an ASCII terminal as
shown below. Gnuplot can also draw figures in xfig (see Section 19.7 and LaTeX(see
Section 2.2) formats.

```
gnuplot> plot sin(x)
    1 ++---------------***--------------+---***-----------+--------**------++
      +                **+ *            +   *   *       +  sin(x) ****** +
  0.8 ++               *    *               *    *            *     *    ++
      |                *    *               *    *            *     *     |
  0.6 ++               *    *               *     *           *      *   ++
      *                *     *              *     *           *      *    |
  0.4 +*               *     *              *     *           *      * ++
      |*               *      *             *      *          *      *   |
  0.2 ++               *      *             *      *          *      * ++
      | *              *      *             *       *         *      * |
    0 +++              *       *            *       *         *      *++
      |  *             *       *            *       *         *      *|
 -0.2 ++  *            *        *           *        *        *      ++
      |   *            *        *           *        *        *      *|
 -0.4 ++   *           *         *          *         *       *      ++
      |    *          *          *          *          *      *       *
 -0.6 ++    *        *            *         *          *      ++
      |      *       *            *          *          *      |
```

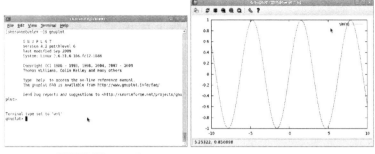

(a) GNUPLOT examples

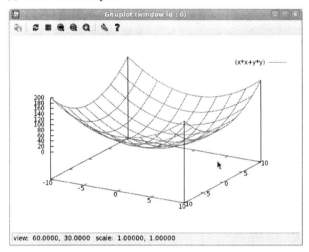

Fig. 19.14 GNUPLOT: scientific plotting software

19.6 Grace/Xmgr

Grace is a graphical WYSIWYG 2D plotting tool for the X-Window system and implemented using Motif. Amongst its many features it includes:

1. Many formats for exporting: includes export to EPS, PDF and SVG format,
2. Graphing flexibility: unlimited number of graphs, unlimited number of curves on graphs, color-fill markers and text annotations,

3. Curve fitting: linear and non-linear least-squares fit, calculation of residuals, and region restrictions,
4. Analysis: including FFT, integration and differentiation, histograms, splines and convolution,
5. Data formats: reads netCDF files,
6. Programmability: built-in programmable library, math functions, and user-defined functions via loadable modules.

An example of XMGR/Grace is shown in Figure 19.15.

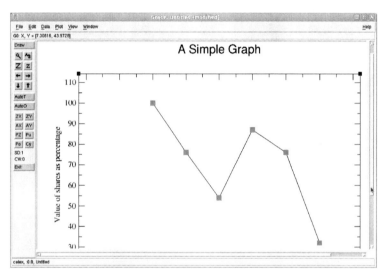

Fig. 19.15 Grace/XMGR plotting software

19.7 Xfig

Xfig is an open-source vector graphics editor running under the X-window system. Xfig is used to draw camera ready pictures. Using xfig it is simple to draw schematics and other technical illustrations containing boxes, lines, circles, ellipses, spline curves and text. Once drawings have been completed an external tool `fig2dev` can be used (or called from within the xfig GUI) to export the image into diverse file formats such as JPEG, EPS and LaTeX.

Like most other UNIX tools, xfig, saves its drawing content as a text-only file (called a '.fig' file). Xfig can also import figures. An example of Xfig generated schematics is shown in Figure 19.16. The text data from Xfig can also be used for non-traditional drawing applications (such as VLSI art work development).

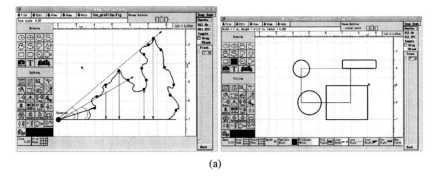

(a)

Fig. 19.16 xfig: vector drawing tool

19.8 Inkscape

Like XFig, Inkscape is also a vector drawing application, but unlike XFig, Inkscape stores its data in XML files, and is fully compliant with XML, SVG and CSS standards. It is designed for artists creating images for the web, and graphic designers, rather than technical illustrators. An example of Inkscape being used to create shapes is shown in Figure 19.17.

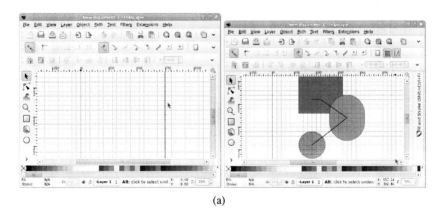

(a)

Fig. 19.17 Inkscape: vector drawing tool

Inkscape supports objects such as (i) boxes, (ii) lines, (iii) Bezier curves, (iv) circles and ellipses, and (v) text. Text can be curved around path, and attributes such as font and size can be changed easily. Inkscape also supports rounded rectangles, with customization for fill, stroke, and opaqueness. It also supports the concepts of layers (similar to a raster editor). Unlike other illustration software, Inkscape has the ability to merge paths together (to perform Boolean AND,OR operations on shapes). Inkscape also has a builtin XML editor to directly edit SVG files document structure, support for LaTeX, and connector tool to create diagrams and flowcharts.

19.9 PovRay : Ray Tracing

POVRAY (Persistence of Vision Ray Tracer) is a *ray tracing* program which can generate photo-realistic renderings of *scenes*. The input to the POVRAY consists of the scene description in its SDL (scene description language). Consider an example as shown below (placed in a file called 'sphere.pov'):

```
/* Example file sphere.pov */
#include "colors.inc"
background { color White }

camera {
  location <0, 2, -3>
  look_at <0, 0, 1>
}

sphere {
  <0, 1, 2>, 1
  texture {
    pigment {
    marble
    turbulence 1
    color_map {
      [0.0 color Gray90]
      [0.8 color Gray60]
      [1.0 color Green]
    }
    }
  }
}
box {
   <-1,0,1.2>, 0.5
  texture {
    pigment {
    marble
    turbulence 1
    color_map {
      [0.0 color Gray90]
      [0.8 color Red]
      [1.0 color Gray20]
    }
   }
  }
}
light_source { <2, 4, -3> color White}
```

This file describes the scene shown in Figure 19.18. The syntax of the POVRAY

Fig. 19.18 Example of scene
rendered using POVRAY

SDL is based on the concept of lights, camera, and objects. Objects can be built up from primitives including, spheres, boxes, torus. Transformations such as translation, scaling and rotation can be applied to objects. Object attributes such as texture, color and size can be specified. The scene also requires the location of a camera and a *look at* direction vector. Moreover, the scene can be illuminated by a light (a light source has a position coordinate, and a color attribute). Figure 19.18 was rendered using POVRAY as:

```
Parsing Options
  Input file: sphere.pov (compatible to version 3.61)
  Remove bounds........On
  Split unions........Off
  Library paths:
    /usr/local/share/povray-3.6
    /usr/local/share/povray-3.6/ini
    /usr/local/share/povray-3.6/include
Output Options
  Image resolution 320 by 240 (rows 1 to 240, columns 1 to 320).
  Output file: /home/skoranne/POVRAY/sphere.png, 24 bpp PNG
  Graphic display......On  (gamma: 2.2)
  Mosaic preview.......Off
  CPU usage histogram..Off
  Continued trace......Off
Tracing Options
  Quality:  9
  Bounding boxes.......On    Bounding threshold: 3
  Light Buffer........On
  Vista Buffer........On    Draw Vista Buffer....Off
  Antialiasing........Off
  Clock value:   0.000  (Animation off)
```

```
 0:00:00 Parsing
 0:00:00 Creating bounding slabs  0:00:00
         Creating bounding slabs 2K tokens
Scene Statistics
 Finite objects:          2
 Infinite objects:        0
 Light sources:           1
 Total:                   3
```

POVRAY can be used as a photo-realistic scene renderer, and is indeed used as the rendering engine for many 3d modeling tools.

19.10 gd (graphics drawing)

GD graphics library is a software library for dynamic generation of graphics. It is used to generate charts and graphics, and is often used as part of report generation system (whether running on a web server, or for document generation). Consider the program listing shown below.

```cpp
// file : dgd.cpp
// Sandeep Koranne, (C) 2010
#include <gd.h>
#include <gdfontl.h>
#include <stdio.h>

int main() {
  gdImagePtr image;
  FILE *fout = fopen( "dgd.png", "wb" );
  image = gdImageCreate(64,64);
  int white = gdImageColorAllocate(image,255,255,255);
  int red = gdImageColorAllocate(image,255,0,0);
  int x=0,y=0;
  for(unsigned int i=0; i < 25; ++i) {
    if( i % 2 ) { // vertical lines
      gdImageLine(image,x,y,x,y+5,red);
      y+=5;
    } else {
      gdImageLine(image,x,y,x+5,y,red);
      x+=5;
    }
  }
  unsigned char *s = (unsigned char*)"Gd";
  gdImageString( image,gdFontGetLarge(),6,30,s,red);
  gdImagePng( image,fout );
  fclose( fout );
  gdImageDestroy( image );
  return 0;
}
```

Compiling this program as:

```
g++ dgd.cpp -lgd -lpng -lm
$./a.out
```

produces the diagram shown in Figure 19.19.

Fig. 19.19 gd : graphics
library

19.11 asymptote

Asymptote is a descriptive vector graphics language to represent and draw coordinate based mathematical drawings of high quality. In this respect it is similar to LATEXdiagram packages, and pictex, but Asymptote has many more features. It supports command-line operation by running the tool asy, and xasy. An example of xasy running in shown in Figure 19.20.

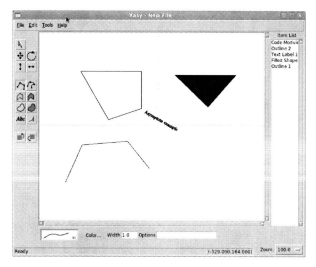

Fig. 19.20 Asymptote drawing language : xasy tool

Since Asymptote is a complete programming language, procedural rendering can be used to draw complex graphs which otherwise would take a long time to draw correctly. Asymptote is also based on and designed for mathematical use, and supports complex numbers as objects to manipulate diagrams. A simple example of vector procedural drawing is:

```
draw((0,0) -- (10,10));
$asy -V line.asy
```

This will create a PostScript file containing a depiction of a line from (0,0) to (10,10), moreover since we invoked asy with the -V option it will call an appropriate PostScript viewer to display this file. The size of the canvas (so to speak) can be specified using the size function in the .asy file. If we want to cap the line with an arrow, we can add:

```
draw((0,0) -- (10,10), Arrow);
```

Similarly, Labels can be added using the label command. Consider the following example:

```
size(5cm);
draw((0,0)--(1,0)--(2.14,1)--
(2.14,2)--(1,2)--(0,2)--cycle);
label("$\sqrt{x}$",(0,0), SW);
label("$\frac{1}{z}$",(1,0), SW);
label("$A$",(2.14,1),SE);
label("$Bz$",(2.14,2),SE);
label("$Az$",(1,2),NE);
label("$\frac{z^2}{x}$",(0,2),NW);
draw((2.14,1){down}..{left}(0,2), Arrow);
```

The output of this file is shown in Figure 19.21.

Fig. 19.21 Mathematical drawing with asymptote

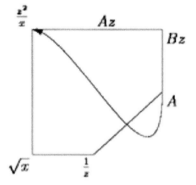

19.12 FreeType : Font Rendering

FreeType is an open-source software library that implements *font rasterization*. Font rasterization is the process of converting font character shapes into bitmaps; a simple enough operation to describe, but which has many subtleties. FreeType simplifies this procedure by providing an easy and uniform interface to access the contents of font description files. Another related project is *Pango*, which is an open-source multi-lingual text rendering engine. An example of using FreeType to render font glyphs as raster spans of pixels is shown in Listing 19.10.

```cpp
// \file freetype_example.cpp
// Example derived from FreeType tutorial
// original (C) Erik M<F6>ller
// \description Example of using FreeType library
#include <iostream>           // program IO
#include <cassert>            // assertion checking
#include <fstream>            // FILE IO
#include <cstdlib>            // exit function
#include <cstring>            // memset function
#include <vector>             // SPAN rep.
#include <ft2build.h>         // FreeType library

// A horizontal pixel span generated by the FreeType renderer.

struct Span
{
  Span() { }
  Span(int _x, int _y, int _width, int _coverage)
    : x(_x), y(_y), width(_width), coverage(_coverage) { }

  int x, y, width, coverage;
};

typedef std::vector<Span> Spans;


void
RasterCallback(const int y,
               const int count,
               const FT_Span * const spans,
               void * const user)
{
  Spans *sptr = (Spans *)user;
  for (int i = 0; i < count; ++i)
    sptr->push_back(Span(spans[i].x, y, spans[i].len, spans[i].coverage));
}
void
RenderSpans(FT_Library &library,
            FT_Outline * const outline,
            Spans *spans)
{
  FT_Raster_Params params;
  memset(&params, 0, sizeof(params));
  params.flags = FT_RASTER_FLAG_AA | FT_RASTER_FLAG_DIRECT;
  params.gray_spans = RasterCallback;
  params.user = spans;

  FT_Outline_Render(library, outline, &params);
}
```

Listing 19.10 Example of using FreeType

19.13 Anti-grain geometry : AGG

Anti-Grain Geometry (AGG) is an open-source graphic library written in C++ which provides rendering engine that produces pixel accurate images in memory from vector data. In addition, AGG also supports anti-aliasing, high-quality, high-performance and numerical stability (important for degenerate input). AGG supports drawing of arbitrary polygons with different stokes, and line styles. It uses the General Polygon Clipper library for polygon Boolean operations.

19.14 Geomview

Geomview is an interactive 3D viewer for UNIX. It can be used as a standalone viewer or as a display engine integrated with other data analysis or computation tool. Figure 19.22 shows the diagram of a 3-dimensional cube with one of its vertices truncated. Using the *polymake* software we can invoke the Geomview visualization as:

```
$polymake t.poly VISUAL geomview
```

Ofcourse, in this instance `polymake` does the conversion of geometry into the format required by Geomview, but afterwards Geomview interacts with the user. The user can change a number of settings in the view including (i) world coordinate, (ii) camera positions, (iii) light settings, (iv) material setting, (v) scaling, and (vi) also perform fly throughs.

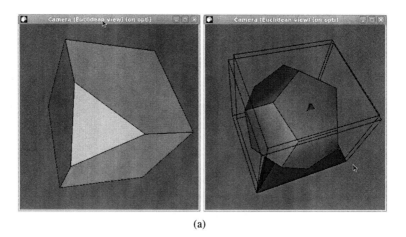

(a)

Fig. 19.22 Geomview visualization software

Most of the parameters controlled by the user have a parallel in OpenGL (see Section 19.2). The file format for Geomview is called Object Oriented Graphics Library (OOGL). Data written in the OOGL syntax and adhering to OOGL specification can be rendered and browsed using Geomview.

Consider the example of a simple quadrilateral in OOGL format:

```
CQUAD
-1 -1  0      1 0 0 1
 1 -1  0      0 1 0 1
 1  1  0      0 1 1 1
-1  1  0      1 0 0 1
```

We save this data in a file "csq.oogl", and invoke Geomview as `geomview csq.oogl`, to get Figure 19.23. In addition to specifying color per vertex, OOGL

Fig. 19.23 Colored square in OOGL format

also allows specifying color per facet. It is also possible to render data made of meshes, vector polylines, and triangulations. In OOGL using the `LIST` command it is possible to refer to other stored geometries. OOGL is also hierarchical, using the `INST` command we can not only refer to another stored geometry but also apply a geometrical transform to the element. The transform is a 4x4 geometrical transform encoding the rotation, translation, scaling, shearing and perspective transformation.

Consider the following example:

```
INST
geom { < csq.oogl }
transforms { =
TLIST
 1 0 0 0
 0 1 0 0
 0 0 1 0
 2 2 0 1

 1 0 0 0
 0 -1 0 0
 0 0 1 0
```

```
5 5 0 1
}
```

In this OOGL file we refer to the previously created colored square "csq.oogl". We create an INST block and specify that the geometry comes from a stored disk file. Next we create a transform list TLIST containing the number of different instantiations we want for the square. In this case we specify two 4x4 matrices. The format of a transformation matrix contains the rotation, scaling and X-flip component in the upper left 3x3 sub-matrix. The last row contains the T_x, T_y translation component (of non-shearing transforms). In the new OOGL file we create an object comprising of two squares placed at (2,2) and (5,5), additionally, the second square is rotated and flipped. See Figure 19.24.

Fig. 19.24 Hiearchical OOGL file with INST

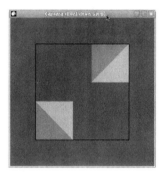

It is thus easy to build complicated models of geometry, and to interact with the 3d model. External tools can also interface with Geomview by generating OOGL files on the fly, and invoking Geomview as their display engine.

19.15 HippoDraw

HippoDraw is a powerful object oriented statistical data analysis package written in C++, with user interaction via a Qt-based GUI and a Python scriptable interface.

19.16 GGobi : multi-dimensional visualization

GGobi is the next generation of XGobi and XGvis, multi-dimensional data visualization tools. It provides dynamic and interactive graphics such as scatter-plot, barchart and parallel coordinate plots. It is best to illustrate the capabilities of `ggobi` with the help of an example. Consider a fictional multi-dimensioned data as given below:

```
Name, Region, Age, Weight, Score, Rank
"Jack", "OR", 12, 76, 88, 4
"Jane", "CA", 11, 43, 54, 32
"Mary", "OR", 14, 76, 66, 23
"Gus", "WA", 11, 56, 82, 7
"Jack2", "WA", 11, 56, 82, 7
"Jane2", "CA", 11, 43, 17, 32
"Mary2", "CA", 11, 43, 45, 32
"Gus2", "OR", 12, 76, 8, 4
"Zack", "WA", 9, 54, 53, 23
"Dave", "CA", 12, 56, 65, 26
"Dawn", "OR", 11, 54, 76, 15
```

This data represents the Age, Weight, Score and Rank of 11 children. We would like to use ggobi to perform and visualize the data. By running ggobi we can load this CSV file and immediately visualize the scatter-plot. Algorithms such as *clustering*, *brush*, *XY-plot* are all included. See Figure 19.25 for an illustration. The axes to be used for the X and Y in the scatter-plot can be chosen using the control panel as shown in Figure 19.25(a). The dimensional dependence (or independence) of data can be visually inferred using the XY plots based on Region-Score, and Region-Weight (as shown in Figure 19.25(a)).

Dynamic display (animated) of the data as the value of the variables in each dimension changes can also highlight a principal axis, or dependence in the data. Using the 2D-Tour facility of ggobi we can analyze this as shown in Figure 19.25(b).

19.17 ParaView and VTK

ParaView is an open-source multi-platform data analysis and visualization software. ParaView was developed to analyze and render extremely large datasets using distributed memory computing resources, and is thus often run on supercomputing clusters to analyze datasets of terascale magnitude. The main window when ParaView is run is shown in Figure 19.17.

The important features of ParaView are:

1. Support for terascale data:
2. Support for structured data: in scientific visualization and analysis, structured data refers to data present and collected on (i) uniform rectilinear grid, (ii) non-uniform rectilinear, and (iii) curvilinear grid,
3. Consistency: processing operations such as filters, themselves produce datasets, allowing for chaining of operations,
4. Contour extraction: isosurfaces and contours can be extracted from the data,

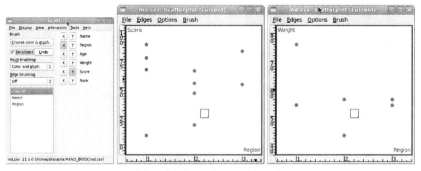

(a) ggobi Scatter Plot with Brush

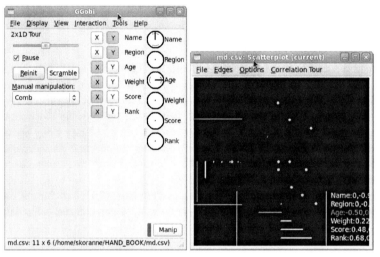

(b) 2d Tour plot

Fig. 19.25 ggobi analyzing CSV data

5. Clipping: a sub-region can be clipped. The clip can be a plane (specified as a threshold), or a *volume-of-interest* (for structured data only),
6. Python integration:

ParaView is implemented using the Qt framework (see Section 19.1.3), and is thus eminently portable. It utilizes MPI (see Section 12.3) and can run on distributed computing platforms and analyze large terascale datasets. It is implemented as a client-server software. Using the Visualization Tool Kit (VTK) it implements a *level-of-detail*(LOD) model to maintain high framerate when displaying large models. We can construct a 3d object in ParaView, by adding geometrical primitives from the "Sources" menu. An object has been constructed comprising of a cylinder, sphere and a box, as shown in Figure 19.26(a). Objects can be *sliced* with a cutting plane as shown in Figure 19.26(b).

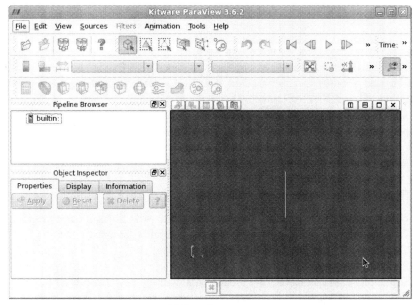

(a) Paraview screen

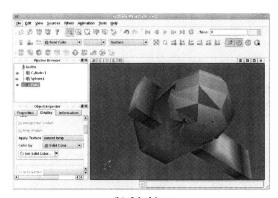

(b) 3d object

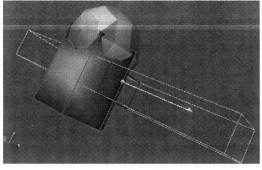

(c) Plane clip

Fig. 19.26 ParaView showing 3d object

19.18 OpenDX

OpenDX (for Open Data Explorer) is a scientific data visualization software written by IBM, but is now available as open-source under IBM Public License. OpenDX can be used not only to visualize data, but also to create user-interaction based applications. OpenDX is implemented using the Motif widget library on X-Windows. Its GUI is oriented around user-interactions; user can interact with the current data display using a number of *interactors*. These interactors can be direct (such as rotation of current view), or indirect (such as applying a filter). Complex interaction sequences and program control can be built up which allow a sophisticated user to use the visual data effectively.

OpenDX has a number of unique features:

- Data prompter: user interface for describing data to be brought into OpenDX,
- Data model: data fields, geometric objects, and rules describing data,
- Data browser: user interface for viewing data file,
- Scripting language:
- Visual program editor: GUI for creating and editing *networks* (visual programs),
- Modules: blocks which constitute visual program network. Each module performs some action,
- Module builder,
- Control panel,
- Main window: rendering window.

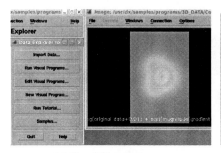 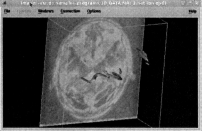

(a) OpenDX control pane   (b) Data from MRI

Fig. 19.27 OpenDX scientific visualization software

As we mentioned above the *visual program* or *network* concept of OpenDX sets it apart from other data visualization software we have discussed. Using networks, the user can visualize both observed and simulated data. *Modules*, many of which implement core transformations and filters, are shipped as part of OpenDX and can be used by the user to create a network quickly. Moreover, the visual program editor can be used to create visual programs even more easily.

19.18.0.1 Data import in OpenDX

OpenDX can import user data in a variety of ways:

1. General Array Importer,
2. NetCDF, CDF, HDF file import,
3. OpenDX ImportSpreadsheet,
4. OpenDX Import module,
5. ReadImage to import TIFF,
6. Read OpenDX native format,
7. Use `gis2dx`.

19.18.0.2 General OpenDX Flow

The steps to visualize data with OpenDX can be broken into the following steps:

1. Collect data,
2. Import data,
3. Define visual form and concept,
4. Define user interaction,
5. Prepare output.

We conclude this section on OpenDX by presenting a small example of analyzing 2d data. Assume that we are given a series of position-dependent 2d data and we would like to develop an interactive environment to analyze and visualize it. The size of the data is moderate, say, 100x100, but could be very large also. We can use OpenDX to quickly put together a visual program which will perform the interactive visualization. The data is in a TEXT file with 100x100 floating point values. We write a data importer using the visual Data Import tool as:

```
file = rainfall.txt
grid = 100 x 100
format = ascii
interleaving = record
majority = row
field = rainfall
structure = scalar
type = float
dependency = positions
positions = regular, regular, 0, 1, 0, 1
end
```

This is the "rain.general" data importer. Next we create a visual program using the VPE (visual program editor) as shown in Figure 19.28(a). The data flows from the Import module, to the AutoColor, to the Collector (which combines the Isosurface pixels), to form the final image (as shown in Figure 19.28(b)).

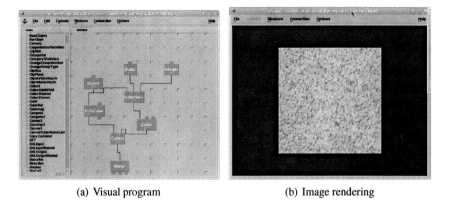

(a) Visual program   (b) Image rendering

Fig. 19.28 Visual program in OpenDX to analyze 2d data

As can be seen in Figure 19.28(a) the controls of the Isosurface module are controlled using input controls which the end-user can control. The rendered image can also be rotated, zoomed, and saved in a disk file.

19.19 Conclusion

In this chapter we have discussed information visualization. This is an important part of any software system, since an image can transmit and convey large amount of information. In some domains (such as robotics), and computational geometry, visualization can lead to insights in problems, which otherwise would be difficult.

While it is true that most high-performance computing environments have excellent batch processing capabilities, more and more users are demanding a graphical user interface (GUI) front-end (atleast to ease in the transition to new features or tools). In this chapter we have discussed the GTK, Qt, Fox toolkit and wxWidgets. Any discussion about visualization would be incomplete without OpenGL, and we discussed OpenGL utility libraries such as GLUT and GLUI.

There are a number of pre-built applications such as ParaView, OpenDX, Geomview which cater to data visualization and rendering. While analysis packages such as HippoDraw and GGobi perform statistical analysis. Visualization has two components, information assimilation and information creation. Tool to create static and animated images from data were also presented in this chapter, including XFig, gnuplot, Inkscape, and Graphviz dot.

Chapter 20
Web and Database Systems

Abstract In this chapter we discuss the use of open-source software to implement Internet software including web servers, virtualization and databases.

Contents

20.1 Web Servers

The Internet has transformed the way information travels across the world, and it has indeed impacted the manner in which information is disseminated, shared, and accessed. Internet based technologies are also experiencing some of the most rapid advances in technology since the space age, and it is not foolhardy to think that this rapid rate of innovation will have significant impact on the disciplines of scientific computing and research. In this chapter we provide an overview of the current open-source solutions for deploying data and computation on the Web. If anything in this book is sure to get obsolete quickly, I fear, this chapter is on the top of the stack, and thus we cover only the most commonly used web technology, or the most promising "new thing", what we expect will last for a technological node or so.

S. Koranne, *Handbook of Open Source Tools*,
DOI 10.1007/978-1-4419-7719-9_20, © Springer Science+Business Media, LLC 2011

20.1.1 HTTP Server : Apache

Apache, is a web server software, which (according to some estimates) is responsible for serving more than 60% of the Internet as of 2010. In sheer volume, Apache is said to server more than 100 million websites on the Internet. Apache supports a number of features, including: (i) compiled module support, (ii) compression support, (iii) virtual hosting, and (iv) high performance. The Apache binary is called 'httpd' and runs as a daemon or service on GNU/Linux.

20.1.2 YAWS: Yet Another Web Server

Yaws is written in Erlang (see Section 1.6.10), and is a robust, efficient and fault-tolerant web server. It is highly scalable, in an experiment Yaws was able to sustain operations during a DDOS (distributed denial of service) simulation, serving 80,000 page requests. It is designed for dynamic web content generation, with arbitrary Erlang code inserted in Web pages, the Yaws framework performs runtime substitution of the Erlang function's output into the web page.

Yaws also has an integrated web applications framework which can be used to deploy applications written in Erlang. The full power of the Erlang language, Mnesia database is available in Yaws.

20.1.3 LAMP Stack

LAMP (Linux, Apache, MySQL, and PHP) is an acronym of the above mentioned tools which are most often associated with a software solution stack to build a general purpose web server and hosting solution. GNU/Linux has already been covered in this book (see Section I), and above we discussed the Apache web server. MySQL is discussed in Section 20.6. Although PHP is a programming language we did not present it in the context of other programming languages as it is mostly (only?) used in the context of web application development. Some have suggested replacing the P in LAMP by Perl and Python as these languages are also used for web development.

20.1.3.1 PHP

PHP: hypertext processor is a scripting language designed for web development, including dynamic web content generation, and user interaction and validation. PHP code is embedded inside HTML documents and is executed by the browser with the help of a PHP module.

20.1.3.2 Ruby and Ruby on Rails (RoR)

Ruby is a general purpose object-oriented programming language that is dynamic and reflective. Its syntax is inspired by Perl, Eiffel and Lisp with Smalltalk like features. Ruby implementation on GNU/Linux is written in C as a single-pass interpreter and this reference implementation is considered the standard as there is no language standard. The programming language, Ruby, is often deployed as part of Ruby on Rails (RoR) which is an open-source web application framework designed for rapid development.

20.2 Hadoop

Hadoop is an Apache project which develops open-source software for reliable, scalable, distributed computing. The main tools within the Hadoop project are:

1. Hadoop Common: utilities common to the Hadoop project,
2. Chukwa: data collection system for managing large distributed systems,
3. HBase: scalable, distributed database that supports structured storage for large dataset tables,
4. HDFS: distributed file system,
5. Hive: data warehousing infrastructure,
6. MapReduce: software framework for distributed processing of large data sets on compute clusters,
7. Pig: high-level data-flow language and execution framework for parallel computation,
8. ZooKeeper: high-performance coordination service for distributed applications.

20.3 Content Management with Joomla

Joomla is a content management system (CMS), which enables one to build Web sites and powerful online applications. Many aspects, including its ease-of-use and extensibility, have made Joomla a popular Web site software available. It helps keeps track of every piece of content on the Web site; content can be simple text, photos, music, video, documents, or just about anything you can think of. A major advantage of using a CMS is that it requires almost no technical skill or knowledge to manage. Joomla is highly extensible and thousands of extensions (most for free under the GPL license) are available in the Joomla Extensions Directory. Joomla is based on PHP and MySQL.

20.4 Virtualization and Cloud Computing

Virtualization is an old computer hardware and systems concept which had been deployed for years on high-end mainframes and server machines. In the last decade, however, virtualization has become a buzzword. The reason for this resurgence (atleast in terms of publicity) can be explained by two factors: (i) rising cost of energy (both utilities of electricity and cooling to run major data-centers) and (ii) rise of multi-core computing, which has placed powerful machines on every desktop.

Once the reason (utility cost) and the physical means (powerful computers) were in place, researches lost no time in taking proven concepts from servers and mainframes, and customizing them for personal computers and their associated processors. The virtualization concept comprises of a powerful *host* computer and one or many *guest* virtual machines which are running on the host. The guest machines are separated from each other using low level (processor based) protection. Now, processor providers also provide special instructions and ring access to virtualization *hypervisors* to increase the performance of the guest machine.

Open-source virtualization systems include Xen, KVM and Qemu. These are described below:

1. Xen: Xen is a virtual-machine monitor for IA-32, X86-64 and PowerPC 970 architecture. Xen uses a hypervisor which runs on the physical machine and can host several guest machines on it. The first guest boots automatically and receives special administration privileges from Xen. Originally, guest kernels had to be modified (paravirtualized) in order to run on Xen, but from version 3.0, unmodified versions of many OS kernels can execute on Xen (when running on Intel-VT or AMD-V hardware assisted virtualization capable processors),
2. KVM (kernel based virtual machine): GNU/Linux has KVM infrastructure which supports the latest Intel VT and AMV-V processor advanced methods of hardware assisted virtualization. KVM by itself does not perform any emulation, instead an user-space program uses the /dev/kvm interface. QEMU also uses the same mechanism.
3. QEMU: in contrast to the above two approaches which don't perform any ISA (instruction set architecture) mapping, QEMU is a processor emulator that relies on dynamic binary translation of instruction sequences to achieve performance, while being capable of virtualizing disparate ISA kernels. QEMU has two modes of operation: (i) user mode emulation, and (ii) complete computer system emulation. QEMU supports the emulation of many architectures, including: (i) x86, X86-64, MIPS R4000, SPARC, ARM, and PowerPC.

Operating system partitioning for load balancing and providing security to different applications can also be done using OpenSolaris zones, GNU/Linux chroot, and BSD jails. By setting appropriate usage caps on applications, rudimentary load balancing can be performed. Nevertheless, this type of OS partitioning is not deemed virtualization, even though it has the potential of reducing system load as the OS kernel is shared between all the guest instances. In some systems (e.g., Open-Solaris zones) the OS partition can even execute binaries of a different OS (although most systems still prefer the ISA to be identical to avoid performance penalty from JIT translation). Also, live system migration of a running guest from one physical machine to another is not supported with OS partitioning.

20.4.1 Cloud computing

Cloud computing has recently been adopted as the new Internet phenomenon. We understand cloud computing to comprise of:

1. Infrastructure as a Service (IaaS): this is the *computing utility* model, where service providers have in place computing infrastructure, and pay-as-you use service model for providing compute resources, storage resources and in some cases complete licensed software on the compute servers as well,
2. Software as a Service (SaaS): in this model SaaS providers includes companies such as RackSpace who have developed OpenStack (see Section 20.4.1.1),
3. Platform as a Service (PaaS): example of PaaS include Gmail, Flickr, etc; since these cloud computing infrastructure provide the complete application as a single platform to the end user. This is the easiest option for the end user since the software is usable out-of-the-box, but has problems of vendor lockin, and reduced customization.

Advantages of cloud computing include:

1. Flexibility of scaling to users: if the demand for computing resources is very dynamic with a large gap between the peaks, then the user can leverage the reduced demand to reduce their infrastructure requirement. Once the application or web service is deployed on the cloud, the cloud computing provider manages the scaling requirement,
2. Reduce fixed hardware cost: following up on the previous theme, the end user can reduce fixed infrastructure cost by deploying the application to the cloud. Since most cloud computing providers use *multi-tenancy*, where a physical machine is shared (using virtualization, for example) between many users, costs can be reduced. The costs include the hardware as well as utility cost of electricity and cooling associated with the datacenter,
3. Replication and Availability: many cloud computing providers support replication of data and compute resources seamlessly; this has an advantage to users who want to deploy the application to a wide audience, and/or have reliable backup options,
4. Reduction in IT maintenance cost: since the cloud computing provider is responsible for maintaining the cloud infrastructure, end user cost of IT maintenance can be reduced.

Most of the cloud computing infrastructure is built upon open-source software such as GNU/Linux, Apache HTTP server, Erlang based CouchDB, load balancing and distributed computing and file systems. Recently, some cloud computing providers have released their software stack as open-source allowing users to deploy the same cloud infrastructure in-house. Ofcourse, deploying a cloud infrastructure requires more than just the access to source code; nevertheless by inspecting the software stack, users can make better informed decisions about deployment of applications, and in some cases can also optimize the application to take advantage of unique features present in the cloud software stack.

20.4.1.1 OpenStack

As discussed above, one such cloud computing software stack is OpenStack developed by RackSpace and released as open-source.

20.4.2 Network and Cluster Monitoring

20.4.3 Ganglia

Ganglia is a scalable distributed system monitor tool for high-performance computing systems such as clusters and grids. It allows the user to remotely view live or historical statistics (such as CPU load averages or network utilization) for all machines that are being monitored.

20.4.4 Nagios

Nagios is a popular open source computer system and network monitoring software application. It watches hosts and services, alerting users when things go wrong and again when they get better.

20.5 PostgreSQL

PostgreSQL is an open-source, object-relational database management (ORDBMS) system. It supports the SQL standard and offers many features:

1. complex queries,
2. foreign keys,
3. triggers,
4. views,
5. transactional integrity,
6. multiversion concurrency control.

PostgreSQL is based on a *client-server* model where a PostgreSQL session consists of a master server process (which manages the underlying database file, connection management and performs database operations). This server program is called `postgres`. The client application which wants access to the database can be diverse in nature. The client can be a command-line based tool (such as the PostgreSQL SQL command-line `psql`), or it can be a Web based client, or it can also be integrated into the application using API. The server and client can be running on different machines (thus the client-server model) communicating over TCP/IP network. The database server itself can handle simultaneous connections from multiple clients (implemented using `fork` system call to create a new process).

Using PostgreSQL to create a database and use it for simple storage of information, retrieval and SQL processing is described below.

20.5.1 *Creating database using PostgreSQL*

Before we actually create a database we have to inform PostgreSQL where it can create the files needed for the database. This is done using the `initdb` command.

```
$initdb /home/skoranne/PGDB/test
```

for small databases we are using the `/home` partition, but for larger databases a specially provisioned filesystem should be used. Now we can launch the `postgres` program and we can inform it of the above file location as:

```
$postgres -D /home/skoranne/PGDB/test > pg.log&
```

We launch the program as a background job, but retain the log output for troubleshooting in case of a problem. Once the process for the database server has launched we can use the `createdb` command to create a database in the above defined file locations.

```
$createdb shopping
```

We can use the PostgreSQL `psql -l` command-line tool to list the active databases.

```
$ psql -l
                              List of databases
   Name    |  Owner   | Encoding |  Collation  |   Ctype     |   Access privileg
es
-----------+----------+----------+-------------+-------------+------------------
-----
 postgres  | skoranne | UTF8     | en_US.UTF-8 | en_US.UTF-8 |
 shopping  | skoranne | UTF8     | en_US.UTF-8 | en_US.UTF-8 |
 template0 | skoranne | UTF8     | en_US.UTF-8 | en_US.UTF-8 | =c/skoranne
                                                             : skoranne=CTc/skor
anne
 template1 | skoranne | UTF8     | en_US.UTF-8 | en_US.UTF-8 | =c/skoranne
                                                             : skoranne=CTc/skor
anne
(4 rows)
```

Once the database has been created we can use it to store data. Data storage can be done using ODBC or using SQL statements. SQL (structured query language) is a standard for database operations.

```
$ psql shopping
psql (8.4.4)
Type "help" for help.

shopping=# SELECT VERSION();
version
---------------------------------
 PostgreSQL 8.4.4 on x86_64-redhat-linux-gnu,
 compiled by GCC gcc (GCC) 4.4.3 20
100127 (Red Hat 4.4.3-4), 64-bit
(1 row)

shopping=#
SELECT (14+22)/SQRT(64);
 ?column?
----------
      4.5
(1 row)
```

PostgreSQL is a *relational database management system* (RDBMS). Databases
can be of many different types, including object-oriented, file/directory based,
property-attribute based, or RDBMS. RDBMS define data storage in terms of *tables*, where each table is a collection of *rows*, and each row of the table has the same
set of columns of fixed type. A table can be created in PostgreSQL using the SQL
CREATE TABLE function as:

```
CREATE TABLE grocery_stores (
shopping(# name varchar(80),
shopping(# items varchar(100),
shopping(# location int,
shopping(# phone varchar(10));
CREATE TABLE
```

As we are entering this using the psql command-line, the prompt changes depending upon the context. Once we enter the semicolon, the system returns CREATE
TABLE denoting success. To remove a table we can use the DROP TABLE command. It should be noted that the above command has created an empty table, to
actually populate it with data we need the INSERT SQL function:

```
INSERT into grocery_stores VALUES
        ('DiscountPlus','Milk,Bread',97070,'5031234567');
INSERT 0 1
```

Bulk data can be imported into PostgreSQL using the COPY function. Consider a
external text file as:

```
'ValueMart'| 'Soup,Veg,Fruits'| 97034| '1234567'
'Fruities'|'Fruits'|97070|'4321123'
\.
```

We are using the | as the *delimiter* and _ denotes end of data. Running this we get

```
#COPY grocery_stores FROM 'grocery.txt' DELIMITER '|';
COPY 2
shopping=# select * from grocery_stores ;
     name      |      items       |loc    | phone
-------------+-----------------+-------+--------
 DiscountPlus|Milk,Bread        | 97070 |'1243657'
 'ValueMart' |'Soup,Veg,Fruits'| 97034 |'1234567'
 'Fruities'  |'Fruits'          | 97070 |'4321123'
(3 rows)

#select name,phone from grocery_stores ;
     name      |    phone
-------------+------------
 DiscountPlus | '1243657'
 'ValueMart'  | '1234567'
 'Fruities'   | '4321123'
(3 rows)
```

Running the SELECT query with name and phone as the columns we can list all the stores. Sorting can be done using:

```
select name,phone from grocery_stores ORDER BY phone;
```

Using PostgreSQL as a remote database server is simplified using SSH (see Section 1.5). Consider the above example where the database was created on machine 'sxde' by the user. On another machine, say HOSTB, we want to use the same database. We can use SSH to create a *tunnel* between host 'sxde:5432' (the notation is host:port) and 'HOSTB:63333'. This is done as:

```
$ssh -L 63333:localhost:5432 skoranne@sxde
```

Once the tunnel has been established, PostgreSQL psql can be used on 'HOSTB' as:

```
$psql -h localhost -p 63333 shopping
psql (8.4.4)
Type "help" for help.

shopping=# select * from grocery_stores ;
     name      |           items           | location |   phone
-------------+--------------------------+----------+-----------
 DiscountPlus | Milk,Bread               |    97070 | 5031234567
 'ValueMart'  | 'Soup,Vegetables,Fruits' |    97034 | '1234567'
 'Fruities'   | 'Fruits'                 |    97070 | '4321123'
(3 rows)
```

To PostgreSQL, the connection appears to come from host 'localhost' on 'sxde' and connects to the same database 'shopping' we created above.

20.6 MySQL

Another open-source DBMS is MySQL. MySQL has become popular with the use
of databases for content serving and managing information presented on websites.
MySQL is efficient, relatively easy to setup and administer and is available on a
number of operating systems and platforms. MySQL is also used as the database for
many other open-source programs such as Bugzilla (see Section 3.5). Using MySQL
on GNU/Linux systems is simple:

```
$mysql
Welcome to the MySQL monitor.  Commands end with ; or \g.
Your MySQL connection id is 5
Server version: 5.1.45 Source distribution

Type 'help;' or '\h' for help. Type '\c' to clear
the current input statement.

mysql> select version();
+-----------+
| version() |
+-----------+
| 5.1.45    |
+-----------+
1 row in set (0.00 sec)
```

The database to be used for the MySQL session can be changed using the USE
function:

```
mysql> select database();
+------------+
| database() |
+------------+
| NULL       |
+------------+
1 row in set (0.02 sec)

mysql> use test;
Database changed
mysql> select database();
+------------+
| database() |
+------------+
| test       |
+------------+
1 row in set (0.00 sec)
```

As with PostgreSQL, SQL statements can be used to create TABLES, INSERT
data into them and perform queries.

```
mysql> CREATE TABLE shopping_list ( name varchar(80),
                                     items varchar(100),
    -> location int, phone varchar(10));
Query OK, 0 rows affected (0.09 sec)
```

```
mysql> INSERT INTO shopping_list VALUES
('PrimeMart', 'Milk,Cookies', 97070, '5031234322');
Query OK, 1 row affected (0.00 sec)

mysql> select * from shopping_list;
+-----------+--------------+----------+------------+
| name      | items        | location | phone      |
+-----------+--------------+----------+------------+
| PrimeMart | Milk,Cookies |    97070 | 5031234322 |
+-----------+--------------+----------+------------+
1 row in set (0.00 sec)
```

In addition to SQL processing, the MySQL command-line also reports status using the \s command:

```
mysql> \s
--------------
mysql  Ver 14.14 Distrib 5.1.45, for redhat-linux-gnu (x86_64)
       using readline 5.1

Connection id: 5
Current database: test
Current user: skoranne@localhost
SSL: Not in use
Current pager: stdout
Using outfile: ''
Using delimiter: ;
Server version: 5.1.45 Source distribution
Protocol version: 10
Connection: Localhost via UNIX socket
Server characterset: latin1
Db      characterset: latin1
Client characterset: latin1
Conn.   characterset: latin1
UNIX socket: /var/lib/mysql/mysql.sock
Uptime: 8 min 3 sec

Threads: 1  Questions: 17  Slow queries: 0
Opens: 17  Flush tables: 1  Open tables: 9
                        Queries per second avg: 0.35
--------------
```

Since a major use of MySQL is in web based HTML generation, MySQL directly supports the creation of reports of queries in XML. To generate XML output, run mysql -X as the command-line:

```
mysql> select name,items from shopping_list;
<?xml version="1.0"?>

<resultset statement="select name,items from shopping_list;"
  xmlns:xsi="http://www.w3.org/2001/XMLSchema-instance">
  <row>
<field name="name">PrimeMart</field>
<field name="items">Milk,Cookies</field>
```

```
  </row>

  <row>
<field name="name">Fruties</field>
<field name="items">Plum</field>
  </row>

  <row>
<field name="name">GardenPlace</field>
<field name="items">Vegetables</field>
  </row>
</resultset>
3 rows in set (0.00 sec)
```

Without the XML flag we get:

```
mysql> select name,items from shopping_list;
+-------------+--------------+
| name        | items        |
+-------------+--------------+
| PrimeMart   | Milk,Cookies |
| Fruties     | Plum         |
| GardenPlace | Vegetables   |
+-------------+--------------+
3 rows in set (0.00 sec)
```

In addition to using the `mysql` command-line MySQL (and PostgreSQL) can also be used with the Python language. Consider the Python program shown in Listing 20.1.

```python
#!/usr/bin/python
# \file mysql_python.py
# \author Sandeep Koranne, (C) 2010
# \description Using MySQL with Python
import sys
import MySQLdb

try:
    connection = MySQLdb.connect( db = "test",
                                  host = "localhost" )
    print "Successful connection"
except:
    print "Unable to connect to MySQL database..\n"
    sys.exit(1)

# Connection is successfull, print ZIP code of stores
cursor = connection.cursor()
cursor.execute("""SELECT name, location from shopping_list""")
while True:
    row = cursor.fetchone()
    if row == None:
        break
    print "Store name = %s, Zip code = %s" % ( row[0], row[1] )

print "Total number of stores = %d\n" % cursor.rowcount
connection.close()
print "Disconnected from server"
sys.exit(0)
```

Listing 20.1 Using MySQL with Python

Running this Python program gets:

```
$python mysql_python.py
Successful connection
Store name = PrimeMart, Zip code = 97070
Store name = Fruties, Zip code = 97068
Store name = GardenPlace, Zip code = 97068
Total number of stores = 3

Disconnected from server
```

Listing 20.1 shows the use of database *cursors* alongwith SQL queries. Care must be taken to ensure that the size of results returned should not exceed the memory space of the Python program, as using `store_results()` method the complete set of all rows from the query are returned. It is preferable to use cursors and fetch one row at a time as shown in Listing 20.1. Python can also be used to populate a RDBMS using SQL `INSERT INTO` statements.

20.7 SQLite

SQLite is a software library that implements a self-contained, server less, zero-configuration, transactional SQL database engine. Although SQLite is available as an API it should not be confused with C database libraries like BerkelyDB (see Section 9.5), or SQL access APIs in ODBC. SQLite is by itself a complete SQL database engine. Its advantages include (i) efficiency, (ii) portability, (iii) low configuration and setup time, (iv) small code footprint and (v) self-contained. SQLite has been used as a substitute for storing XML property based meta-data in files, storing information in embedded systems such as PDAs and settop boxes, and website database for small to medium sized websites. Using SQLite as a database is simple:

```
$sqlite3 mydb
SQLite version 3.6.20
Enter ".help" for instructions
Enter SQL statements terminated with a ";"
sqlite> select date();
2010-03-17
sqlite> create table shopping_list
        (name varchar(80), location int);
sqlite> insert into shopping_list values
        ('PrimeMart', 97070);
sqlite> insert into shopping_list values
        ('Fruties', 97068);
sqlite> select name from shopping_list;
PrimeMart
Fruties
```

The current database, table and table schema can be printed using:

```
sqlite> .databases
seq   name                    file

0     main                    /home/skoranne/HAND_BOOK/mydb
sqlite> .tables
shopping_list
sqlite> .schema
CREATE TABLE shopping_list(name varchar(80),location int);
```

The main advantage of SQLite is the ease of setting up the database, as well as the ability of using the database from C/C++ programs. Bindings for many other languages are also available. Consider a C++ program to access the database we have created above, the code is shown in Listing 20.2.

```cpp
// \file sqlite_db.cpp
// \author Sandeep Koranne (C) 2010
// \description Example of SQLite API
#include <iostream>          // program IO
#include <cstdlib>           // exit
#include <cstdio>            // C File
#include <sqlite3.h>         // SQLite

int main( int argc, char *argv [] ) {
  if( argc != 2 ) {
    std::cerr << "Usage: ./sqlite_db <db>...\n";
    exit(1);
  }
  // if db does not exist, it is created.
  sqlite3 *handle = NULL;
  int status = sqlite3_open( argv[1], &handle );
  if( status ) {
    std::cerr << "Connection to db " << argv[1] << "failed.\n";
    exit(1);
  }
  sqlite3_stmt *stmt;
  char query[] = "SELECT name,location from shopping_list;";
  status = sqlite3_prepare_v2( handle, query, -1, &stmt, 0 );
  if( status ) {
    std::cerr << "Db retrieval failed.\n";
    exit(1);
  }
  int cols = sqlite3_column_count( stmt );
  std::cout << "Total of " << cols << " columns.\n";
  while( true ) {
    // get a single row from the db
    status = sqlite3_step( stmt );
    if( status == SQLITE_ROW ) { // valid data
      std::cout << "Name = " << (const char*) sqlite3_column_text( stmt, 0 )
      << " Zip = " << (const char*) sqlite3_column_text( stmt, 1 )
          << std::endl;
    } else if( status == SQLITE_DONE )
      break; // all done
    else {
      std::cerr << "Error in SELECT query.\n";
      exit(1);
    }
  }
  sqlite3_close( handle ); // close the database
  std::cout << std::endl;
  return (0);
}
```

Listing 20.2 Example of SQLite API in C++

Compiling and running this program as:

```
g++ sqlite_db.cpp -o sqlite_db -lsqlite3
$ ./sqlite_db mydb
Total of 2 columns.
Name = PrimeMart Zip = 97070
Name = Fruties Zip = 97068
```

Looking at the code in Listing 20.2 we can appreciate the brevity of the API, which after-all does implement a complete SQL engine. Comparing SQLite with BerkelyDB 9.5 is instructive at this time. For application which need database access in C/C++ program alongwith SQL support, SQLite is clearly a good candidate.

20.8 CouchDB

Before we close this chapter, let us discuss a new class of database management systems which is designed to be a fault-tolerant and scalable database. CouchDB is based on the concept of data as documents, and as such for application which are document-based, using CouchDB feels almost natural. CouchDB is written in Erlang (see Section 1.6.10 for more information on the Erlang programming language), and uses the HTTP protocol. It includes many powerful ways to query, map, and combine data. Running CouchDB on a GNU/Linux system is straightforward, and once running we can use any HTTP client to verify its status:

```
$curl http://localhost:5984
{"couchdb":"Welcome","version":"0.10.2"}
```

The above example shows that indeed, CouchDB is alive and running on port 5984. To list all databases CouchDB is aware of:

```
$curl -X GET http://127.0.0.1:5984/_all_dbs
[]
```

It simply returns the empty set, as no databases have been created. To create a database we can use the PUT method of curl to send information to the CouchDB instance which is running:

```
$curl -X PUT http://127.0.0.1:5984/shopping_list
{"ok":true}
$curl -X GET http://127.0.0.1:5984/_all_dbs
["shopping_list"]
```

True to its web based design, CouchDB has an integrated web-browser based administration interface (called Futon), as shown in Figure 20.1, which is accessible by going to localhost:5984/_utils/.

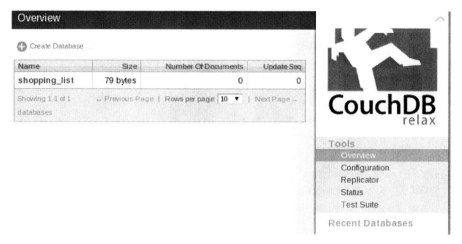

Fig. 20.1 Futon: CouchDB web-based administration

Using CouchDB and Futon, databases can be created and documents can be added to the database. We added the 'shopping_list' database, as well as adding information for three stores containing their name, location (zip-code) and phone number, as shown in Figure 20.2.

To use the collection of documents we can write JSON (Java Script Object Notation) programs, as (shop.json) :

```
{
  "_id" : "_design/example",
  "views" : {
  "name" : {
  "map" : "function(doc)
       { emit( doc.Name, doc.Phone)}"
     }
   }
}
```

We have to place this document in CouchDB using PUT and access the view location:

```
$curl -X PUT
 http://localhost:5984/shopping_list/_design/example \
     -d @shop.json
{"ok":true,"id":"_design/example",
 "rev":"1-27f9c1f60960da2589ebfc27d6cc0c0a"}
$ curl http://localhost:5984/shopping_list/
    _design/example/_view/name
{"total_rows":3,"offset":0,"rows":[
{"id":"9034968401851e4997a08c13aa435101",
 "key":"Fruties","value":12345678},
```

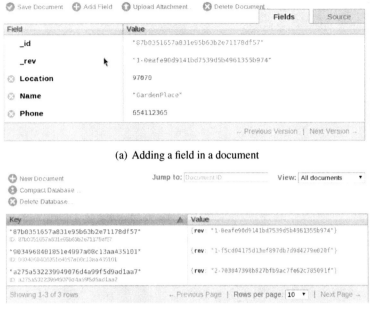

(a) Adding a field in a document

(b) Adding multiple documents

Fig. 20.2 Creating databases and documents in CouchDB

```
{"id":"87b0351657a831e95b63b2e71178df57",
 "key":"GardenPlace","value":654112365},
{"id":"a275a532239949076d4a99f5d9ad1aa7",
 "key":"PrimeMart","value":765112343}
]}
```

The JSON script code we wrote uses the `emit` function of CouchDB. `emit` takes two arguments, the *key* and the *value*. On access, CouchDB performs a *map* of the list of documents against this function (which can also have conditionals on the document fields).

Above, we have given a very brief overview of the capabilities of CouchDB. It is instructive to compare CouchDB to MemCache library (see Section 9.6) which also does object based caching over the network.

20.9 Conclusion

In this chapter we discussed the use of open-source software to implement and use Internet software, including web servers, virtualization and databases. The database systems included PostgreSQL, MySQL, SQLite and CouchDB.

Chapter 21
Conclusion

As we stated at the outset of this book (in the Preface), the moral of this book is "stand on the shoulders of giants", and in the modern software development parlance, "don't reinvent the wheel", which means, use standard library functions, and libraries which use standardized interfaces.

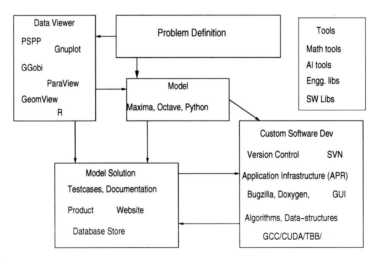

Fig. 21.1 Design process for a solution using Open-source tools

In my view the ideal software application now develops on GNU/Linux, using GCC as the main compiler, with GNU binutils as the underlying system tools. GDB/Insight is used as the debugger. GCC with $-Wall$ was used as the static code checker. Source code is written on Emacs/Eclipse/Kdevelop and version controlled using CVS/SVN/git. Source navigator is used as the project browser. Regular backups, system administration tasks are automated with Bash scripting using UNIX tools such as tar, $sort$ and $grep$. The source code is well organized, follows a

S. Koranne, *Handbook of Open Source Tools*,
DOI 10.1007/978-1-4419-7719-9_21, © Springer Science+Business Media, LLC 2011

uniform style guide and has been extensively cross-referenced using CTAGS, and documented using `doxygen`. Class diagrams, ancillary design documents are available in open-source documentation formats, SGML, Open Office, or LaTeX. Dependency diagrams are drawn using `dot`, schematics using `xfig`.

Automatic software builds using GNU make or SCons are done, and Bonsai, Tinderbox and Bugzilla manage the software release activities. Platform differences are mitigated by using GNU buildtools such as `autoconf` and `automake`. The software itself is regressed and profiled using `gprof`, `gcov`, and `valgrind` and static code checking.

The code itself uses standard C library functions, system calls, and C++ library (including streams, algorithms, templates and dynamic memory). For complex algorithms garbage collection is implemented using Boehm GC. Parallel algorithms are implemented in OpenMP, and distributed using MPI. Persistent storage is provided using Berkeley DB, while user data comes in HDF5. Internal data is compressed using zlib and checksummed using MD5. Object serialization is performed using standardized XML protocol.

The algorithms have been prototyped using scientific software as discussed in Chapter 14. Computations involving mathematics libraries are addressed using one of the fine open-source libraries mentioned in Chapter 16 including linear algebra, graph theory, matrices, and FFT. Logic applications are implemented using BDD. The command language for the application is compiled using a compiler written in lex, yacc and optimized using LLVM.

The bulk of the software system is written as a module with SWIG generated interfaces which generate Python bindings, allowing testing and reuse of individual components. The Python interface also gives an interactive batch mode, while an attractive graphical user interface written in Qt is available for users.

Graphics used by the application are written in OpenGL and diagrams for hard copy are generated using Cairo, `gd` or `asymptote`, while plots are generated using `gnuplot`. Database access is implemented by Postgres. A pictorial depiction of this whole process is shown in Figure 21.1.

Agreed, such a combination may never arise in the first software you write after reading this book, or in any single software system. But the goal of this book is to introduce the many available tools for efficient work flow. This may mean not having to write any code, but to use one of the tools mentioned above. Or, better yet, perhaps you could enhance one of the tools to add the missing functionality. I hope this book has introduced you to the world of open-source programs in general, and scientific software in particular.

Appendix A
Websites of Open-Source Applications

In this chapter we describe the websites of many of the open-source tools we have presented in this book. The website often contains links to download and install the software. As I stated in the Preface, the fact that the software considered for inclusion in this book has to be Open Source does not imply that it is free to use, especially in a commercial product. While the code is available to look at and learn, I would advise the reader to contact the author of the software, and read the license carefully to determine the responsibility the user has prior to including the software library, or using the software system.

Table A.1 Open-source software websites

Name	Website
ACL2 (theorem prover)	www.cs.utexas.edu/users/moore/acl2/
ASA (adaptive simulated annealing)	www.ingber.com/
ATLAS (automatically tuned linear algebra)	math-atlas.sourceforge.net/
Alliance (VLSI CAD)	www-asim.lip6.fr/recherche/alliance/
Anti-grain geometry (graphics rendering)	www.antigrain.com/
Antlr (parser generator)	www.antlr.org
Apache Portable Runtime (APR)	apr.apache.org
Apache Webserver	www.apache.org
Asymptote (graphics drawing)	asymptote.sourceforge.net/
Audacity (sound processing)	audacity.sourceforge.net
Axiom (math software)	www.open-axiom.org
BLAS (linear algebra)	www.netlib.org/blas/
BRL-CAD (3d CAD)	www.brlcad.org
BZIP2 Compression	www.bzip.org
Blender (solid modeling)	www.blender.org

Table A.2 Open-source software websites

Name	Website
Boost C++ Library	www.boost.org
Bugzilla (defect tracking)	www.bugzilla.org
CGAL (computational geometry)	www.cgal.org
CImg (image processing)	cimg.sourceforge.net
CLIPS (expert system)	clipsrules.sourceforge.net/
CMake (dependency tool)	www.cmake.org
CTags	ctags.sourceforge.net
CVS Version Control	www.nongnu.org/cvs
Cairo Graphics Library	cairographics.org
CoCoA (commutative algebra)	cocoa.dima.unige.it/
Comp. Infra. for OR (COIN-OR)	www.coin-or.org
CouchDB Database	couchdb.apache.org
Doxygen (documentation generator)	www.doxygen.org
Eclipse (editor and IDE)	www.eclipse.org
Emacs (editor)	www.gnu.org/software/emacs/
Erlang Programming Language	www.erlang.org
Expat (XML processing)	expat.sourceforge.net
FFTW (FFT library)	www.fftw.org
FWTools (GIS tools)	fwtools.maptools.org/
Fast Artificial Neural Networks (FANN)	eenissen.dk/
Folding At Home (protein folding)	folding.stanford.edu
Fox Toolkit (GUI library)	www.fox-toolkit.org
FreeType (font rendering)	www.freetype.org
GAUL (genetic algorithm)	gaul.sourceforge.net/
GDAL	www.gdal.org
GGobi (multi-dimensional viewing)	www.ggobi.org
GLUI (OpenGL UI)	www.cs.unc.edu/ rademach/glui/
GLUT (OpenGL)	www.opengl.org/resources/libraries/glut/
GNU Autoconf (build tool)	www.gnu.org/software/autoconf/
GNU Automake (build tool)	www.gnu.org/software/automake/
GNU Binutils	www.gnu.org/software/binutils
GNU Bison (parser generator)	www.gnu.org/software/bison/
GNU Compiler Collection (GCC)	gcc.gnu.org
GNU Image Manipulation Program (GIMP)	www.gimp.org
GNU Libtool (build tool)	www.gnu.org/software/libtool/
GNU Linear Programming (GLPK)	www.gnu.org/software/glpk/
GNU Make (dependency tool)	www.gnu.org/software/make/
GNU Octave	www.gnu.org/software/octave/
GNU PSPP (statistical)	www.gnu.org/software/pspp
GNU Scientific Library (GSL)	www.gnu.org/software/gsl/
GNU debugger (gdb)	www.gnu.org/software/gdb
GNU flex (lexical analysis)	www.gnu.org/software/flex/
GNU multi-precision library (GMP)	www.gmplib.org
GNUPlot (plotting software)	www.gnuplot.info

Table A.3 Open-source software websites

Name	Website
GNUs Not UNIX (GNU)	www.gnu.org
GRASS (GIS)	www.grass.itc.it/
GROMACS	www.gromacs.org
GTK (GUI library)	www.gtk.org
GTKWave (waveform viewer)	gtkwave.sourceforge.net/
GeomView (visualization)	www.geomview.org
Go Programming Language	golang.org
Grace/Xmgr (plotting)	plasma-gate.weizmann.ac.il/Grace/
GraphicsMagick Image Processing	www.graphicsmagick.org
Graphviz (graph layout)	www.graphviz.org/
HDF5 (high-performance data format)	www.hdfgroup.org/HDF5/
Hadoop (distributed computing framework)	hadoop.apache.org
Icarus (verilog simulator)	www.icarus.com/eda/verilog/
Inkscape (vector drawing)	www.inkscape.org
Insight (debugger)	sourceware.org/insight/
JMol (molecule viewer)	jmol.sourceforge.net/
KDevelop (IDE)	www.kdevelop.org
LAPACK (linear algebra)	www.netlib.org/lapack/
LIBSVM (machine learning)	www.csie.ntu.edu.tw/ cjlin/libsvm/
LLVM (low level virtual machine)	www.llvm.org
LZMA (compression)	www.7-zip.org/sdk.html
LibGD (graphics drawing)	www.libgd.org
Linux Kernel	www.kernel.org
Lout (document typesetting)	www.qtrac.eu/lout.html
Lua (programming language)	www.lua.org
MPFR (multi-precision floating library)	www.mpfr.org
Macaulay2 (commutative algebra)	www.math.uiuc.edu/Macaulay2/
Magiv (VLSI layout editor)	opencircuitdesign.com/magic
MapServer	mapserver.gis.umn.edu/
Maxima (computer algebra)	maxima.sourceforge.net/
Memcached Library	www.memcached.org
Message Passing Interface (MPI)	www.mcs.anl.gov/mpi/
NAMD (molecular dynamics)	www.ks.uiuc.edu/Research/namd/
NTL (number theory library)	www.shoup.net/ntl/
NVIDIA CUDA	www.nvidia.com/object/cuda_home.html
Nauty (graph isomorphism)	cs.anu.edu.au/ bdm/nauty/
NgSPICE (circuit simulator)	ngspice.sourceforge.net/
OGDI	ogdi.sourceforge.net/
OGRE (3d graphics)	www.ogre3d.org
OpenCL (parallel programming)	www.khronos.org/opencl/
OpenCV (computer vision)	opencv.willowgarage.com

Table A.4 Open-source software websites

Name	Website
OpenDX (data visualization)	www.opendx.org
OpenEV (map viewer)	OpenEV: openev.sf.net/
OpenFOAM (CFD tool)	www.openfoam.com
OpenGL (3d graphics)	www.opengl.org
OpenMP	www.openmp.org
OpenOffice suite	www.openoffice.org
OpenSSL	www.openssl.org
PARI/GP (group theory software)	pari.math.u-bordeaux.fr/
POVRay (ray tracer)	www.povray.org
PROJ4	www.remotesensing.org/proj
ParaView (high-performance 3d viewer)	www.paraview.org
Perftools	http://code.google.com/p/google-perftools/
Polymake (polytope)	www.opt.tu-darmstadt.de/polymake/
PostgreSQL database	www.postgresql.org
PyCUDA	mathema.tician.de/software/pycuda
QCAD (2d CAD drawing)	www.qcad.org
QUCS (simulator)	qucs.sourceforge.net/
Qt (GUI and application development)	qt.nokia.com
Quantum GIS	www.qgis.org
R (statistical software)	www.r-project.org
REDUCE (algebra)	reduce-algebra.sourceforge.net/
SAGE (math software)	www.sagemath.org
SCons (dependency tool)	www.scons.org
SGML (markup language)	www.w3.org/MarkUp/SGML/
SOX (sound processing)	sox.sourceforge.net
SQLite database	www.sqlite.org
SWIG (wrapper generator)	www.swig.org
Scala (programming language)	www.scala-lang.org
Scheme (programming language)	www.schemers.org
Scientific Python (SciPy)	numpy.scipy.org
Scribus (desktop publishing)	www.scribus.net
Singular (computer algebra)	www.singular.uni-kl.de/
SourceNavigator	sourcenav.sourceforge.net/
Subversion Version Control	subversion.tigris.org

Table A.5 Open-source software websites

Name	Website
Tcl/Tk (programming language)	www.tcl.tk
TeXMacs (math editor)	www.texmacs.org
Thread Building Block (TBB)	www.threadingbuildingblocks.org/
Tinderbox	www.mozilla.org/projects/tinderbox/
TkCVS	kcvs.sourceforge.net/
VNC (remote computing)	www.realvnc.com
Valgrind (profiler)	www.valgrind.org
WxWidget (GUI library)	www.wxwidgets.org
X Window System	www.x.org
X10 (programming language)	x10.codehaus.org
XCircuit	opencircuitdesign.com/xcircuit/
XFig (vector drawing)	www.xfig.org
ZLIB Compression	www.zlib.org
LaTeX(document typesetting)	www.latex-project.org
TeX(document typesetting)	www.tug.org

Index